Chemistry of High-Temperature Superconductors

A C S   S Y M P O S I U M   S E R I E S **351**

Chemistry of High-Temperature Superconductors

David L. Nelson, EDITOR
Office of Naval Research

M. Stanley Whittingham, EDITOR
Schlumberger-Doll Research

Thomas F. George, EDITOR
State University of New York at Buffalo

Developed for symposia sponsored
by the Divisions of Inorganic Chemistry
and Physical Chemistry
at the 194th Meeting
of the American Chemical Society,
New Orleans, Louisiana,
August 30–September 4, 1987

American Chemical Society, Washington, DC 1987

Library of Congress Cataloging-in-Publication Data

Chemistry of high-temperature superconductors.
   (ACS symposium series, ISSN 0097-6156; 351)

   "Developed from a symposium sponsored by the
Divisions of Physical Chemistry and Inorganic
Chemistry at the 194th Meeting of the American
Chemical Society, New Orleans, Louisiana, August 30–
September 4, 1987."

   Includes bibliographies and indexes.

   1. Superconductors—Chemistry—Congresses.
2. Materials at high temperatures—Congresses.

   I. Nelson, David L., 1942-   . II. Whittingham,
M. Stanley (Michael Stanley), 1941-   . III. George,
Thomas F., 1947-   . IV. American Chemical Society.
Division of Physical Chemistry. V. American Chemical
Society. Division of Inorganic Chemistry. VI. American
Chemical Society. Meeting (194th: 1987: New Orleans,
La.) VII. Series.

QD473.C488   1987   537.6'23   87-19314
ISBN 0-8412-1431-X

PRINTED IN THE UNITED STATES OF AMERICA

Second printing 1988

Foreword

The ACS SYMPOSIUM SERIES was founded in 1974 to provide a medium for publishing symposia quickly in book form. The format of the Series parallels that of the continuing ADVANCES IN CHEMISTRY SERIES except that, in order to save time, the papers are not typeset but are reproduced as they are submitted by the authors in camera-ready form. Papers are reviewed under the supervision of the Editors with the assistance of the Series Advisory Board and are selected to maintain the integrity of the symposia; however, verbatim reproductions of previously published papers are not accepted. Both reviews and reports of research are acceptable, because symposia may embrace both types of presentation.

Contents

Preface

THE BREAKTHROUGH IN RAISING THE CRITICAL TEMPERATURE (T_c) of superconductivity is approaching its first anniversary $(1,2)$. Since the breakthrough, there has been a tremendous flurry of activity to investigate this phenomenon and to search for new classes of materials in an effort to continue to raise the critical temperature as high as possible. After what has been largely an Edisonian approach in this exploding area, we now feel a compelling need to provide a snapshot of current progress and understanding from the point of view of solid-state chemistry. We also want to define future directions for research in basic materials so as to accelerate the achievement of leading technologies for the commercial exploitation of high-temperature superconductivity.

Although this area is highly multidisciplinary, chemistry plays a prominent role throughout, from materials preparation through processing for small- and large-scale applications. In this volume, we present the following major areas: theory, materials preparation and characterization, structure–property relationships, surfaces and interfaces, processing and fabrication, applications, and research needs and opportunities. Our intention here is to further involve the chemistry community in this scientifically exciting research area of such vital importance to our nation.

DAVID L. NELSON
Office of Naval Research
Arlington, VA 22217-5000

M. STANLEY WHITTINGHAM
Schlumberger-Doll Research
Ridgefield, CT 06877-4108

THOMAS F. GEORGE
State University of New York at Buffalo
Buffalo, NY 14260

June 30, 1987

References

1. Bednorz, J. G.; Mueller, K. A. *Z. Phys. B.* **1986,** *64,* 189.
2. Wu, M. K.; Ashburn, J. R.; Torng, C. J.; Hor, P. H.; Meng, R. L.; Gao, L.; Huang, Z. J.; Wang, Y. Q.; Chu, C. W. *Phys. Rev. Lett.* **1987,** *58,* 908.

THEORY

Chapter 1

High-Temperature Superconductivity in Oxides

Arthur W. Sleight

Central Research and Development Department, E. I. du Pont de Nemours and Company, Experimental Station, Building 356, Wilmington, DE 19898

The common structural feature of oxide superconductors is the presence of $+$(M-O)$+$ chains which are linear or nearly so. In addition, mixed valency and high covalency are key factors for high temperature superconductivity in both the $BaPb_{1-x}Bi_xO_3$ system and the new oxide superconductors based on copper. The disproportionation reactions of Bi^{IV} and Cu^{II} are especially crucial. Highly electropositive cations, such as Ba^{2+}, are required to sufficiently stabilize oxygen $2p$ levels so that the Cu^{III} and Bi^V states can in turn be stabilized. High-temperature superconductors are normally expected to be metastable materials; the evidence for metastability in these oxide superconductors is presented.

The discovery in 1975 of high-temperature superconductivity in the $BaPb_{1-x}Bi_xO_3$ system ([1]) was widely hailed as a novel type of superconductivity, and new mechanisms have been proposed ([2]). The recently discovered oxide superconductors based on copper ([3-6]) are considered by many to be new examples of this type of superconductivity ([4,5]). The higher critical temperatures (T_c) in the copper systems are presumably due to the fact that states near the Fermi level (E_f) are primarily $3d$ in the copper system whereas they are primarily $6s$ in the case of the $BaPb_{1-x}Bi_xO_3$ system. Since d bands are generally narrower than s bands, a higher density of states at E_f is expected in the copper systems. A higher density of states at E_f has been observed experimentally and, according to BCS theory ([7]), this could account for higher T_c's in the copper systems.

0097–6156/87/0351–0002$06.00/0
© 1987 American Chemical Society

It should be emphasized, however, that all the oxide systems mentioned have a much lower density of states at E_f than expected for such high T_c's. The T_c of 13 K found in the $BaPb_{1-x}Bi_xO_3$ system still stands as the record for any material not containing a transition metal. There are many materials with higher density of states at E_f but with lower T_c's than in the new copper oxide superconductors.

In this paper I point out the similarities and differences between the $BaPb_{1-x}Bi_xO_3$ and the copper systems. This will be done from a chemistry viewpoint considering such issues as covalency, mixed valency, structure, and metastability.

In an attempt to be consistent with accepted chemistry nomenclature, Roman numeral superscripts to the right of element symbols (e.g. Bi^V) will frequently be used to indicate oxidation number and to emphasize that this number does not suggest any particular degree of ionization. This paper contains some discussion of the meaning of these superscripts.

Structure

The oxides so far mentioned in this paper can all be considered to have perovskite-related structures. In fact, the structure for superconductors in the $BaPb_{1-x}Bi_xO_3$ system is very close to the ideal cubic perovskite structure. The stoichiometry is essentially AMO_3, and there is just a small tetragonal distortion from cubic symmetry. The structure of the new oxide superconductors based on copper (i.e. $La_{2-x}A_x^{II}CuO_4$ and $RBa_2Cu_3O_{7-y}$) deviate much more from the ideal perovskite structure. The La_2CuO_4 or K_2NiF_4 structure may be viewed as an intergrowth between the perovskite and sodium chloride structures, i.e. $AO \cdot AMO_3 = A_2MO_4$. However, this results in a layered structure with MO_2 sheets and no M-O-M bonding perpendicular to the sheets. For the 1-2-3 compound, e.g. YBa_2CuO_{7-x}, there is extensive oxygen deficiency, i.e. $AMO_{2.3}$ instead of AMO_3. The oxygen vacancies created are mostly ordered. This causes a severe departure from the perovskite structure, and viewing $RBa_2Cu_3O_7$ as perovskite-related is of dubious value. Nonetheless, all of the cations in the 1-2-3 structure have the perovskite arrangement; R and Ba are ordered on A cation sites, and Cu occupies the M sites. The oxygen vacancies drop the cation coordination numbers: R, 12→8, Ba; 12→10; Cu(1), 6→5 and Cu(2), 6→4.

In the ideal perovskite structure, there are infinite linear -M-O-M-O- chains along all axes of the unit cell. These chains intersect at the M cations to give octahedral coordination around M. It should be noted that the structure of previous T_c record holders, e.g. Nb_3Ge, has strongly bonded Nb chains along each of the axes of its unit cell; however, in this case the

chains do not intersect. Thus, a general structural
feature of superconductors is strongly bonded chains that
are linear or nearly so. These chains may or may not
intersect. In the case of the tetragonal K_2NiF_4
structure, there are intersecting linear $+Ni-F+$ chains
along two axes only. In the case of the orthorhombic
La_2CuO_4 structure, these chains are slightly bent at the
oxygen site. For the $RBa_2Cu_3O_7$ structure, there are
infinite linear non intersecting chains along one axis
and there are nonlinear intersecting chains but only
along two axes. The importance of the chains in Nb_3Ge
has been debated for many years, and undoubtedly the
importance of chains in the oxide superconductors will be
debated for years to come.

Oxides of Cu, Ag and Au

Cu_2O and Ag_2O are isostructural and show the tendency of
Cu^I and Ag^I to prefer two-fold linear coordination to
oxygen. The band gap narrows on going from Cu_2O to Ag_2O
and presumably narrows so much on going to Au_2O that this
compound does not exist. This trend from Cu to Ag to Au
can be taken as a reflection of increased covalency of
the metal-oxygen bonds going down this row of the
periodic table.
    CuO and AgO are also well known whereas AuO is not.
Although CuO contains Cu^{II}, AgO contains a mixture of Ag^I
and Ag^{III}. This tendency for the divalent state to
disproportionate increases going down the periodic table,
$Cu{\rightarrow}Ag{\rightarrow}Au$, to such an extent that compounds which might
appear to contain Au^{II} actually always contain a mixture
of Au^I and Au^{III}. On the other hand Cu^{II} has never been
observed to disproportionate in oxide systems. I propose
that as one increases the covalency of the $Cu^{II}-O$ bond,
Cu^{II} will eventually disproportionate into Cu^I and Cu^{III}
states. I further propose that there can be an
intermediate range where Cu^{II} in oxides exhibits
delocalized electron behavior and superconductivity. I
am thus suggesting a metallic region sandwiched between
two semiconducting regions (Figure 1).
    The covalency of $Cu^{II}-O$ bonds can be controlled
inductively. In series such as A_2CuO_4, the covalency of
the Cu-O bonds increases as the A cation becomes more
electropositive. Thus for a A-O-Cu linkage, Cu-O bonds
are forced more covalent as the A-O bond becomes more
ionic. This can be thought of as an issue of competition
between A and Cu for overlap of the oxygen orbitals.
Perhaps a better view is that when a highly
electropositive cation ionizes, the electron density
shifts to the rest of the system, e.g. oxygen *and copper*.
    When the Cu-O bond is in the ionic limit, the Cu^{2+}
ion will of course exhibit the behavior of a localized d^9
ion. However, as the covalency increases, the Cu-O
distance decreases, the mixing with oxygen orbitals

increases and in a concentrated system the levels broaden
into bands which can support metallic behavior. The
highest occupied state for Cu^{II} in the tetragonal
distorted octahedron ($d^1_{x^2-y^2}$) is sigma antibonding. The
fact that this state is strongly antibonding from
symmetry considerations is of little consequence in the
ionic limit. However, as we increase the covalency, it
is this antibonding situation that would cause Cu^{II} to
disproportionate into Cu^I and Cu^{III}. Stabilization
occurs for Cu^I because the $3d$ shell fills completely and
thus shrinks. Stabilization occurs for Cu^{III} because the
offending σ^* electron has disappeared.

   The π antibonding states for Cu^{II} are also filled,
and there are no possibilities for relief by depopulating
$3d$ levels because that would require a copper oxidation
state higher than three which cannot be achieved. Again
this is not an issue in the ionic limit. However, as
Cu-O covalency increases and Cu-O distances decrease, the
π^* problem increases. The system responds by bending
Cu-O-Cu bonds away from 180° as in La_2CuO_4. This
structural feature is also present in the $RBa_2Cu_3O_{7-y}$
phases. Bending this bond away from 180° is coupled with
oxygen s-p rehybridization which pushes electron density
out of the bond. Nature's way of dealing with
antibonding π situations is to bend bonds and convert
antibonding states into nonbonding states. The reason
that Cu-O-Cu bonds bend in La_2CuO_4 is precisely the same
reason that C-O-C bonds are bent in ethers. Thus we find
linear Cu-O-Cu bonds in the ionic limit (e.g. A_2CuO_4
where A is Pr, Sm, Nd and Gd), but the bond bends in the
covalent limit prior to the disproportionation limit.

   The intermediate metallic region (Figure 1) is not
expected in all Cu^{II}/O systems. We could go directly
from the ionic Cu^{2+} limit to the covalent Cu^I + Cu^{III}
limit. Indeed, this is to be expected in dilute systems
where there are no Cu-O-Cu linkages. In the $La_{2-x}A^{II}_xCuO_4$
systems, the disproportionation reaction may be prevented
by frustration. It is easy to draw a La_2CuO_4 derived
Cu-O layer which would contain "isolated" $Cu^{III}O_4$ and
linear Cu^IO_2 units. However, such ordering might not be
cooperative from one layer to the next due to small
interlayer interactions. More importantly there is no
good way to stack such layers, and this leads to
frustration of the disproportionation process.
Furthermore, the substitution of divalent cations leads
to disorder on the A cation sites as well as either
oxygen vacancies or small levels of Cu^{III}. These defects
could also inhibit the disproportionation reaction which
requres long range cooperative structural reorganization.

Ba(Pb,Bi)O$_3$ System

Since many have considered superconductivity in this
system to be the first example of a new type of

superconductivity more recently discovered in mixed
oxides of copper, it is important to review the
$Ba(Pb,Bi)O_3$ system for analogies to the superconducting
systems based on copper oxides. The disproportionation
reaction $2Bi^{IV}(6s^1) \rightarrow Bi^{III}(6s^2) + Bi^V(6s^o)$ is a crucial
feature of the $BaPb_{1-x}Bi_xO_3$ system (Figure 2). In the
metallic region, Bi is diluted in a $BaPbO_3$ matrix. The
$6s$ electron from Bi^{IV} is delocalized in a band derived
mostly from Pb $6s$ and Bi $6s$ states but with strong
admixture of oxygen $2p$ states. [Although paramagnetic s^1
cations such as Bi^{4+} are not observed in concentrated
systems, they are frequently found in dilute systems such
as glasses.] As the Bi concentration in the Pb matrix
increases, the Bi atoms begin to sense each other and the
tendency to disproportionate into Bi^{III} and Bi^V
increases. At about 30% Bi, the disproportionation
reaction occurs. The Bi^{III} and Bi^V do not take on long
range order at this point because they are still diluted
in a Pb matrix. However, as one approaches pure $BaBiO_3$,
Bi^{III} and Bi^V are ordered onto discrete lattice sites
(8).

Two other questions should be addressed in this
system. Why is $BaPbO_3$ a metal and why is the system
semiconducting above $x \sim 0.3$? We know that $BaPbO_3$ is best
viewed as a zero gap semiconductor, i.e. the conduction
and valence bands just slightly overlap. By this
explanation we might expect that $BaPb_{1-x}Bi_xO_3$
compositions should remain metallic even if Bi^{IV}
disproportionates to remove Bi $6s$ states from E_f.
However, as x increases in the $BaPb_{1-x}Bi_xO_3$ system, the
lattice expands. Thus Pb-O overlap decreases and a gap
opens up.

Occurrence of Superconductivity

Superconductivity in the $BaPb_{1-x}Bi_xO_3$ system and the
recently discovered copper oxide systems is presumably
related to the disproportionation reactions of Bi^{IV} or
Cu^{II}. In the Bi^{IV} system, we are able to push into the
region where disproportionation acutally occurs. One
could say that superconductivity occurs when the
situation $2Bi^{IV}$ vs. $Bi^{III} + Bi^V$ becomes degenerate. It is
even tempting to suggest that lone pairs of Bi^{III} and
Cooper pairs are somehow related. In the case of copper
oxide superconductors, we can increase the covalency of
Cu-O bonds enough to push into the metallic region but we
have not yet increased it enough to see the predicted
disproportionation reaction.

La_2CuO_4 sits right on the boundary between the ionic
semiconducting region and the covalent metallic region.
As normally prepared, La_2CuO_4 is metallic at room
temperature but becomes semiconducting at low
temperature. Since there has been no symmetry change, we
know that Cu^{II} disproportionation reaction has not

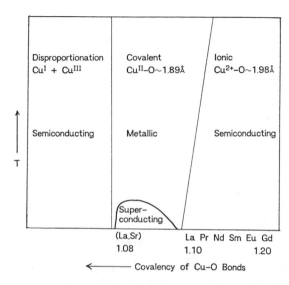

Figure 1. Schematic phase diagram for A_2CuO_4 phases:
temperature vs. Pauling electronegativity of A. The
value for La,Sr of 1.08 is the weighted average for
$La_{1.8}Sr_{0.2}CuO_4$.

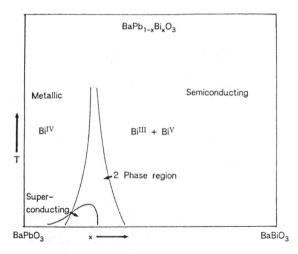

Figure 2. Schematic phase diagram for the
$BaPb_{1-x}Bi_xO_3$ perovskite system: temperature vs. x.

occurred. However, this compound is nonstoichiometric and is best represented as $La_{2-x}CuO_{4-y}$ where x and y can each reach values of about 0.05. At the higher values of y, this material becomes semiconducting at all temperatures (9). For finite values of x and very low values of y, parts of "La_2CuO_4" samples become superconducting below about 40K (10). Although most samples of "La_2CuO_4" are antiferromagnetic, apparently, others are not (9).

Since La_2CuO_4 sits right on the boundary (Figure 1), a more electropositive A cation is required to push the material into the desirable region for superconductivity. There are no trivalent cations which are more electropositive than La. Thus, we must resort to the use of alkaline earths or alkalis. The alkaline earth cations Ba, Sr, and Ca are all significantly more electropositive than La and can be used to push into the metallic region. The divalent cations of Pb and Cd apparently can substitute for La (11) but do not increase the covalency of Cu-O bonds because Pb and Cd are less electropositive than La.

When Ba, Sr, or Ca are substituted for La in La_2CuO_4, the Cu-O distance steadily decreases. This happens despite the fact that Ba is much larger than La and thus causes a significant increase in the unit cell volume. It is apparently the inductive effect of the electropositive alkaline earth cation that causes increased covalency of Cu-O bonds and therefore shorter Cu-O bonds. (One could argue that A^{II} substitution causes decreased population of σ^* states and thus decreased Cu-O distances. Although this is true, the net effect is still that all Cu-O distances decrease.) The shorter Cu-O bonds in turn results in increased T_c's. The reason that T_c's are highest in the Sr system is because Cu-O distances become shortest in this system through a combination of geometric and inductive effects. The T_c's correlate very well with the Cu-O distance (12). High pressure can be used to further decrease the Cu-O distance, and T_c continues to increase (4).

The disproportionation reactions of Bi^{IV} and Cu^{II} relate in a straight forward manner to soft phonons and the electron-phonon coupling mechanism required by BCS theory. The cation-oxygen distances for the various ions Bi^{III}, Bi^{IV}, Bi^V or Cu^I, Cu^{II}, Cu^{III} differ from each other. Therefore, there will tend to be oxygen displacements as the conduction electron moves between cation sites. Thus, the disproportionation reaction creates breathing mode soft phonons which are strongly coupled to the conduction electrons. This mechanism has been accepted by many for the $BaPb_{1-x}Bi_xO_3$ system. Furthermore, structural refinements for $YBa_2Cu_3O_7$ show very large vibrations for the oxygen in the apparently very important linear -(Cu-O)- chains. Thus the breathing mode soft phonon model has great appeal for the oxide superconductors of copper. It could be further suggested that larger oxygen displacements are required for Cu^{II} disproportionation than for Bi^{IV} disproportionation. Thus T_c's will be higher for the copper systems. A breathing mode soft phonon model predicts that T_c should change on substitution of ^{18}O for ^{16}O. Such an isotope effect has very recently been found for $(La,Sr)_2CuO_4$ (13). This would seem to confirm the importance of the breathing mode for this system. In the $YBa_2Cu_3O_{7-y}$ system, an isotope effect has not yet been found (14,15). However, the disproportionation tendency does not necessarily have to be simply related to a soft breathing mode to promote high T_c superconductivity. There are other structural instabilities in the copper oxide and $BaPb_{1-x}Bi_xO_3$ systems. Bi^{III} tends to be unstable at a center of symmetry, and cation-oxygen-cation angles of 180° are unstable in both systems when the covalency is high. Although coupled to atomic displacements, these instabilities result in electronic polarizations much larger than the atomic displacements.

Covalency and High Oxidation States
Real Charge vs. Formal Oxidation States

It is no accident that Ba (or some other highly electropositive cations) is found in these high T_c superconductors: $Ba(Pb,Bi)O_3$, $(La,Ba)_2CuO_4$ and $RBa_2Cu_3O_{7-y}$. I have already discussed the importance of electropositive cations in promoting high covalency of Cu-O bonds. There is another important aspect of the presence of highly electropositive cations such as Ba^{2+}. In mixed oxides of transition metals, highly electropositive cations stabilize high oxidation states. It is essential that ions such as Ba^{2+} be present to stabilize Cu^{III} or Bi^V. Otherwise the Cu^I or Bi^{IV} disproportionation reaction does not become favorable.

The O^{2-} ion is not stable in the gas phase but is stabilized in solids where it is surrounded by cations. The more electropositive the cations, the greater the stabilization of the O^{2-} levels. These levels can be so effectively stabilized that unusually high oxidation states (Cu^{III} or Bi^V) are obtained.

We also know that Ba^{2+} and other highly electropostive cations stabilize peroxide ions and other oxygen anions intermediate between O^{2-} and O_2. Indeed, the covalency of the $Cu^{III}-O$ or Bi^V-O bonds are so high, some suggest that we should formulate these as $Cu^{2+}-O^{1-}$ and $Bi^{4+}-O^{1-}$. When we are dealing with metallic compounds, formulating as $Cu^{2+}-O^{1-}$ vs. $Cu^{III}-O$ makes little difference. In fact, it seems acceptable to consider the metallic state as a situation where the two formulations are degenerate. However, once we have a semiconductor, we can see the folly of formulating

$Ba_2Bi^{III}Bi^VO_6$ as for example $Ba_2Bi^{3+}Bi^{4+}O_5^{2-}O^{1-}$. The
latter formulation implies unpaired electrons which are
not present.

Some studies of $BaBiO_3$ accept two different charges
on the two crystallographically distinct bismuth cations.
However, they prefer to write $Ba_2Bi^{4-\delta}Bi^{4+\delta}$ and speak of
a charge density wave instead of disproportionation.
This shows a fundamental lack of understanding of the
meaning of formal oxidation states. It makes no more
sense to write $Ba_2Bi^{4-\delta}Bi^{4+\delta}O_6$ than for example
$Ba_2Bi^{2-\delta}Bi^{2+\delta}O_6$ because the framework of formal oxidation
states has been ignored and the superscripts no longer
have meaning. There is a commensurate charge density
wave in $BaBiO_3$, and this is precisely equivalent to a
$Ba_2Bi^{III}Bi^VO_6$ formulation.

There continues to be great confusion about the
meaning of formal oxidation states vs. real charges on
ions. As expected this is more of a problem in the
physics community, but it is unfortunately also a problem
in parts of the chemistry community. When one writes
$Ba_2Bi^{III}Bi^VO_6$, there is no implication that the real
charges on bismuth cations should resemble +3 and +5.
Also, there is no implication that the charge difference
between Bi^{III} and Bi^V is two. Real charges depend on
electronegativity and covalency. There have been many
proposals to calculate real charges of ions based on
electronegativity factors. For example, Sanderson's
method (16) would give charges on Bi^{III} and Bi^V of about
+0.4 and +0.5, respectively. This important issue is
worth pursuing somewhat further. To a chemist there is
no doubt but that CuO should be formulated as $Cu^{II}O$, AgO
as $Ag^IAg^{III}O_2$, CuS as $Cu_4^ICu_2^{II}(S_2)_2^{-II}S_2^{-II}$, CuS_2 as
$Cu^{II}(S_2)^{-II}$, BaO_2 as $Ba^{II}(O_2)$ and OsO_4 as $Os^{VIII}O_4$. In
the last case, the real charge on Os^{VIII} is less than
one. Some refer to formal oxidation states as a
chemist's simple-minded way of electron accounting.
However, such a system of formal oxidation states is
highly useful in predicting the number and character of
unpaired electrons in a system.

It is no surprise when number crunching operations
in a computer show that the conduction bands in
$Ba(Pb,Bi)O_3$ and the copper oxide superconductors are very
high in oxygen $2p$ character. This is entirely expected
on the chemical grounds of very high covalency. There is
no reason to pervert the system of formal oxidation
states by suggesting that O^{-I} is present in these
systems.

Metastability

A rule for high temperature superconductors has been that
they must be metastable materials. This rule follows
from BCS theory in the sense that pushing the various
parameters involved to their limits will naturally cause

materials to become thermodynamically unstable. Metastability is in fact observed in the $BaPb_{1-x}Bi_xO_3$ system (Figure 2). The best superconductors are quenched from high temperature. If equilibrium were achieved, two phases with different values of x would coexist. The phase with lower x would be metallic; the one with higher x would be semiconducting. Thus, the most interesting region for superconductivity does not exist at equilibrium.

The superconducting 1-2-3 phases are also apparently metastable. There has been no success preparing these compounds below about 800°. This in itself is very suggestive of metastability considering the intense worldwide effort to find low temperature routes to these exciting materials. Presumably the 1-2-3 compound is thermodynamically stable in air in a temperature range of about 800 to 950°; however, its oxygen content is then too low (e.g. $RBa_2Cu_3O_{6.3}$) to produce a superconducting material. Attempts to prepare high oxygen content mateials (e.g. $RBa_2Cu_3O_{6.8}$) directly at high temperature under high oxygen pressure have so far failed. Thus it appears that 1-2-3 compounds with sufficient oxygen to become superconducting are not stable at any combination of temperature and oxygen pressure. The 1-2-3 phases are first prepared with low oxygen content; they are then annealed under oxygen at lower temperatures where they oxidize. However, this oxidation presumably is not an equilibrium process. At equilibrium there would be phase separation. This conclusion has recently been verified by work at the University of Tokyo showing that the superconductivity of 1-2-3 phases is destroyed by long anneals at low temperature ($\sim$400°) and fixed oxygen stoichiometry (17).

Literature Cited

1. Sleight, A. W.; Gillson, J. L.; Bierstedt, P. E. Solid State Commun. 1975, 17, 299.
2. for example, Rice, T. M.; Sneddon, L. Phys. Rev. Lett. 1981, 47, 689.
3. Bednorz, J. G.; Müller, K. A. Z. Phys. 1986, B64, 189.
4. Chu, C. W.; Hor, P. H.; Meng, R. L.; Gao., L.; Huang, Z. J. Science 1987, 235, 567.
5. Cava, R. J.; van Dover, R. B.; Batlogg, B.; Rietman, E. A. Phys Rev. Lett. 1987, 58, 408.
6. Wu, M. K.; Ashburn, J. R.; Torng, C. J.; Hor, P. H.; Meng, R. L.; Gao, L.; Huang, Z. J.; Wang, Y. Q.; Chu, C. W. Phys. Rev. Lett. 1987, 58, 908.
7. Bardeen, J.; Cooper, L. N.; Schrieffer, J. R. Phys. Rev. 1957, 108, 1175.
8. Cox, D. E.; Sleight, A. W. Acta Cryst. 1979, B35, 1.

9.   Johnston, D. C.; Goshorn, D.; Thomann, H.; Alvarez,
     M.; Bhattacharya, S.; Higging, M.; Jacobson, A.;
     Keweshen, C.; Ping, H.; Kwiapek, P.; Moncton, D.;
     Sinha, S.; Stokes, J.; Pindall, P.; Vaknin, D. Mat.
     Res. Soc. Meeting, Anaheim, 1987.
10. Grant, T. M.; Parkin, S.S.P.; Lee, V. Y.; Engler, E.
     M.; Ramirez, M. L.; Vazquez, J. E.; Lim, G.;
     Jacowitz, R. D.; Green, R. L. Phys. Rev. Lett. 1987,
     58, 2482.
11. Gubser, D. V.; Hein, R. A.; Lawrence, S. H.;
     Osofsky, M. S.; Schrodt, D. J.; Toth, L. E.; Wolf,
     S. A. Phys. Rev. 1987, B35, 5350.
12. Kishio, K.; Kitazawa, K.; Sugii, N.; Kanbe, S.;
     Fueki, K.; Takagi, H.; Tanaka, S. Chem. Lett., 1987,
     635.
13. Stacy, A. M.; Zettl, A., private communication.
14. Battlogg, B.; Cava, R. J.; Jayaraman, A.; van Dover,
     R. B.; Kourauklis, G. A.; Sunshine, S.; Murphy, D.
     W.; Rupp, L. W.; Chen, H. S.; White, A.; Short, K.
     T.; Myisce, A. M.; Rietman, E. A. Phys. Rev. Lett.
     1987, 58, 2333.
15. Bourne, L. C.,; Crommie, M. F.; Zettl, A.; Zur loye,
     H.-C.; Keller, S. W.; Leary, K. L.; Stacy, A. M.;
     Chang, K. J.; Cohen, M. L.; Morris, D. E. Phys Rev.
     Lett. 1987, 58 2337.
16. Sanderson, R. T. Chemical Bonds and Bond Energy;
     Academic Press, New York, 1976.
17. Kitazawa, K., private communication.

RECEIVED July 6, 1987

Chapter 2

Valence Bond Theory and Superconductivity

Richard P. Messmer[1,2] and Robert B. Murphy[2]

[1]**General Electric Corporate Research and Development, Schenectady, NY 12301**
[2]**Department of Physics, University of Pennsylvania, Philadelphia, PA 19104**

Two necessary criteria for a theory of superconductivity are a phase coherence of the wave function and an attractive electron-electron interaction. We review how the BCS theory achieves these criteria and then show how a valence bond wave function can also meet these conditions. The energy scale of the latter approach has a larger range in principle than that possible *via* the electron-phonon interaction.

The recent discovery of ceramic high-T_c superconductors has forced a re-examination of the basic concepts and physical assumptions employed in current theoretical approaches. In re-examining basic concepts, it is well to remember that the true N-electron wave function may be expanded in terms of components each of which is made up of N single particle functions and that this expansion can be made in (at least) two different ways:

$$\Psi = \sum_\nu c_\nu \; \Phi_\nu \tag{1}$$

(*molecular orbitals/Bloch orbitals/delocalized basis*)

$$\Psi = \sum_\nu d_\nu \; \Phi_\nu^\dagger \tag{2}$$

(*valence bond orbitals/localized basis*)

The former expansion is the one typically assumed both in molecular and solid state work. The ease with which the single particle basis can be obtained in this case is certainly a significant advantage. Furthermore, it might be argued that either approach is equivalent *in the end*, and hence it makes sense to choose the mathematically more straightforward approach. In fact, one *always* considers only a small fraction of the terms in either expansion and the more relevant question is which is more rapidly convergent and/or more physically

0097–6156/87/0351–0013$06.00/0

motivated. The concepts derived by the two approaches may be quite different. This has been illustrated recently in a series of molecules and clusters, (1-5) for which calculations using Equation (2) were carried out. For metal clusters it was found (1,2) that electrons became localized into interstitial regions. And for double (3), triple (4) and conjugated bonds (C_6H_6) (5), it has been found that bent-bonds made up of essentially tetrahedral hybrids describe the bonding. These conclusions are quite different than those based on MO theory.

A key factor in superconductivity is the presence of an energy gap which separates the ground state (superconducting state) from the continuous single particle spectrum characteristic of the normal state of a metal. Let us write the many electron wave function as

$$\Psi = \sum_{\nu} a_\nu \ \Phi_\nu' \tag{3}$$

without specifying the representation. Assume Φ_ν are solutions of H_o ($H_o \Phi_\nu' = E_\nu \Phi_\nu'$) and the full Hamiltanian is $H = H_o + U$. The question is what forms do U and Ψ need to have in order to produce a significant energy lowering of the ground state with respect to the E_ν? That is, how can an energy gap be produced? The total energy is

$$W = \sum_{\nu} E_\nu \ |a_\nu^2| \ + \ \sum_{\mu\nu} U_{\mu\nu} a_\mu^* a_\nu \tag{4}$$

A significant lowering in energy can be achieved, as is well known (6), if the E_ν are all nearly equal and the $U_{\mu\nu}$ are all nearly equal. In this case, if the $E_\nu \simeq \bar{E}_o$ and the $U_{\mu\nu} \simeq -V$, the energy is

$$W = E_o - V \sum_{\mu\nu} a_\mu^* \ a_\nu \tag{5}$$

with the lowest energy obtained when all the a_ν are the same. If there are m terms contributing in the expansion of the wave function, the result is

$$W = E_o - mV \tag{6}$$

with the energy gap

$$\Delta = (E_o - W) = V \sum_{\mu\nu} a_\mu^* a_\nu = mV \tag{7}$$

Hence to produce a gap, one has to devise a physically meaningful wave function that has phase coherence and equal amplitudes for the Φ_ν' and a potential U which is attractive (i.e., leads to matrix elements $U_{\mu\nu} = -V$). In the next section, we review how Bardeen, Cooper and Schrieffer (7) (BCS) met this challenge.

Review of the BCS Theory of Superconductivity

Here we present a brief review of the essential aspects of the BCS theory of superconductivity which was introduced in 1957 and is still the most successful and complete theory of superconductivity which exists. The purpose of this section is to stress the unique features of the superconducting wave function which should be preserved in any new theory which involves a different mechanism for obtaining the ground state wave function.

In the BCS theory, the normal (non-superconducting) state of the metal is described by Bloch single particle eigenstates (8) |k> labelled by a wavevector **k**,

$$h \mid k> \ = \ \varepsilon_k \mid k> \tag{8}$$

where h is the single electron Hamiltonian and ε_k is the single particle energy of state $\mid k>$. The wave function of the metal is described by the occupation of single particle states $\mid k>$ and in the ground state all single particle levels are filled up to the Fermi energy E_F with states above E_F being unoccupied. This Bloch model does not include correlations between electrons due to Coulomb forces nor the interaction of the electrons with the lattice vibrations (phonons). In the superconducting state the phonon interaction is accounted for, but the electronic correlation is neglected except for effects produced by the electron-phonon coupling. The argument for ignoring other electronic correlation effects is that they are characteristic of both the normal and superconducting phases and therefore cannot be responsible for producing a gap.

The BCS superconducting state is characterized by an attractive potential between electrons arising from an electron-phonon interaction. We will now review how this interaction is derived (9). The direct electron-phonon interaction is a result of the potential between the electrons and the nuclear charges when the nuclei vibrate about their equilibrium positions. The electron-ion (e-i) interaction is expanded in terms of displacements Q_j of the nuclear coordinates from their equilibrium positions R_j,

$$V_{e\text{-}i}\,(\mathbf{r}_\gamma - \mathbf{R}_j - \mathbf{Q}_j) = V_{e\text{-}i}\,(\mathbf{r}_\gamma - \mathbf{R}_j) + \mathbf{Q}_j \cdot \nabla V_{e\text{-}i}\,(\mathbf{r}_\gamma - \mathbf{R}_j) + \cdots \tag{9}$$

$$\mathbf{r}_\gamma = \gamma^{th}\ electron\ coordinate\,;\ V_{e\text{-}i}\,(\mathbf{r}_\gamma - \mathbf{R}_j) = \frac{-Z}{\mid \mathbf{r}_\gamma - \mathbf{R}_j \mid}$$

By making a Fourier transformation of $V_{e\text{-}i}$ and using a plane wave basis $\mid k>$ for the electrons, we can consider the electron-ion interaction in the usual scattering terms in which we consider the amplitude M_q to transfer momentum q to the electron state $\mid k>$ "scattering" it into state $\mid k + q>$. The corresponding diagram for this process in Figure 1.

Fröhlich (10) showed how second order perturbation theory could be applied to derive an effective interaction between electrons from the direct electron-ion interactions. The physical idea is that as one electron scatters from a nuclear center it distorts the lattice, this distortion is felt by another electron, and thus the electrons experience an indirect interaction. The result is that we can think of the electrons as exchanging phonon momentum q in an electron-electron scattering process shown in Figure 2. The effective potential of interaction between the electrons for a scattering involving a change in momentum q is (11),

$$V_{e\text{-}ph} = \frac{\mid M_q \mid^2}{(\varepsilon_k - \varepsilon_{k+q})^2 - (\hbar\omega_q)^2} \tag{10}$$

where M_q is the amplitude for the electron to directly absorb a phonon of momentum q; ε_k and ε_{k+q} are the single particle energies before and after scattering, and $\hbar\,\omega_q$ is the energy of the phonon. The important point is that for $\mid \varepsilon_k - \varepsilon_{k+q} \mid < \hbar\,\omega_q$, the potential is attractive and it is this attractive interaction which leads to the superconducting state. Counterbalancing this attractive interaction is the repulsive Coulomb interaction which in momentum space is

$$V_C = \frac{4\pi e^2}{\mid \mathbf{k}_1 - \mathbf{k}_2 \mid^2} \tag{11}$$

creating the potential for the scattering of an electron from $\mid k_1>$ to $\mid k_2>$ in the Coulomb field of another electron.

The object of the BCS theory was to maximize this attractive electron-phonon (e-ph) interaction since it could possibly lead to a physically different state, lower in energy than the normal metal. To maximize the interaction BCS considered only electron-phonon interactions for which V_{e-ph} is negative which requires scattering of electrons from states $|k>$ to $|k'>$ such that $|\varepsilon_k - \varepsilon_{k'}| < \not{h} \, \omega_c$ where $\not{h} \, \omega_c$ is some average phonon frequency of the order of a Debye frequency. In other words only $|k>$ states within a region $\pm \not{h} \, \omega_c$ of the Fermi energy can contribute to the scattering process. To simplify matters further an average constant attractive potential -V was defined as $-V = <V_{e-ph} + V_c>$ which is an average of the electron phonon potential and the Coulomb potential over the k region defined for attractive V_{e-ph}.

The next crucial step in the theory was the development of a superconducting wave function that both optimizes the attractive phonon interaction and minimizes the Coulomb repulsion. This means specifying a particular occupation of $|k>$ states. Considering the matrix elements of the attractive potential, BCS showed that the optimal wave function involved an occupation of k states such that if state $|k\sigma>$ with momentum k and spin σ is occupied then the state with opposite momentum and spin, $|-k-\sigma>$ is also occupied. This condition is referred to as electron pairing and is responsible for the coherence of the wave function. The scattering from V_{e-ph} only occurs from the pair state $|k,-k>$ to another pair state $|k',-k'>$ where $k' = k+q$ as shown in Figure 2 and again k and k' are in the restricted region near the Fermi level. The paired wave function is written as,

$$\Psi_{BCS} = \prod_k (v_k + u_k \, b_k^{\dagger}) \, | 0> \qquad (12)$$

where u_k is the amplitude to have pair state $|k,-k>$ occupied, u_k^2 is the probability to occupy pair state $|k,-k>$, v_k is the amplitude to have pair state $|k,-k>$ unoccupied, $b_k^{\dagger} = c_{k\sigma}^{\dagger} \, c_{-k-\sigma}^{\dagger}$ is the pair creation operator in second quantized form, and $|0>$ is the vacuum.

In the ground state at 0 K of the normal metal u_k^2, the probability to occupy pair state $|k,-k>$, is unity up to the Fermi energy after which it is zero. In the superconducting state u_k^2 differs from the Fermi distribution by the excitation of some pair states above the Fermi level. These pairs interact via the attractive potential which more than compensates for the excitation energy above the Fermi level. Thus a rounded Fermi distribution is obtained (Figure 3).

The Hamiltonian which describes these interactions is simply,

$$H_{BCS} = \sum_{k,\sigma} \varepsilon_k \, c_{k\sigma}^{\dagger} \, c_{k\sigma} - \sum_{k,k'} V_{k,k'} b_{k'}^{\dagger} \, b_k \qquad (13)$$

$V_{k,k'}$ is the potential obtained for each scattering of pair state $|k,-k>$ to $|k',-k'>$ for k,k' in the region near the Fermi level. The energy is simply

$$E_{BCS} = 2 \sum_k u_k^2 \, \varepsilon_k - \sum_{k,k'} V_{k,k'} \, u_k \, v_k \, u_{k'} \, v_{k'} \qquad (14)$$

The coefficients u_k and v_k are determined variationally giving (see Figure 3),

$$u_k^2 = 1/2 \, (1 - \varepsilon_k/E_k), \quad v_k^2 = 1/2 \, (1 + \varepsilon_k/E_k) \qquad (15)$$

It is assumed that the matrix element $V_{k,k'} \simeq V$ (i.e., are independent of k and are non-zero only in the vicinity of ε_F). This leads to a gap parameter

$$\Delta = V \sum_{k'} u_{k'} \, v_{k'} \qquad (16)$$

$|\vec{k}+\vec{q}\rangle$

$\vec{q}$

$|\vec{k}\rangle$

Figure 1. Electron of momentum **k** being scattered by phonon into state with momentum **k** + **q**.

$|\vec{k_1}+\vec{q}\rangle$

$|\vec{k_2}-\vec{q}\rangle$

$\vec{q}$

$|\vec{k_1}\rangle$

$|\vec{k_2}\rangle$

Figure 2. Two electrons with momenta k_1 and k_2 exchanging a phonon of momentum **q** in an electron-electron scattering process.

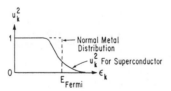

Figure 3. Occupancy of **k**-states for the normal metal and in the superconducting state.

which may be compared to Equation (7) and a new quasiparticle energy given by $E_k = \sqrt{\varepsilon_k^2 + \Delta^2}$ (12).

That a set of nonzero u_k has been obtained for $k > k_F$ means that a new many-electron state has formed with a lower energy than the normal state. The key feature of the new state is its stability to single particle excitations. In the normal metal, single particle excitations above the Fermi level can be made with vanishingly small energies. In the superconducting state there is not only an energy lowering of the whole system relative to the normal metal but there is also a finite energy gap to single particle-like excitations of the order of the gap parameter Δ. This gap to single particle excitations is responsible for most of the physically observable properties of a superconductor including a specific heat which grows as exp(-Δ/kT) and perfect diamagnetism to be discussed below.

When a magnetic field is applied to a superconductor, a current is induced in the material which creates a new field opposed to the applied field such that $B = 0$ in the material (Meissner effect). The quantum mechanical current density j in the material is given by,

$$j = - \frac{-ie\hbar}{2mc} \, [\Psi^* \nabla \Psi - (\nabla \Psi)^* \Psi] - \frac{e^2}{mc} A \Psi^* \Psi \qquad (17)$$

where A is the vector potential defined by $B = \nabla \times A$. The first term is the paramagnetic contribution to the current and the second is the diamagnetic term. Ψ is the wave function in the presence of the field A which differs from the ground state superconducting function Ψ_0 to first order in A by,

$$\Psi = \Psi_0 + \sum_{n \neq 0} \frac{<n \mid A \cdot p \mid 0>}{(E_n - E_0)} \mid n > \qquad (18)$$

where $\mid n >$ are excited states above Ψ_0. $E_n - E_o$ is at least as large as the gap parameter Δ which makes the perturbation to Ψ relatively small; thus we can use $\Psi = \Psi_0$ in the calculation of j. In the ground state Ψ_0, the paramagnetic term in the current is zero, hence the only response of the system is purely diamagnetic, $j = -(e^2/mc) A \Psi^* \Psi$. Since $\nabla^2 A = -(4\pi n/c) j$, $A = A_0 \exp(-\sqrt{(4\pi n e^2/mc^2)} \cdot z)$, which shows, in the one-dimensional case, the exponential drop of the field as z increases into the material from the surface. This shows that diamagnetism is a direct consequence of the stability of the superconducting state to single particle like excitations.

Finite temperature effects are easily included in the theory by modifying all the matrix elements by including Fermi-Dirac factors, $1/(\exp(-\beta E_k) + 1)$. Variationally minimizing the free energy $G = U - TS$ shows the gap function decreasing as a function of temperature until it reaches zero at the critical temperature T_c, at which the transition to the normal state occurs. The expression for T_c is, $kT_c \sim \hbar\omega_c \exp(-1 / N(0)V)$, for $N(0)V < 1$. $N(0)$ is the density of states of the normal metal at the Fermi level.

To summarize, the BCS theory of superconductivity provides an energy gap and a physical mechanism to achieve: (1) phase coherence in the wave function and (2) an attractive interaction. Both of these are required to produce a gap as seen in the discussion of the Introduction. The BCS mechanism of pairing electrons with k and $-k$ momenta is the key to accomplishing these goals.

Why, then, are there so many discussions of new mechanisms to explain the high T_c superconductors? First, there is a limitation to the magnitude of V if it arises from the electron-phonon interactions; second, the density of states for these oxide materials is quite low. Taken together, these observations make it difficult to understand a transition temperature of 90°K. Recently, the band structure of the tetragonal La_2CuO_4 compound has been calculated (13,14). The results indicate a partially filled band at the Fermi level which can be well described in a tight binding model as Cu $d_{x^2-y^2}$ and O p orbitals in the plane hybridizing to form the four Cu-O bonds in the plane. The Fermi surface of the undistorted tetragonal lattice in 2-D is just a square. The lattice vibrations associated with a transition from the tetragonal phase to an orthorhombic structure have wave vectors that exactly span the Fermi square and hence can cause a Peierls distortion (15) to the orthorhombic phase and opens a gap at the Fermi level, creating a semiconductor. The Ba substituents are believed to change the Fermi level thus changing the shape of the Fermi surface and destroying the Peierls instability, bringing the symmetric structure back without the semiconducting band gap. Unfortunately, new experiments (16) seem to indicate that the lattice distortion can occur in the superconducting state.

Calculations of the phonon modes and BCS electron phonon coupling constants (17) suggest that a high T_c is the result of the high frequency of a Cu-O(1) in-plane bond-stretching mode (ω_c) rather than a strong electron phonon coupling parameter. The highest T_c calculated was 40K, limited by the highest Cu-O frequency. Unfortunately replacement of ^{16}O with ^{18}O in this system (18) shows no change in T_c casting doubts on this mechanism.

The first conclusion to be drawn from this work is that calculations based on the BCS model cannot presently explain the T_c's of order 90 K given the phonon frequencies and band structures calculated. This points out that perhaps a mechanism other than an electron-phonon mechanism is involved, which depends on parameters with larger energy scales. A possibility is an electronic mechanism, many of which have been proposed. We will discuss one of these in the next section.

The second point is that the oxide electronic structure is discussed in a real space local bonding framework although traditionally band structure/k-space methods have been used. The simple picture of the k,-k pairing in BCS seems to be lost in these complicated band structure calculations. It should be noted also that there have been serious difficulties associated with applying traditional band structure *concepts* for predicting the *normal* behavior of metallic oxides. As a typical example, simple band filling models for MnO cannot avoid attaining a partially filled band indicating that it should be a conductor (19,20); however, MnO is one of the best insulators known in nature. Such difficulties led to the use of the Hubbard model (21) of metals, which includes strong atomic-like correlation effects in a model Hamiltonian form to rationalize some of these band problems. More of the history of the importance of correlation effects in the oxides can be found in References 19,20,22. The point is that correlation effects have been known to be very important for predicting the properties of oxides, and this should be kept in mind when using the results or concepts of mean field theories which neglect these very important effects.

Valence Bonds and Superconductivity
_____

One of the authors (22), has proposed a model of superconductivity which is based on a highly correlated description of metals, closer to the Wigner lattice limit of electron localization (23) than to the free electron limit usually assumed. For small metal clusters of Li atoms McAdon and Goddard (1,2) have shown using correlated *ab-initio* methods that electron pairs localize in tetrahedral interstices of the lattice. It also has been shown (24) that one can obtain a very good approximation to the experimental charge density of Be using this description. Furthermore, studies on molecular systems (3-5) have shown an energetic preference in sp valence atoms for bent bonds formed from *tetrahedral hybrids* rather than the traditional σ,π bonds based on MO theory. The use of tetrahedral hybrids for Be metal suggests that the electron pairs can localize into tetrahedra in two separate ways, as shown in Figure 4. The fully symmetric wave function is a coherent superposition of the two structures. However, each of these structures is only a shorthand way of describing a large number of valence bond structures with alternative hybrids forming bonds. But, what does this have to do with superconductivity?

Let us return to the discussion in the Introduction and recall that to obtain a theory of superconductivity it is necessary (see Equations 4-7) to have an attractive potential, -V, and to have the a_ν of Equation (3) be equal. In the BCS theory, Equation (1) was used as the basis of the model. Here, we use Equation (2) as our starting point. The question, as it was above in the discussion of BCS theory, is how do we obtain a physically meaningful wave function and potential to satisfy these criteria?

Returning, then, to the expansion of Equation (2), we note that the terms represent different valence bond structures. Why should they all have the same amplitude and phase? This situation is very similar to the problem of determining the "resonance energy" of benzenoid molecules (25,26,27). In that case, of all the possible valence bond structures which might contribute, *only* the Kekule´ structures are used. For large benzenoid systems this is only a small fraction of the total number of structures. Furthermore, it is assumed that they all enter with equal expansion coefficients (*i.e.,* equal amplitude and phase). In addition, the matrix elements which convert one structure into another are set equal to a common value, determined empirically. Thus, the energy lowering associated with "resonance" in benzenoid molecules has a mathematical structure which maps onto the discussion in the Introduction. However, there are some important differences.

A necessary, but not sufficient, condition for producing a superconducting ground state is that the number of available orbitals exceeds the number of electrons. In order to describe a superconducting metal, the lowering of the ground state produces a gap with respect to a continuum of single particle excitations. From a valence bond viewpoint such a continuum is easily achieved when the number of orbitals (hybrids) is significantly larger than the number of electrons (e.g., the ratio $\frac{\# \ of \ orbitals}{\# \ of \ electrons}$ may be in the range $\sim 1.1 - 4.0$). For the oxide materials it is the ratio of orbitals to holes which is important. *However,* for the benzenoid molecules this ratio is unity and the ground state is an insulator; the lowest lying excited states will have significant excitonic and polaronic effects. Thus, in spite of the coherent superposition of alternative bonding structures ("spatial resonance") in these molecules, which is one of the necessary conditions for superconductivity, there are insufficient orbitals for the number of electrons, therefore failing another of the requirements.

For superconducting metals, the non-Kekule´ valence bond structures (analogs of Dewar and "long-bond" structures) provide the basis for constructing the continuum of single particle excitations. Raising the temperature in this model will eventually destroy the stability gained from the resonance because the entropic term, -ST, in the free energy will increase when the electron pairs take on the many other possible configurations that are available to them than just those which maximize the resonance energy. A more quantitative discussion will be given elsewhere (28).

In order to discuss the new superconductors with this model, it must be shown that such a coherent spatial resonance can occur in these materials as a consequence of their local bonding. As a first step, we have calculated the electronic structure of the SF_6 molecule to gain some insight into the bonding occurring in octahedral complexes which are an important part of the environment in the new superconductors.

Bonding in a Octahedral Environment

The molecule sulfur hexafluoride (SF_6) has recently challenged both molecular spectroscopy with its unexpected rotational spectra (29) and electronic structure theories with novel correlation effects (30,31,5). The electronic structure must explain the molecule's high stability, octahedral symmetry, and, most importantly, provide a simple picture of the bonding. At first glance, the traditional chemical models do not appear to be appropriate because sulfur seemingly forms six bonds to fluorines, yet the sulfur s^2p^4 valence configuration allows for at most two covalent bonds.

In this section we report on some preliminary results which suggest a novel interpretation of the electronic structure of SF_6 by considering the wave function as a coherent superposition of low symmetry generalized valence bond structures involving ionic bonding and little sulfur d-orbital character. This coherent superposition provides a significant fraction of the correlation energy by including both intra- and inter-pair correlation while retaining a local picture of the bonding.

At the lowest level, SF_6 is described by the molecular orbital Hartree-Fock wave function with doubly occupied orbitals ϕ_i which are determined self consistently in the mean field of the other pairs,

$$\Psi_{HF} = \det [\phi_1 \alpha \, \phi_1 \, \beta \, \phi_2 \alpha \, \phi_2 \, \beta \, ...] \tag{19}$$

The Hartree-Fock calculations were performed at the experimental (32) octahedral geometry with a S-F bond length of 1.564Å. A standard double zeta basis was used for the fluorine atoms (33), while the sulfur was described by an effective potential (34) with a valence double zeta s-p basis. In a separate calculation a single d polarization function with exponent .532 was added to the sulfur basis to assess the importance of d functions.

The most striking feature of the results is that SF_6 is not bound with respect to the separated atoms when d functions are not included in the sulfur basis while the introduction of the d function on sulfur lowers the energy by 10.5 eV making SF_6 bound by 5.6 eV. Similar results were obtained by Reed (30). Without further evidence this result would suggest that sulfur has a large $sp^3 d^2$ component from which six equivalent S-F bonds can be made.

From the Hartree-Fock calculation, the zeroth order description of SF_6 is of fluorine forming partially ionic bonds to sulfur, with a small population of the sulfur d function energetically very important for bonding. Reed has concluded that the d orbital occupation is small because the d orbitals are high in energy. The Hartree-Fock binding energy falls far short of the experimental value of 20.1 eV indicating that there are major correlation effects neglected, which we will discuss below.

In order to gain both a more local and a more accurate description of the bonding in SF_6 we introduced intra-pair correlations *via* the perfect pairing generalized valence bond method (GVB-PP) (35). The valence electrons are described by generalized Heitler-London pairs,

$$(\phi_a \, \phi_b + \phi_b \, \phi_a) \ (\alpha\beta - \beta\alpha) \tag{20}$$

where the singlet coupled overlapping spatial orbitals ϕ_a, ϕ_b forming a local valence bond are variationally determined. The GVB-PP wave function is the antisymmetrized product of the pair functions and perfect pairing refers to the orthogonality of the pair functions and the nature of the spin coupling:

$$\Phi_{GVB\text{-}PP} = \det [\phi_a(1) \, \phi_b(2) \, \phi_c(3) \, \phi_d(4) \ \cdots \ \Theta_{PP}] \tag{21}$$

with

$$\Theta_{PP} = (\alpha(1)\beta(2) - \beta(1)\alpha(2)) \ (\alpha(3)\beta(4) - \beta(3)\alpha(4)) \ \cdots \tag{22}$$

Six pairs of electrons were described as correlated pairs (Equation 20) while the other pairs were treated at the Hartree-Fock level. No symmetry restraints were placed on the wave function for reasons that are explained below. The basis set included the sulfur d functions. The correlation energy obtained relative to Hartree-Fock is 3.0 eV indicating the extent of intra-pair correlation.

These GVB results suggest that we may think of SF_6 forming in a hypothetical sequence in which the axial fluorines first form largely ionic bonds to sulfur, thus promoting an effective sp^3 valence configuration of sulfur. The four sulfur electrons are left in a tetrahedral orientation on sulfur available for the bonds with the equatorial fluorines.

The equatorial bonds have small z components on the sulfur in the GVB results. The lack of z character in the orbitals is a result of the high electronegativity of the fluorines, causing highly polarized bonds, the lack of inter-pair correlations in the perfect pairing method, stemming from the orthogonality constraints between the pairs, and the neglect of the correlation effects described next.

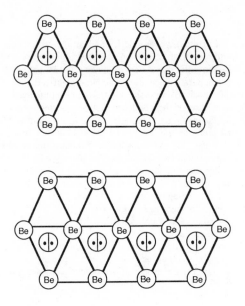

Figure 4. Schematic representation of electrons localized in tetrahedral interstices of hcp Be metal. The electron pairs are distributed among four hybrids at each site. Only a representative number of electron pairs are shown.

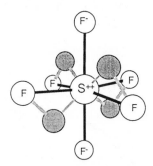

Figure 5. Schematic representation of bonding in SF_6 for one of the *spatial resonance* structures. There are six equivalent structures, each has two "ionic" bonds and four "covalent" bonds. The shaded spheres represent the positions of electron pairs; the light connecting lines represent approximate tetrahedral hybrid orbitals. The dark connecting lines merely show the octahedral geometry.

Inter-pair correlations introduced by the tetrahedral sp^3 orbitals bonding to the equatorial fluorines can be seen in Figure 5, where we have shown schematically what the equatorial bonds would look like with more sulfur p_z character in the bonds. This figure shows the inter-pair correlation occurring by pulling two pairs above the equatorial plane and two below, thus reducing the Pauli repulsion relative to having all pairs in the plane. The choice of the axial direction for the ionic components is not unique. To restore the symmetry in this scheme we must also include the degenerate configuration in which this sp^3 state is reflected about the equatorial plane and consider the four other orientations obtained from the other equivalent "axial" positions. Thus we suggest that the SF_6 wave function is described by a coherent superposition of these six structures,

$$\Psi_{SF_6} = C\,(\,\Phi_1 + \Phi_2 + \ldots + \Phi_6\,) \tag{23}$$

Such a coherent superposition of degenerate states leads to the well known "resonance" energy defined as the difference in energy between the energy of Φ_1 alone and Ψ_{SF_6} (33,5). This bonding scheme for introducing inter-pair correlations beyond the GVB-PP method using a superposition of resonance structures emphasizing the atomic hybridization has been shown to be successful for benzene (5).

Much remains to be done to implement the ideas outlined here about valence bond wave functions, in order to address the many questions about new high-T_c materials. However the fact that it is formulated in *real space* and is based on chemical bonds, should allow much more direct contact with the *chemical aspects* than has been previously possible.

Acknowledgment

This work was supported in part by the Office of Naval Research.

Literature Cited

[1] McAdon, M.H.; Goddard, III, W.A. *Phys. Rev. Lett.* 1985, *55* 2563.

[2] McAdon, M.H.; Goddard, III, W.A. *J. Phys. Chem.* 1987, *91* 2607.

[3] Messmer, R.P.; Schultz, P.A.; Tatar, R.C.; Freund, H.-J. *Chem. Phys. Lett.* 1986, *126* 176.

[4] Messmer, R.P.; Schultz, P.A. *Phys. Rev. Lett.* 1986, *57*, 2653.

[5] Schultz, P.A.; Messmer, R.P. *Phys. Rev. Lett.* 1987, *58*, 2416.

[6] Kittel, C. "Introduction to Solid State Physics", Third Edition, p. 624, Willey, 1966, See Reference 11, also.

[7] Bardeen, J.; Cooper, L.N.; Schrieffer, J.R. *Phys. Rev.* 1957, *108*, 1175.

[8] Bloch, F. *Z. Phys.* 1928. *52*, 555.

[9] Mahan, G.D. " Many Particle Physics", Plenum, 1981.

[10] Fröhlich, H. *Phys. Rev.* 1950, *79*, 845.

[11] Feynman, R.P. "Statistical Mechanics, A Set of Lectures", Ch. 10, Benjamin Press, 1972.

[12] Reference [7] and an alternative derivation by Bogoliubov, N.N.; Nuovo cimento, 1958, *7*, 794.

[13] Mattheiss, L.F. *Phys. Rev. Lett.* 1987, *58*, 1028.

[14] Yu, J.; Freeman, A.J.; Xu, J.H. *Phys. Rev. Lett.* 1987, *58*, 1035.

[15] Peierls, R. "Quantum Theory of Solids", Clarendon, 1956.

[16] Fleming, R.M.; Batlogg, B.; Cava, R.J.; Rietman, E.A. to be published.

[17] Weber, W. *Phys. Rev. Lett.* 1987, *58*, 1371.

[18] Batlogg, B;, Cava, R.J.; *et al. Phys. Rev. Lett.* 1987, *58*, 2333.

[19] Adler, D. *Solid State Physics*, 1968, *21*.

[20] Adler, D.; Feinleib, J. *Phys. Rev. B* 1970, *2*, 3112.

[21] Hubbard, J. *Proc. Royal Soc.*, *A276* 238 (1968); *ibid. A277* 237 (1964); *ibid. A281* 401 (1964); *ibid. A285* 542 (1965); *ibid. 296* 82 (1966); *ibid. A296* 100 (1966).

[22] Messmer, R.P. *Solid State Commun.* 1987, *63*, 405.

[23] Wigner, E. *Faraday Discussions* 1934, *34*, 678.

[24] Tatar, R.C.; Messmer, R.P. to be submitted for publication.

[25] Pauling, L. "The Nature of the Chemical Bond", Third Edition, Chapter 6, Cornell University Press, 1960.

[26] Wheland, G.W. "Resonance in Organic Chemistry", Wiley, 1955.

[27] For recent work, see for example: Gutman, I.; Cyvin, S.J. *Chem. Phys. Lett.* 1987, *136*, 137.

[28] Murphy, R.B.; Messmer, R.P. to be published.

[29] "Physics Today", July 1984, 17.

[30] Reed, A.E.; Weinhold, F. *J. Amer. Chem. Soc.* 1986, *108*, 3587.

[31] Hay, P.J. *J. Amer. Chem. Soc.* 1977, *99*, 1003.

[32] Herzberg, G. "Molecular Spectra and Molecular Structure", Vol. III, Polyatomic Molecules, Van Nostrand, NY 1966.

[33] Dunning, Jr., T.H.; Hay, P.J. "Methods of Electronic Structure Theory" Vol. 3 of Modern Theoretical Chemistry, edited by H.F. Schaefer III, (Plenum, New York, 1977), p. 1.

[34] Wadt, W.R.; Hay, P.J. *J. Chem. Phys.* 1984, *82*, 284.

[35] Hunt, W.J.; Hay, P.J.; Goddard, III, W.A. *J. Chem. Phys.* 1972, *57*, 738; Goddard, III, W.A.; Dunning, Jr., T.H.; Hunt, W.J.; Hay, P.J. *Acc. Chem. Res.* 1973, *6*, 368; Bobrowicz, F.W.; Goddard, III, W.A. "Methods in Electronic Structure Theory", Vol. 3 of "Modern Theoretical Chemistry", edited by H.F. Schaefer III (Plenum, New York, 1977), p. 79.

[36] Voter, A.F.; Goddard, III, W.A. *Chem. Phys.* 1982, *57*, 253.

RECEIVED July 17, 1987

Chapter 3

Interpretation of the Temperature Dependence of Conductivity in Superconducting Oxides

R. J. Thorn

Chemistry and Material Science Divisions, Argonne National Laboratory, Argonne, IL 60439

The diverse variations of the conductivities vs. temperature observed in oxide superconductors can be interpreted with a model based on microcanonical ensembles of a semiconductor and a semimetal originally derived for organic radical cation salts. The equation for the model transverses, through the variations of the Fermi and gap energies and a scattering factor, the range from a semiconductor to a semimetal and defines quantitatively the boundary between the two corresponding to the optimum superconducting state. Therein it describes quantitatively the roles of composition and pressure in determining the onset of the superconducting state. An examination of the published resistivity curves vs. temperature for $La_{2-x}A_xCuO_{4-\delta}$ and $YBa_2Cu_3O_{6.5+\delta}$ yield information consistent with the model and relevant to the defects and range of homogeneity needed for phase diagrams.

Prior to the advent of superconducting oxides, the superconducting state was approached and conceptionally viewed from the perspective of the metallic state. In that case, the nature of the variation of conductivity with temperature is essentially monotonic. Beginning with the first superconducting oxide, $SrTiO_{3-\delta}$ ([1]), and especially with $BaPb_{1-x}Bi_xO_3$ ([2,3]), more diverse, nonmonotonic curves were observed. The measured resistance vs. temperature consists of two types: (1) In one, the resistance decreases monotonically, almost linearly, with decreasing temperature, to the abrupt decrease at the attainment of the superconducting state. (2) In others, the

0097–6156/87/0351–0025$06.00/0
© 1987 American Chemical Society

resistance increases logarithmically or parabolic-like with decreasing temperature to a peaked maximum and thereafter decreases abruptly to what appears to be a superconducting state. Investigations with $BaPb_{1-x}Bi_xO_3$ are particularly significant because mixtures of $BaPbO_3$, a conductor, and of $BaBiO_3$, a nonconductor or a semiconductor, form a continuous solid solution. Hence, the complete set includes a third type: (3) In the extreme, the resistance vs. temperature displays a broad minimum usually identified as a metal-insulator transition. Similar variations are observed with the recently discovered high temperature superconducting oxides, $La_{1.85}Ba_{0.15}CuO_4$ (4,5) and $YBa_2Cu_3O_{6.5\pm\delta}$ (6,7), and the semiconductor La_2CuO_4 (8).

The approach to the superconducting state from the semiconducting state has thus added a new dimension to the conceptionalization of the nature of the superconducting state, and therein it will lead to a more complete understanding of the state, particularly relative to the semiconducting and conducting states. However, to date this understanding has been limited by the absence of a quantitative description or model which predicts these diverse variations.

In a study of the nature of the conductivity vs. temperature of the organic charge transfer compounds, such as tetramethyltetraselenafulvenium perchlorate and perrhenate, $(TMTSF)_2ClO_4$ and $(TMTSF)_2ReO_4$, a model (9) was derived which describes quantitatively the normal state immediately preceding the onset where superconductivity occurs, "nearly" occurs, or can be made to occur. The model contains parameters which are physically significant. Examinations of the resistance curves and the chemical variables for the oxides reveal that they provide a better application of the model to yield some significant insights to understanding the evolution of the superconducting state.

Measured Resistance vs. Temperature and Oxygen Pressure

Whereas with the organic charge transfer conductors, the variables available are rather incremental in the form of a limited number of anions, with the oxides a nearly continuous range of compositions exist, especially as cited above with $BaPb_{1-x}Ba_xCuO_3$, but also in some other cases. In both cases, effects of hydrostatic pressure occurs, probably attributable to the same cause. The evolving literature on the oxides is now so large that an extensive review cannot be presented herein. However, for the intended purpose such is not needed. Rather, some of the more pertinent measurements on $La_{2-x}A_xCuO_4$ and $YBa_2Cu_3O_{6.5\pm\delta}$ have been selected below to identify the nature of the variations of resistance R(T).

The variations of R(T)/R(300) obtained by Tarascon et al. (10) when oxygen is removed from $La_{1.94}Sr_{0.16}CuO_{4-\delta}$ are shown in Figure 1. With successive annealing at approximately 500°C in a vacuum of 10^{-3} torr, the character of R(T)/R(300) changes from one in which the values decrease almost linearly to the abrupt decrease (the first type cited above) to one in which R(T)/R(300) increases somewhat parabolically to a sharp peak at the onset of the abrupt decrease (the second type). After the sample was annealed in oxygen at approximately 500°C, the original state was nearly restored. Similar variations undoubtedly occur when the trivalent lanthanum cation is replaced with a divalent barium or strontium cation. If the alkaline earth cation concentration is reduced to zero or near zero, the R(T)/R(300) curve has only the broad minimum (the third type cited above (5,8)).

Similar variations are observed with $YBa_2Cu_3O_{6.5\pm\delta}$ (11). In Figure 2 are plotted values for R(T)/R(300) for the following annealing conditions: (1) A sample was annealed at 900°–945°C for 17 to 18 hours in oxygen, then cooled from 900° to 600° C over 4 hours, and finally from 600° to 25° C over 2 to 3 hours in oxygen. (2) Other samples were annealed at 900°–945°C for 17 to 18 hours in an atmosphere of oxygen, and then they were rapidly cooled by removing them from the furnace. Some variabilities occurred in the R(T)/R(300) curves for the three samples shown in Figure 2, but in general they are either types 2 or 3 cited above, whereas that for the sample slowly cooled in oxygen is type 1. Undoubtedly the first sample continued its equilibration as the temperature was slowly reduced, so that it contained more oxygen than did the quenched samples.

From a neutron diffraction investigation of the structure of $YBa_2Cu_3O_{6.5\pm\delta}$ at various temperatures between 400° and 900°C and various oxygen pressures, in situ, Jorgensen et al. (12) have derived a composition–temperature diagram at three pressures of oxygen. (See Figure 3). At a temperature near 700°C in pure oxygen, the diffraction measurements reveal what appears as a tetragonal-to-orthorhombic transition as the temperature is decreased. However, this is not a first order phase transition; as the variation in lattice parameters reveals (12), no discontinuity in volume occurs; and no discontinuity in partial pressure of oxygen occurs. The change is primarily in the oxygen–copper network. The tetragonal structure has partially occupied, nearly octahedral Cu–O arrangements; the orthorhombic structure has one-dimensional Cu–O chains. This is not an unusual situation in a nonstoichiometric phase. The data contained in Figure 3 demonstrates that the variations observed by Kini with the two kinds of samples are attributable to the oxygen content of essentially one phase. For the formula $YBa_2[(Cu^{2+})_{1-x}(Cu^{3+})_x]_3O_{6.5+\delta}$ one finds that x = 2/3 δ. At δ = o, x = o and the phase contains essentially all 2+. At δ = 0.35(400°C and 1 atm of oxygen), x = 0.23, i.e., approximately one-fourth of the copper sites are occupied with 3+. Because only one-third of the copper "cations" are in the conducting network, this value of x implies that the nominal ratio of 3+ to 2+ for this situation of preparation is 2.3; the fraction of the sites in the network which is in the 3+ state is 0.70. For negative values of δ, the couple is Cu^{1+}/Cu^{2+}. Thus for δ = -0.2 at 900°C and one atmosphere of oxygen, the fraction of 1+ is 0.13, or 40 percent of the sites in the conducting network are monovalent copper "cations", formally stated.

The types of curves shown in Figures 1 and 2 are also observed with the organic charge transfer conductors containing the organic radical cations $(TMTSF)_2^+$. For instance, with slowly cooled $(TMTSF)_2ClO_4$, a monotonic curve with a slight plateau before the onset of the abrupt decrease is obtained (13). With quenched $(TMTSF)_2ClO_4$, a curve having a broad minimum and a rather sharp maximum followed by the abrupt decrease (with decreasing

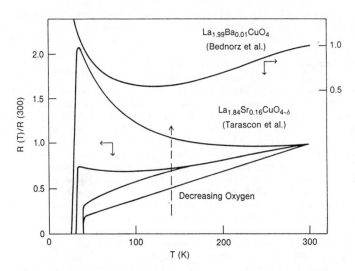

Figure 1.   For $La_{1.84}Sr_{0.16}CuO_{4-\delta}$ and $La_{1.99}Ba_{0.01}CuO_{4-\delta}$, variation of ratio of resistance at T and at 300 K vs. temperature at various annealing times at 10^{-3} torr at 500°C.   Data from Refs. 10 and 5 respectively.

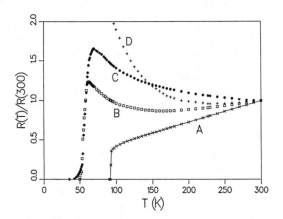

Figure 2.   For $YBa_2Cu_3O_4$ variation of ratio of resistance at T and at 300 K vs. temperature for samples: A-annealed at approximately 925°C for 17 hours and slowly cooled in an atmosphere of oxygen. B-, C-, D-annealed at approximately 925°C for 17 hours in oxygen and quenched in air.   Reproduced from Ref. 11.

temperature) is observed (the second type cited above). With $(TMTSF)_2ReO_4$, a curve of the second type is obtained at ambient pressure, but with increasing hydrostatic pressure a curve of type 1 is eventually obtained (14).

Although the quantitative dependencies of the precise shape of $R(T)/R(300)$ on the equilibrium partial pressure of oxygen have not been determined yet for any of the superconducting oxides, the data presented above and those existing in the literature demonstrate that such dependencies exist and that formally, at least, in $YBa_2Cu_3O_{6.5\pm\delta}$ the ratio of Cu^{3+}/Cu^{2+} or Cu^{1+}/Cu^{2+} determine the shapes. One must recognize that such a rationalization is only formal, because undoubtedly the copper-oxygen network in which conduction occurs is covalently bonded, a situation which is made possible in part because the O^{2-} ion is generally unstable and exists only in a highly ionic environment. Thus the conducting network consists of covalently bonded $Cu-O$'s in which charge is transferred formally through the $Cu^{3+}-O^{2-}$ $Cu^{2+}O^{1-}$ network. Whatever is the precise description, it is apparent that the single, most decisive variable which determines the characteristics of $R(T)/R(300)$ is the density of carriers.

It is risky, of course, to generalize from one oxide to another, but in the absence of adequately completed studies with anyone of the oxides, the assumption concerning the density of carriers seems justified. All of the superconducting oxides display $R(T)$'s similar to those cited above. In the case of strontium titanate, $SrTiO_{3-\delta}$, Schooley et al. (15) have shown that the critical superconducting temperature determined from the midpoints of the abrupt decreases in the resistance vs temperature and the magnetic susceptibility vs temperature depend on the density of carriers determined from measurements of Hall coefficients. Thus the results of T_c vs density of carriers is shown in Figure 4. Therein, one observes that the T_c's from both $R(T)$ and $\chi(T)$ increase to maxima near 10^{20} carriers cm^{-3} and thereafter they decrease.

Model for Conductivity in a Semiconducting and Semimetallic Micro-composite

Although numerous variations of resistance vs temperature similar to those given in Figures 1 and 2 have been published, the information which has been extracted from these measurements has been qualitative at the most. Much of the quantitative information contained in them has been unavailable because of the absence of quantitative expressions which describe them. The solution of the physically detailed description of the variations are evasive because of the complexity of the conductive process in terms of band structure and the wealth of other phenomena, such as charge density and spin density waves, which occur but the significances of which are unknown. Before all these are placed in appropriate perspective and indeed as a guide for doing such, it may be productive to search for the minimum essential parameters which describe quantitatively the diverse variations which have been and are being observed and to do so in such a way that the parameters have physical meaning connected to the complex physical situation and that the detailed structures can be incorporated as necessary. Thus a model has been constructed

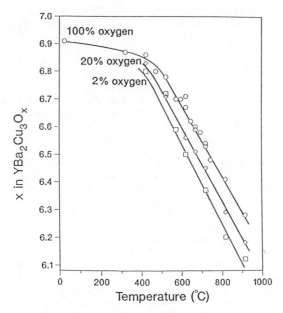

Figure 3. For YBa$_2$Cu$_3$O$_x$ variation of oxygen content with
temperature and three oxygen partial pressures (100 %, 20%, and 2%)
obtained from neutron diffraction study, by Jorgensen et al. (Ref.
12).

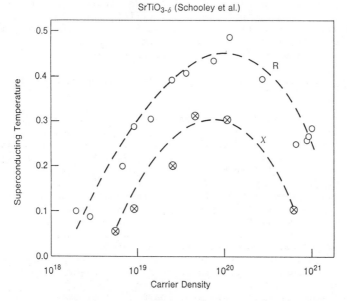

Figure 4. For SrTiO$_{3-\delta}$ variation of superconducting temperature
from half-height of resistance R and magnetic susceptibility χ vs.
carrier density determined from Hall coefficient. Data from Ref.
15.

starting from the basic fact that, phenomenologically at least, the conductivity depends on the thermal variation of the density of carriers, $N(T)$, and of the scattering in terms of the mean-free-path, $\lambda(T)$, or relaxation time between collisions, $\tau(T)$:

$$\sigma(T) = \frac{e^2 N(T) \, \lambda(T)}{m \, v_f} = \frac{e^2 N(T) \, \tau(T)}{m} \qquad (1)$$

in which e is the electronic charge, m its mass, and v_f is the velocity at the Fermi level. This expression is somewhat phenomenological in that the more correct formulation requires an integration over the Fermi surface (16). However, if the concentration of carriers is sufficiently small, Equation 1 can be adequate (16), and the derived results may be useful. For a given density of states, $g(\varepsilon)$, and the required Fermi-Dirac distribution, $f(\varepsilon)$, the density of carriers is

$$N(T) = \int_0^\infty g(\varepsilon) f(\varepsilon) d\varepsilon \qquad (2)$$

Without the loss of generality, the integral can be split into three parts convenient for the model desired. If E_f is the Fermi energy and $(E_g - E_f)$ is the energy of the semiconducting gap, then one writes:

$$N(T) = \int_0^{E_f} g_1(\varepsilon) f(\varepsilon) d\varepsilon + \int_{E_f}^{E_g} g_2(\varepsilon) f(\varepsilon) d\varepsilon + \int_{E_g}^\infty g_3(\varepsilon) f(\varepsilon) d\varepsilon \qquad (3)$$

This separation has been useful in defining metals and semi-conductors in terms of the first and third integrals respectively and exclusively. In the first case, $N(0) > 0$ and is constant, because F_f is sufficiently large compared with kT; in the second case, $N(0) = 0$ and $N(T)$ never reaches saturation. Formally at least, other kinds of materials can be defined such as semimetals in which $N(0) \neq 0$, but $N(T)$ varies slowly. And microcomposite materials in which all or only two of the integrals are significant can be added as a new category. Such can be decided of course by distinct regions for $g_1(\varepsilon)$, $g_2(\varepsilon)$, and $g_3(\varepsilon)$. And the existence or the presence of any combination can be dictated by the measurements.

Examination of the structure of the organic charge transfer conductors and the oxide superconductors reveals that, in general, they are anisotropic In each case the structures consist of two parts. In the organic radical cation salts, the parts are:

1. The cation stacks containing essentially a conjugated system in which there is a species (usually sulfur or selenium) which has two valence states separated by small energies in the solid.

2. Donor anions, usually, which serve as a source of charge which can be transferred into the conjugated system to adjust its conductivity.

In the oxides, the parts are:

1. A network of aliovalent cations and oxygen ions, actually covalently bonded, in which the dual valence states in both species (copper and oxygen) are sufficiently close energetically to establish conductivity.

2. The remainder of the structure within which substitution at certain sites can have indirect effects, such as polarization, in the conductive network.

The existence of these two parts suggest that these materials are neither metallic or semiconductive, exclusively, but rather are microcomposites of the two. Thus one defines a material which is not one of those usually classified in solid state physics.

The fact of anisotropy and the parts identified above require that one recognize dimentionality in the conductive network. In an elementary way such can be done by finding the forms of $g(\varepsilon)$ from the number of modes of an elastic system in the usual way except for one, two, or three dimensions. The result is that $g(\varepsilon)$ is proportional to $\varepsilon^{(d-2)/2}$ in which $d = 1$, 2, or 3 for the three cases. Substituting these expressions for $g(\varepsilon)$ into Equation 3 and assuming that a semiconductive gap exists, one finds (9) with $\beta = (kT)^{-1}$ that

$$N(\beta; E_f, E_g) = C_d \{A_d E_f^{d/2} f_d + \beta^{-d/2} e^{-\beta(E_g - E_f)} g_d\} \qquad (4)$$

In this expression C_d and A_d are constants which depend on dimensionality, and f_d and g_d are series expansions which are nearly unity but vary extremely slowly with temperature (9). For the two dimensional case, the integrations can be made without the expansions to yield:

$$N(\beta; E_f, E_g) = C\beta^{-1} \{\ln(e^{\beta E_f} + 1) + \ln(\frac{e^{-\beta(E_g - E_f)} + 1}{2})\} \qquad (5)$$

for a gap in which the middle integral in Equation 3 is zero, and

$$N(\beta; E_f, E_g) = C_{\beta-1} \ln(e^{\beta E_f} + 1) \qquad (6)$$

with no gap.

Several investigations of the dependence of the mean-free-path or relaxation time on temperature have found a T^{-n} dependence with n having a value of 1 or 2. For phonon scattering a value of $n = 1$ has been found (17); for electron scattering, $n = 2$ (18, 19). Hale and Ratner (20) found $n = 1$ for phonon scattering and $n = 2$ for librational (2 phonon) scattering. Thus one writes that,

$$\sigma(T) = KN(T) \, T^{-n} \text{ (with } 1 \leqslant n \leqslant 2). \qquad (7)$$

The structure of this equation and its curves with $N(T)$ given either by Equation 4 or 5 do not depend generally on dimensionality, d; the values for the parameters for the three dimensionalities are of course different, but the evolution of the curves from the semiconductive state to the semimetallic state is the same for all three dimensions. Hence, Equation 5 can be used in Equation 7 to illustrate the predictions of the model. Such is shown in Figure 5 wherein $\sigma(T) / \sigma(200)$ is plotted for $E_g = 10^{-2}$eV, $n = 1$, and E_f is varied from 0 to 2×10^{-3} eV. In this sequence, the curves vary from that for a semiconductor at a to that for a semimetal (a degenerate semiconductor) at d. If curve a were extended to higher temperatures, then it would display a broad maximum, or minimum in the resistivity. The maximum is attributable to the competition

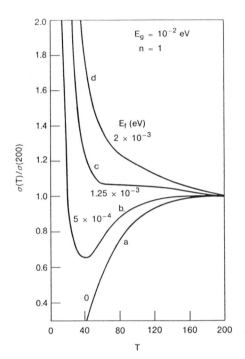

Figure 5. Variation of ratio of conductivity at T and at 200 K vs. temperature for various values of Fermi energy, 0 to 2 x 10^{-3}eV and fixed values of E_g = 1.5 x 10^{-2}eV and of the scattering exponent, n = 1 calculated from Equation 5 substituted into Equation 7. Temperature range is slightly above that for $La_{1.85}Sr_{0.15}CuO_4$.

between N(T) and τ(T). At low temperatures, the increase is dominated by N(T); at high temperature τ(T) dominates. Thus the equation describes what is usually identified as a metal—insulator transition and therein serves as an explanation of it. As E_f is increased from zero, curves like _b_ with E_f = 5 x 10^{-4} eV in Figure 5 results.

Curves a and b, or more correctly their reciprocals, resemble
the uppermost curves in Figures 1 and 2; the variation from a to b
or from top downwards in Figures 1 and 2 are caused in both cases by
an increase in the density of carriers. As E_f is increased further
the density of carriers increases and the sequence passes through
curves like c before the semimetallic curve d is attained. At c the
peaked minimum has disappeared, and consequently it corresponds to
the lowest curves in Figures 1 and 2 of the "optimum" superconducting state.

Although both the semiconductive and semimetallic components
(the second and first terms in the curly brackets of Equations 4 and
5) contribute to values of $\sigma(T)/\sigma(200)$ or the curves in Figure 5,
exclusive of curve a where $E_f = 0$, one can still define a boundary
between the two extremes as the place where the mimimum and the
broad maximum merge, and these disappear at curve c (9). This
situation defines a set of values for n, E_f, and E_g, which can
correspond to those for the superconducting state (9). Hence, the
model enables one to identify the meaning of the statement by Chu et
al. (21) that in the $BaPb_{1-x}Bi_xO_3$ system "(1) superconductivity
occurs only over a limited range of x, and the superconducting phase
[state] usually is sandwiched between a conducting and a semiconducting phase [state], (2) T_c increases rapidly as x approaches
the semiconducting phase boundary and disappears abruptly beyond the
phase boundary---."

It must be emphasized, of course, that the model presented
above is intended to describe only part of the evolution from the
semiconductive state to the superconductive state and beyond into
the semimetallic state. It is intended to describe only the normal
state at the onset of the evolution of the superconductive state
dictated by the density of carriers. To complete the description it
is necessary to recognize the generation of the paired electron
state and the resulting strong diamagnetic susceptibility which
approaches $-\frac{1}{4\pi}$. A thermochemical model describing the equilibria
among single electrons, paired electrons, and plasmons has been
derived (22). The derived expressions show that the concentration
of pairs varies sigmoidally with temperature, as does the magnetic
susceptibility generally, and is proportional to the density of
carriers. The superconducting temperature, defined as the half-
height of the magnetic susceptibility vs. temperature, is given by
the expression

$$T_o = \frac{\Delta H}{\Delta S - R\ln[(N-ne_n)/_N o]} = \frac{\Delta_o^2/V}{\Delta S - R\ln[(N-ne_n)/_N o]} \tag{8}$$

in which ΔH and ΔS are the dissociation enthalpy and entropy of the
electron pair, N is the total concentration of carriers of which ne_n
are plasmons, N^o is the usual standard state factor, Δ_o is the BCS
gap at 0 K, and V is the averaged interaction potential between
electrons to form the pair. This relation predicts that T_o vs.
concentration of carriers has a maximum in agreement with T_c vs.
carrier density (Figure 4). As the temperature above T_o is
decreased to near T_o, the system changes from one in which the

electrons are nearly all in the single state to one in which pairs form, and then the conductivity is through the pairs with apparently an extremely long mean-free-path. The two models, that for $\sigma(T)$ and that for $\chi(T)$ have not yet been adequately connected to identify precisely where the onset of the pair formation occurs on the $\sigma(T)$ plots shown in Figure 5, but it is obvious that they are connected by the carrier density and it is likely that the pair formation occurs near the abrupt increases shown in Figure 5. If so, then the abrupt increase for the optimum superconductive case, curve c in Figure 5, is made more abrupt and that for a curve like b is modified slightly because at the lower density of carriers only a relative small concentration of pairs is formed. (See Tarascon et al. (10) for the effect of oxygen on the magnetic susceptibility in $La_{1.85}Sr_{0.15}CuO_4$).

Acknowledgments

I express my appreciation for the opportunities extended to me by W. T. Carnall and J. M. Williams in this study and I thank Mrs. V. Bowman and Mrs. C. Cervenka for typing and processing the manuscript.

This work was performed under the auspices of the Office of Basic Energy Sciences, Division of Material Science, U.S. DOE, under contract number W-31-109 ENG 38.

Literature Cited

1. Schooley, J. F.; Hosler, W. R.; and Cohen, M. L. Phys. Rev. Lett. 1064 17 474.
2. Suzuki, M.; Murakami, T.; and Inamura, T. Jap. J. Appl. Phys. 1980 19 L72 .
3. Wu, M. K.; Meng, R. L.; Huang, S. Z.; and Chu, C. W. Phys. Rev. B 1981 24 4075.
4. Bednorz, J. G. and Müller, K. A. Z. Phys. B 1986 64 189.
5. Bednorz, J. G.; Takashige, M.; and Müller, K. A. "Susceptibility Measurements Support High T_c Superconductivity in Ba-La-Cu-O System", IBM Research Division, Zurich Research Laboratory, preprint.
6. Wu, M. K.; Ashburn, J. R.; Torng, C. J.; Hor, P. N.; Meng. R. L.; Gao, L.; Huang, Z. J.; Wang, Y. G.; and Chu, C. W. Phys. Rev. Lett. 1987 58 908.
7. Cava, R. J.; Batlogg, B.; van Dover, R. B.; Murphy, D. W.; Sunshine, S.; Siegrist, T.; Remeika, J. P.; Rietman, E. A.; Zahurak, S.; and Espinosa, G. P. "Bulk Superconductivity at 91 K in Single Phase Oxygen Deficient Perovskite $Ba_2YCu_3O_{9-\delta}$". AT&T Bell Laboratories, Murray Hill, NJ 07974, preprint.
8. Jorgensen, J. D.; Hinks, D. G.; Capone II, D. W.; Zhang, K.; Schuttler, H. B.; and Brodsky, M. B. "Lattice Stability and High-T_c Superconductivity in $La_{2-x}Ba_xCuO_4$", Argonne National Laboratory, Argonne, IL 60439, preprint.
9. Thorn, R. J. J. Phys. Chem. Solids 1987 48 355.
10. Tarascon, J. M.; McKinnon, W. R.; Green, L. H.; Hull, G. W.; Bagley, B. G.; Vogel, E. M. Mat. Res. Soc.

11. Kini, A. M.; Geiser, U.; Kao, H-C. I.; Carlson, K. D.; Wang, H.
     H.; Monaghan, M. R.; and Williams, J. M. Inorg. Chem. 1987 26
     1834.
12. Jorgensen, J. D.; Beno, M. A.; Hinks, D. G.; Soderholm, L.;
     Volin, K. J.; Hitterman, R. L.; Grace, J. D.; Schuller, I. K.;
     Segre, C. V. Zhang, K.; and Kleefisch, M. S. preprint submitted
     to Phys. Rev. B.
13. Ishiguro, T.; Murata, K.; Kajimura, K.; Kinoshita, N.;
     Tokumoto, N.; Tokumoto, M.; Uhachi, T.; Anzai, H.; and Sato, G.
     J. de Physique 1983 44 C3-831.
14. Parkin, S. S. P.; Jerome, D.; and Bechgaard, K. Molec. Cryst.
     Liq. Cryst. 1982 79 213.
15. Schooley, J. F.; Hosler, W. R.; Ambler, E.; Becker, J. H.;
     Cohen, M. L.; and Koonce, C. S. Phys. Rev. Lett. 1965 14 305.
16. Ziman, J. M. Principles of Theory of Solids, University Press,
     Cambridge, U.K. (1972).
17. Sham, L. J. and Ziman, J. M. Solid State Physics (Edited by F.
     Seitz and D. Turnbull) Vol. 15 p. 289 Academic Press, New York
     (1963).
18. Abraham, E. Phys. Rev. 1954 95 839.
19. Pines, D. Solid State Physics (Edited by F. Seitz and D.
     Turnbull) Academic Press, New York (1955).
20. Hale, P. D. and Ratner, M. A., J. Chem. Phys. 1985 83 5277.
21. Chu, C. W.; Huang, S.; and Sleight, A. W. Solid State Commun.
     1976 18 977.
22. Thorn, R. J. "Thermochemistry of the Electron-Paired,
     Diamagnetic State of Superconductivity," Argonne National
     Laboratory, Argonne, IL 60439, preprint.

Note Added in Preparation

After this manuscript was completed, another preprint arrived. In
this study, Mohanran et al.* found for $YBa_2Cu_3O_7$ and for
$Y_{1.05}Ba_{1.95}Cu_3O_7$ R(T) curves and type 1 (optimum superconductivity)
and type 2 (presence of the semiconductor component) respectively
and additionally for $Y_{0.95}Ba_2Cu_3O_7$ they found a variation in which
R(T) decreases monotonically with decreasing temperature towards
zero with no abrupt decrease. This behavior, which Mohanran et al.
found to be curious and unique, is part of the model described
herein. The variation in R(T) corresponds to that shown by curve d
in Figure 5, i.e., the curve dominated by the semimetallic
component. Hence, for the Y-Ba-Cu-O system, curves which span the
range from that for a semiconductor (curve a) to that dominated by
the semimetal (curve d) have been observed. The observation of the
existence of the two end members in which pair formation is
negligible demonstrates the applicability of the model, and thereby
makes the existence of the intermediate behavior (curves b and c)
plausible, even though the behavior is not observable because it is
accompanied by a more abrupt increase in $\sigma(T)$ through the
conductivity attributable to the pairs.

*Mohanran, R. A.; Seedhar, K.; Raychaudhuri, A. K.; Ganguly, P.; and
Rao, C. N. R. "High T_c Superconductivity in Perovskite Oxides of the
Y-Ba-Cu-O System," submitted to Phil. Mag.

RECEIVED July 6, 1987

MATERIALS PREPARATION AND CHARACTERIZATION

Chapter 4

Chemical Aspects of High-Temperature Superconducting Oxides

Kazuo Fueki[1], Koichi Kitazawa[1], Kohji Kishio[1], Tetsuya Hasegawa[1], Shin-ichi Uchida[2], Hidenori Takagi[2], and Shoji Tanaka[2]

[1]Department of Industrial Chemistry, University of Tokyo, Hongo, Bunkyo-ku, Tokyo 113, Japan
[2]Department of Applied Physics, University of Tokyo, Hongo, Bunkyo-ku, Tokyo 113, Japan

The composition, structure and superconducting properties of high T_c oxide superconductors, La–Sr–Cu–O and
Y–Ba–Cu–O systems have been investigated. From com-
parative studies of effects of partial replacements of
metal sites by other elements, a low dimensional
nature of conduction path was suggested in both oxide
systems. Critical relevance of oxygen deficiency and
the conditions of heat treatments to the evolution of
superconductivity, is discussed in terms of the relatively large nonstoichiometry found in these systems.

Discovery of high T_c oxide superconductors was initiated by a pio-
neering work of Bednorz and Muller (1–2) on a Ba–La–Cu–O system. The
succeeding reconfirmation by Uchida et al. (3) and the identification of the superconducting phase as $(La_{1-x}Ba_x)_2CuO_4$ of the K_2NiF_4
structure by Takagi et al. (4) at the end of 1986 firmly established
the record-breaking T_c of 30 K for the first time in 13 years.
Intensive efforts were then initiated by many researchers to substitute the component ions with similar elements on both of the La and
Ba sites. The first success was reported by Kishio et al. (5) with a
$(La_{1-x}Sr_x)_2CuO_4$ system which exhibited T_c higher than the former by
about 7 K. Other groups (6–7) independently reached the same material shortly afterwards.

Then the substitution on the La sites led Wu et al. (8) to find
another superconductor, Y–Ba–Cu–O system, and finally to overcome a
long-waited technological barrier of liquid nitrogen temperature in
February 1987. Independent discoveries of the same system were
also reported (9–10) in a short period. The identification of the
superconducting phase as $Ba_2YCu_3O_7$, a new crystal structure, was
performed by Siegrist et al.(11) and further refined through neutron
diffraction studies (12–13).

The present paper is mainly concerned with chemical aspects of
these oxide superconductors. By chemically modifying the composition, changes in structural parameters and superconducting properties of $(La_{1-x}Sr_x)_2CuO_4$ and $Ba_2YCu_3O_7$ will be investigated in de-

0097–6156/87/0351–0038$06.00/0

tail. This is accomplished by partial replacements of cation sites by other elements, or by changing the oxygen stoichiometry through various heat treatment conditions. The effect of residual water on superconductivity is also presented. These results will serve to clarify the effects of impurities and to show the necessity of precise control of oxygen composition in optimizing the super-conducting properties of these oxides. From a physical point of view, a low dimensional nature of conduction paths inferred from the present experimental results and the compatibility of supercon-ductivity with magnetic impurities are discussed. These information should be quite helpful in understanding the superconductivity of oxide superconductors realized at these extraordinarily high tem-peratures.

Effects of partial replacements of component cations

It is now well known that all systems of $(La_{1-x}M_x)_2CuO_4$ where M=Ba, Sr or Ca at x = about 0.1 are high-T_c superconductors with the same crystal structure, the tetragonal K_2NiF_4-type. After the dis-covery of these three systems, it became of immediate interest whether these compounds were soluble to each other and how the superconductivity would be affected by mixing of alkaline earth elements. A quasi-ternary composition of $(La_{1-x}(Ba,Sr,Ca)_x)_2CuO_4$ at x=0.1 was thus prepared and it was shown that the entire composition was a homogeneous solid solution and actually superconducting (14). Moreover, T_c was found to vary smoothly with composition. The maximum T_c was found to be 37.0 K and a minimum, 18.2 K, both of which were found at corner compositions of Sr and Ca, respectively. In order to find out factors which determine the superconducting properties, the tetragonal lattice parameters, a_0 and c_0, of various compositions are plotted, in Fig. 1, as a function of onset T_c determined from a.c. susceptibility measurements. It is immediately noticed that a_0 is rather well correlated with T_c, while c_0 is not. Close check of composition reveale that c_0 seems to reflect simply the difference in the ionic sizes of alkaline earth elements. On the other hand, the strong correlation of a_0 and T_c was also supported by high pressure studies of the Sr system (15-16).

Replacement of Ba by Sr in the case of the La-Ba-Cu-O system was thus favorable in decreasing a_0 and increasing T_c. However the same is not true in the case of the Y-Ba-Cu-O system. Figure 2 shows the orthorhombic lattice parameters of a_0, b_0 and c_0 for $(Ba_{1-x}Sr_x)_2YCu_3O_7$ for x=0 to 0.7. It is clearly seen that these parame-ters all decrease with Sr composition, x. Figure 3 shows the resis-tivity vs. temperature of these specimens. Quite unexpectedly from the discussions up to this point, T_c in this particular system decreases with increase in x. Although the difference in these two examples is not clear yet, it should be quite essential in under-standing the mechanism of the superconductivity in the both systems.

Partial replacements of the rare earth sites were also exa-mined. Figure 4 shows the effect of 5% substitution of 4f ions on La sites for the La-Sr-Cu-O system (17). The lattice parameters and T_c are plotted in the order of atomic number. Since most of these ions possess unclosed 4f orbitals, they are considered to be magne-tic impurities. It is interesting to see that T_c first decreases, passes through a minimum for Gd and increases again. This suggests

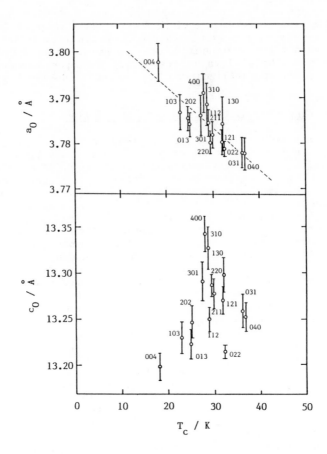

Fig. 1 : Tetragonal lattice parameters as a function of onset
         T_c (ac susceptibility) for solid solution system
         $(La_{0.9}(Ba,Sr,Ca)_{0.1})_2CuO_4$. Indices attached to each
         data point express the ratio of alkaline earth ion
         compositions in the order of Ba, Sr and Ca; e.g.
         [013] represents $(La_{0.9}Sr_{0.025}Ca_{0.075})_2CuO_4$.
         Reproduced with permission from Ref. 14. Copyright
         1987, The Chemical Society of Japan.

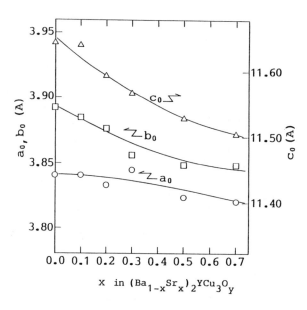

Fig. 2 : Lattice parameters of $(Ba_{1-x}Sr_x)_2YCu_3O_7$ as a
function of x.

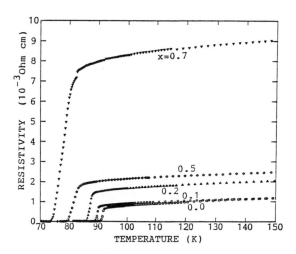

Fig. 3 : Resistivity vs. temperature of $(Ba_{1-x}Sr_x)_2YCu_3O_7$.

interactions of conduction electrons with the 4f shell spins on the La site, although the effect should be considered to be quite small in terms of the destruction of the superconductivity.

In the case of the Y–Ba–Cu–O system, 4f magnetic ions may not be considered to be impurities as far as the superconductivity is concerned. As many investigators have discovered(18–24), most of the rare earth elements give 90 K superconductors without any appreciable difference in T_c. Kishio et al. (25) have shown that 90 K conductors can be also successfully prepared even from certain compositions of unseparated mixture of rare earths as starting materials.

Although the replacements of La or Y sites by 4f magnetic ions have been shown to be rather insensitive to superconducting properties of both La–Sr–Cu–O and Y–Ba–Cu–O systems, similar replacements of Cu sites are found to be more significant. Hasegawa et al. (26) have shown that incorporation of 3d elements, or Zn on the Cu site of La–Sr–Cu–O at the level of 5 atomic percent was enough to nearly destroy the superconductivity. Figure 5 shows the effect of replacement of Cu with Fe in the Y–Ba–Cu–O system. In this case, it is seen that 3% doping reduces T_c to almost a half value and 10% doping loses zero-resistivity state.

All of the effects of replacements of each metal site so far described can be understood within a framework of a low dimensional nature of conduction mechanisms. More specifically, superconduction takes place through the O–Cu–O bonds of the interconnected CuO_6 octahedra or CuO_4 squares in the La–Sr–Cu–O and Y–Ba–Cu–O systems, respectively. The La or Y sites are less sensitive for the presence of magnetic impurities and seem to be rather away from the superconduction path. This tendency is more significant in the Y–Ba–Cu–O system, because the conduction path would be along the one dimensional chains of O–Cu–O bonds which are most apart from the Y sites.

Oxygen nonstoichiometry and significance of heat treatment

It has been well recognized that the superconducting properties of both La–Sr–Cu–O and Y–Ba–Cu–O systems are quite sensitive to the oxygen composition which is easily altered by various heat treatments. Examples are shown in Figs. 6 and 7. Fig. 6 shows the effect of post-annealing on the resistivity of $(La_{0.92}Sr_{0.08})_2CuO_4$. As is seen , the specimen prepared at 1100°C and quickly cooled shows a clear onset at about 38 K but it exhibits a significantly broad transition and a tailing structure supressing the zero resistivity to as low as 24 K. This tailing structure is presumed to be due to formation of oxygen depleted region in the vicinity of grain boundaries of the polycrystalline structure formed during the cooling process. When the specimens are annealed in oxygen at lower temperatures, the transition becomes appreciably sharper and the resistivity decreases. The best result has been obtained with annealing at 600°C for 50 h, while lower temperatures seem to be too low in sufficient incorporation of oxygen within a limited time of heat treatment. Figure 7 presents the similar effect of heat treatment in the Y–Ba–Cu–O system. Specimens of $Ba_2YCu_3O_7$ were quickly cooled from various temperatures after annealing in pure O_2. It is seen in Fig. 7 that the effect of annealing is much more pronounced in the Y–Ba–Cu–O system.

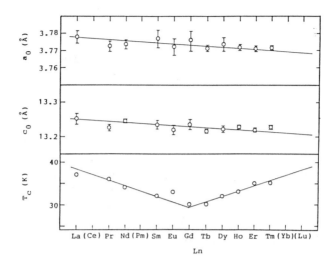

Fig. 4 : Variations of tetragonal lattice parameters a_0 and c_0 and T_C for the solid solution systems of $(La_{0.85}Ln_{0.05}Sr_{0.1})_2CuO_4$ with different lanthanide ions. Reproduced with permission from Ref. 17. Copyright 1987, Publication Board, Japanese Journal of Applied Physics.

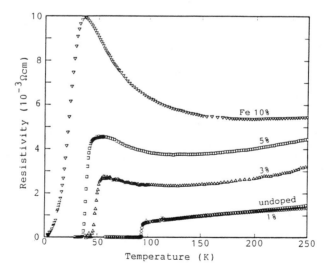

Fig. 5 : Resistivity vs. temperature of $Ba_2YCu_{3(1-x)}Fe_{3x}O_7$.

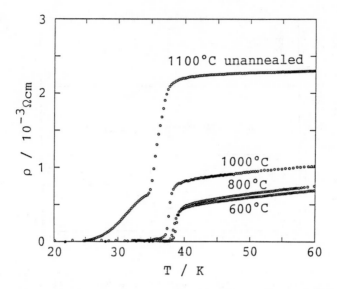

Fig. 6 : Effect of O_2 annealing in $(La_{0.92}Sr_{0.08})_2CuO_4$ at various temperatures.

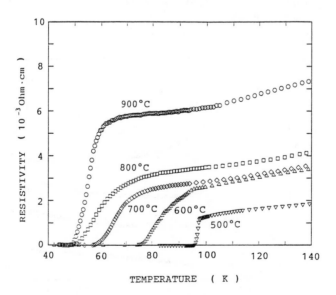

Fig. 7 : Effect of O_2 annealing in $Ba_2YCu_3O_7$ at various temperatures.

Since these remarkable changes in superconducting behaviors of high-T_c oxides are associated with the oxygen composition of the specimens, it is essential to determine equilibrium values of oxygen composition under various conditions. In this paper we describe the result of oxygen nonstoichiometry measurements of the $Ba_2YCu_3O_{7-\delta}$. The measurement was made by a thermogravimetric technique with the aid of chemical analysis using iodometric titration to determine the absolute values of oxygen compositions. As details are found elsewhere (27), only the observed result is presented in Fig. 8. The nonstoichiometry is found to vary continuously as a function of temperature and oxygen partial pressure. The maximum deviation from the tentatively assigned stoichiometric composition 7 is about -0.9 amounting to 13% of the total oxygen content, suggesting the nonstoichiometry is exceptionally large in Y-Ba-Cu-O system. The result suggests that the annealing in an appropriate atmosphere and at relatively low temperatures below 500°C is important to minimize the oxygen deficiency.

However, prolonged annealing in the oxygen atmosphere may not be necessarily optimum, according to our preliminary results. Appearance of highly resistive grain boundary layer has been suggested to form, if the specimen is subjected to prolonged annealing at low temperatures. This indicates the necessity of carefully designed annealing program for the practical fabrication process of the oxide superconducting materials.

Effect of water in superconductivity in La-Sr-Cu-O system

In addition to metallic ions, some other impurities may affect the superconductivity of the high-T_c oxide superconductors. Water was found to be one of such species in the early stage of our investigation. Figure 9 shows an effect of residual water(28) on the volume fraction of superconductivity in $(La_{1-x}Sr_x)_2CuO_4$ at x=0.08. The samples were prepared in presence of water intentionally added during the mixing procedure of starting materials. The samples had been fired at 1100°C for 22 h. As the expected amounts of residual water in the samples A to E' decrease, the volume fractions are seen to increase in Fig. 9. The resistivity measurements of these samples exhibited almost same T_c but the temperature dependence of the resistivity (28) was well correlated again with the amount of residual water. Although X-ray diffraction analyses could not detect formation of any extra phase in these samples, the effect of residual water in deteriorating the superconductivity is quite clear in this oxide.

Conclusion

Some chemical aspects such as the effects of comosition and impurities on the manifestation of superconductivity in high-T_c oxides have been described. They are shown to be quite significant in determining superconducting behavior and probably many other physical properties of these materials as well. As the developments of oxide superconductors with higher T_c and better quality continue, it should be essential to take these aspects fully into account.

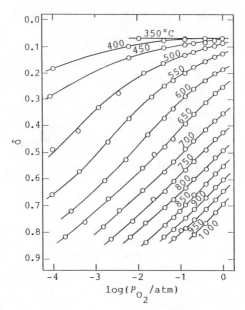

Fig. 8 : Nonstoichiometry of $Ba_2YCu_3O_{7-\delta}$ as a function of
oxygen partial pressure. Reproduced with permission
from Ref. 27. Copyright 1987, Publication Board,
Japanese Journal of Applied Physics.

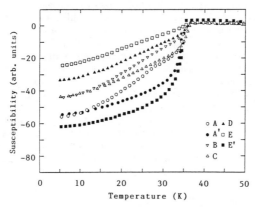

Fig. 9 : Ac susceptibility signals as a function of
temperature for $(La_{0.92}Sr_{0.08})_2CuO_{4-\delta}$ prepared under
various conditions. ([A]: no medium for mixing,
[A']: same as A plus vacuum anneal, [B]: ethanol,
[C]: ethanol+0.5ml H_2O, [D] ethanol+1.5ml H_2O, [E]:
water, [E']: same as E plus vacuum anneal, see text
for details). Signal intensities are normalized to
the sample weights. Reproduced with permission from
Ref. 28. Copyright 1987, Publication Board, Japanese
Journal of Applied Physics.

Literature Cited

1. Bednorz, J. G.; Muller, K. A. Z. Phys. 1986, B64, 189.
2. Bednorz, J. G.; Takashige, M.; Muller, K. A. Europhys. Lett. 1987, 3, 379.
3. Uchida, S.; Takagi, H.; Kitazawa, K.; Tanaka, S. Jpn. J. Appl. Phys. 1987, 26, L1.
4. Takagi, H.; Uchida, S.; Kitazawa, K.; Tanaka, S. Jpn. J. Appl. Phys. 1987, 26, L123.
5. Kishio, K.; Kitazawa, K.; Kanbe, S.; Yasuda, I.; Sugii, N.; Takagi, H.; Uchida, S.; Fueki K. : Tanaka, S. Chem. Lett. 1987, 429.
6. Cava, R. J.; van Dover, R. B.; Batlogg, B.; Rietman, A. Phys. Rev. Lett. 1987, 58, 408.
7. Tarascon, J. M.; Greene, L. H.; Mckinnon, W. R.; Hull G. W.; Geballe T. H. Science 1987, 235, 1373.
8. Wu, M. K.; Ashburn, J. R.; Torng, C. J.; Hor, P. H.; Gao, R. L. ; Huang, Z. J.; Wang Y. Q.; Chu C. W. Phys. Rev. Lett. 1987, 58, 908.
9. Hikami, S.; Hirai, T.; Kagoshima, S. Jpn. J. Appl. Phys. 1987, 26, L314.
10. Zhao, Z.; Chen, L.; Yang, Q.; Huang, Y.; Chen, G.; Tang, R.; Liu, G.; Cui, C.; Chen, L.; Wang, L.; Guo, S.; Li, S.; Bi, J. to be published in Kexue Tongbao 1987, No.6.
11. Siegrist, T.; Sunshine, S.; Murphy, D. W.; Cava, R. J.; Zahurak, S. M. Phys. Rev. 1987, B35, 7137.
12. Beno, M. A.; Soderholm, L.; Capone, D. W.; Hinks, D. J.; Jorgensen, J. D.; Schuller, I. K. to be published in Appl. Phys. Lett.
13. Izumi, F.; Asano, H.; Ishigaki, T.; Takayama-Muromachi E.; Uchida, Y.; Watanabe, N.; Nishikawa T. Jpn. J. Appl. Phys. 1987, 26, L649.
14. Kishio, K.; Kitazawa, K.; Sugii, N.; Kanbe, S.; Fueki, K.; Takagi H.; Tanaka, S. Chem. Lett. 1987, 635.
15. Takahashi, H.; Murayama, C.; Yomo, S.; Mori, N.; Kishio, K.; Kitazawa, K.; Fueki, K. Jpn. J. Appl. Phys. 1987, 26, L504.
16. Yomo, S.; Murayama, C.; Takahashi, H.; Mori, N.; Kishio, K.; Kitazawa, K.; Fueki, K. Jpn. J. Appl. Phys. 1987, 26, L603.
17. Kishio, K.; Kitazawa, K.; Hasegawa, T.; Aoki, M.; Fueki, K.; Uchida, S.; Tanaka, S. Jpn. J. Appl. Phys. 1987, 26, L391.
18. Moodenbaugh, A. R.; Suenaga, M.; Asano, T.; Shelton, R. N.; Ku, H. C.; McCallum, R. W.; Klavina, P. Phys. Rev. Lett. 1987, 58, 1185.
19. Hosoya, S.; Shamoto, S.; Onoda, M.; Sato, M. Jpn. J. Appl. Phys. 1987, 26, L325.
20. Kitazawa, K.; Kishio, K.; Takagi, H.; Hasegawa, T.; Kanbe, S.; Uchida, S.; Tanaka, S.; Fueki, K. Jpn. J. Appl. Phys. 1987, 26, L339.
21. Hikami, S.; Kagoshima, S.; Komiyama, H.; Hirai, T.; Minami, H.; Masumi, T. Jpn. J. Appl. Phys. 1987, 26 L347.
22. Fisk, Z.; Thompson, J. D.; Zirngiebl, E.; Smith, J. L.; Cheong S-W. Solid State Commun., in press.
23. Takagi, H.; Uchida, S.; Sato, H.; Ishii, H.; Kishio, K.; Kitazawa, K.; Fueki, K.; Tanaka, S. Jpn. J. Appl. Phys. 1987, 26, L601.

24. Kanbe, S.; Hasegawa, T.; Aoki, M.; Nakamura, O.; Koinuma, H.;
     Kishio, K. ; Kitazawa, K.; Takagi, H.; Uchida, S.; Tanaka, S.;
     Fueki, K. Jpn. J. Appl. Phys. 1987, 26, L613.
25. Kishio, K.; Nakamura, O.; Kuwahara, K.; Kitazawa, K.; Fueki, K.
     Jpn. J. Appl. Phys. 1987, 26, L694.
26. Hasegawa, T.; Kishio, K.; Aoki, M.; Ooba, O.; Kitazawa, K.;
     Fueki, K.; Uchida, S.; Tanaka, S. Jpn. J. Appl. Phys. 1987, 26,
     L337.
27. Kishio, K.; Shimoyama, J.; Hasegawa, T.; Kitazawa, K.; Fueki,
     K. To be published in Jpn. J. Appl. Phys. 1987, 26, No.7.
28. Kishio, K.; Sugii, N.; Kitazawa, K.; Fueki, K. Jpn. J. Appl.
     Phys. 1987, 26, L466.

RECEIVED July 8, 1987

Chapter 5

Synthesis of New Classes of High-Temperature Superconducting Materials

Francis J. DiSalvo

Department of Chemistry, Baker Laboratory, Cornell University, Ithaca, NY 14853

At present (May 1987), there are several known oxide compounds that are superconducting at temperatures above 30K, with some approaching 100K, and numerous reports of metastable drops in resistance in some materials at temperatures as high as room temperature. There are three types of copper oxides in this class with different but related crystal structures: $La_{2-x}M_xCuO_4$ with M = Ca, Sr, or Ba [1,2]; $MBa_2Cu_3O_7$ with M = trivalent metals [3-6]; and $La_{3-x}Ba_{3+x}Cu_6O_{14}$ [7]. The structures are all derivatives of perovskite, which has the composition $MM'O_3$, and are variously described as having square planar coordinated copper-oxygen sheets or copper in highly distorted octahedral or square pentagonal oxygen coordination. Another oxide, $BaPb_{1-x}Bi_xO3$, should probably also be included in this class, even though its T_c is only 13K, since on the basis of its density of states, its T_c should only be a few degrees [8]. Perhaps coincidently, it also has the perovskite structure.

Since the end of 1986, there has been an enormous amount of synthetic work on related oxides obtained by substitution of the cations or anions or by an Edisonian variation of the composition. These approaches are an important process in the search for new high T_c materials. If general characteristics of the high T_c phases can be used as a guide to future synthesis strategies, it is hoped that a more rapid route to the discovery of other high T_c phases can result. Such a search presupposes a "faith" that the known oxide superconductors are not unique but that other compounds with similar or enhanced properties must exist. In this paper I outline some ideas on the unusual features of the oxide superconductors and some thoughts on what other compounds might show similar characteristics.

I start with a "disclaimer": while my students and I have prepared some of the high T_c materials by following others recipes and are just now starting to attempt to prepare new materials, I have not been one of the contributors to the large body of information that exists concerning these oxides and from which my thoughts arise. Further, I have been fortunate to receive many of the preprints that active groups have been circulating to avoid

0097–6156/87/0351–0049$06.00/0

the "slowness" of the normal publication procedures. Without the
openness of others, I would have little to say. I have also
somewhat arbitrarily chosen the particular references to cite,
choosing to pick one that illustrates the point, rather than
trying to give an exhaustive list of all the preprints that I have
seen on the topic.

The first and most surprising feature of these materials is
apparent in band structure calculations. Such calculations may in
fact not be correct in detail, since these materials appear to be
close to the composition at which a Mott transition to the
insulating state takes place [9]. This suggests that Coulomb
correlations must be included in a realistic description of the
properties. However, the observation based on the band structure
calculations [10-12] that the states near the Fermi level have a
strongly mixed character (of Cu d and O p states) will also
likely survive in a "correct" description of the electronic
structure. This admixture arises from the closeness in energy of
the corresponding atomic energy levels and from the fact that the
band at the Fermi level is derived from a sigma antibonding state.
This is indeed rather unusual for a metallic compound. The band
structure results suggest that 50% or more of the wavefunction
amplitudes are based on the oxygen! Perhaps, then, it is better to
describe these materials as "metallic oxygen". In the vast
majority of other metallic compounds the wavefunctions at the
Fermi level are predominantly of cation (metal) character,
typically containing 10% or less anion character. In other
conducting oxides, particularly of early transition metals (such
as $LiTi_2O_4$), the states at the Fermi level are derived from non-
bonding or weakly pi bonding levels, further reducing the already
low oxygen character resulting from the large energy difference
between the oxygen p and titanium d states.

There are several ways to describe the electronic state of
the materials. Band structure is one way and formal valence states
is another. While the latter is oversimplified, it is also very
easy and widely used. But it can be a little misleading. For
example, the formal valence description of $YBa_2Cu_3O_7$ is
$Y^{+3}Ba^{+2}_2Cu^{+2}_2Cu^{+3}O^{-2}_7$. This implies that the oxygen p levels are
fully occupied (i.e., that the oxygen is fully divalent). Since
the oxygen and copper levels each make a considerable contribution
to the wavefunctions at the Fermi level, it is impossible to
oxidize the copper to +3 without also oxidizing the oxygen. In
fact some recent experiments can detect only Cu^{+2}, suggesting that
only the oxygen is oxidized. In that case the fraction of carrier
density that is based on oxygen is even higher than the band
structure calculations would suggest. Therefore, when authors
speak of needing "mixed valence" Cu^{+2}/Cu^{+3} in the oxides to
produce superconductivity, a broader interpretation of what
species are being oxidized should be kept in mind.

The calculations also suggest that the electropositive
cations (alkaline earth or rare earth metals) have little
influence on the electronic structure near the Fermi level. I
believe that their role is two fold. First, they help "enforce" a
particular structure; that is, the large cations are responsible
for the compounds adopting the perovskite structure. Second, the
electropositive metals effectively increase the oxidizing power of

the oxygen. In binary copper oxides, the maximum formal valence that can be obtained using the usual high temperature solid state preparation techniques is +2. Whereas in ternary copper oxides containing, for example, alkali metals, copper can be formally +3, as in $NaCuO_2$.

The next unusual feature is that they are lousy normal state metals. Their resistivity at 300K is between 10^{-3} and 10^{-4} Ohm-cm, two to three orders of magnitude higher than "good" metals and at least a few factors higher than intermetallic transition metal compounds [4,9,13]. This implies rather short mean free paths for the conduction electrons, perhaps as short as a lattice parameter. Yet the resistivity is quite temperature dependent, typically varying linearly with temperature. It is likely that the unusual behavior of the resistivity is related to the nearness to the Mott transition and may be an important key in determining the detailed mechanism that leads to the high T_c. At the same time the mechanical properties are more like those of ceramics (brittle) than metals (ductile). Recall, however, that the intermetallic compounds are also often brittle, and that some techniques have been invented to allow flexible wires to be fabricated from them.

These materials are also fast ion conductors; at least oxygen is able to diffuse rather readily through the bulk of the compound at temperatures as low as 300C. This implies a bonding potential vs position for oxygen in the lattice that is rather flat, or at least barriers that are rather small in the direction of diffusion. This in turn implies that oxygen vibrations at lower temperatures will have a large amplitude, perhaps being rather anharmonic. Recent neutron diffraction measurements indeed show that the crystallographically unique oxygen atoms in the "one dimensional" chains in $YBa_2Cu_3O_7$ have large thermal amplitudes of motion [5] and are the sites that are reduced in occupancy when oxygen is removed from the lattice [14]. (Some others have also suggested that the fact that Cu^{+2} is a Jahn-Teller ion will also make its motion rather anharmonic. However, the copper is already in a rather distorted site, and the neutron diffraction measurements show only a normal vibrational amplitude.) This ready diffusion also has important consequences for the synthesis and processing of the material.

Some attention has been drawn to the low dimensional aspects of these materials. While the high T_c materials have two dimensional sheets of Cu-O, and the 90K superconductor has in addition one dimensional Cu-O chains, $Ba(Pb/Bi)O_3$ has an almost cubic structure. Consequently, low dimensionality seems not to be a necessary condition for materials in this class to be superconducting. In any case low dimensional structures will likely result in other materials from low coordination number of the transition metal and by including large electropositive cations.

Before discussing some rather straightforward generalizations of the above properties that might be used as a guide for new synthesis, I want to say a few words about the theories that have been proposed to explain the high T_c's observed. While no theory has been developed to the point that it explains in microscopic detail all the observed phenomena, it might be useful to take a brief look at them to extract essential

features. Then using these features and the above observations, we
can try to predict what new compounds might be synthesized that
reproduce some or even all of the desired aspects of the known
high T_c materials.

I will limit this discussion, both since I am not a theorist
and because other such reviews exist [15]. The theories can be
divided into three types, each characterized by the principal
interaction responsible for the high T_c: phonon, magnetic, or
exciton. The phonon mechanisms might be divided into two
categories: enhanced electron—phonon interaction of the BCS type
[10] and local distortions about carriers to produce bipolarons
[16]. Both rely on the large coupling of the electron energy to
lattice motions that result from the large mixing of the cation
and anion orbitals and the antibonding character of the
wavefunctions at the Fermi level. The magnetic models may also be
grouped into two categories: interaction with diffuse magnon modes
and interaction on a more local level to produce localized singlet
pairs (the RVB state, Resonating Valence Bond) [17,18]. These
theories emphasize the nearness of the Mott insulating state to
produce localized spin 1/2 Cu^{2+} sites. Some theories emphasize the
large oxygen contribution to the conductivity and the possibly
local nature of the electrons on copper [19]. Finally, the exciton
theories are based on a direct coupling of the electrons to an
electronic excitation. These might include plasmons [20] or charge
transfer excitations [21]. The latter mechanism also relies on the
small energy difference between the cation d and anion p states
and on a large oscillator strength of the transition. Each theory
then picks one aspect of the many unusual properties and attempts
to explain the high T_c based on that feature.

Now we can try to put this all together to try to predict
where else to look for new superconducting phases. The main themes
that arise from the preceding discussion can be condensed into
four general characteristics: 1) A large cation-anion mixing of
the wavefunctions near the Fermi level, 2) metallic conductor, but
close to a Mott transition, 3)fast anion conductor, 4) the
electropositive cations do not play an essential role in the
electronic properties. For the sake of discussion, we can
arbitrarily break these materials into oxides and other anion
compounds.

<u>Oxides</u>

The atomic d states of copper are about 1eV above the p
states of oxygen. On moving to the left in the periodic table, the
d states rise by about 3eV when at Ti. The energy of these levels
is shifted when the atom is incorporated in a compound. The anion
levels rise and the cation levels fall in energy, due to the
charge transfer and configuration mixing inherent in compound
formation. The charge transfer and energy shifts depend upon the
electronegativity difference between the anion and cation and upon
the cation-anion ratio. However, the real charge transfer is
usually considerably smaller than that suggested by the formal
valence. Unfortunately, these shifts are difficult to predict
before a compound is even prepared. We can rely on general trends
to "guess" what will happen if the cation is different than

copper. If compounds are prepared from 3d elements to the left of copper the cation-mixing will tend to decrease, unless the transition metal is in a higher average oxidation state than that obtained in the copper compounds (which is about +2.3). Perhaps if nickel compounds with an average valence of greater than +3 could be prepared, the mixing would again be strong. Using the same reasoning it is possible that other late transition elements will also be suitable. These include Ag, Pd, Co, and Pt. For each element it may require special preparation conditions to obtain the "correct" phases. That is, unusual conditions such as preparation under high pressure or at low temperatures using solution techniques may be necessary.

In oxides, condition 2 (near a Mott transition) is likely to be met, if the compound is conducting at all. Most oxides are in fact not metallic conductors, and, of those that are, many exhibit metal insulator transitions with changing temperature or pressure. It may also be necessary that in the insulating state the cation has a spin 1/2 configuration (probably necessary for the RVB mechanism). The occurrence of the state depends upon the valence of the cation and upon its local coordination, with highly distorted near-neighbor environments favoring non degenerate states and spin 1/2 configuration. Alternatively, for a given environment the cation valence necessary to produce a spin 1/2 ion can easily be determined in the low and high crystal field strength limits.

It is not clear to me how to design a material to be a fast ion (anion) conductor, nor is it clear that this is a necessary condition for high T_c superconductivity. If the large oxygen vibrations that are a consequence of the flat elastic potential for vibration produce an enhanced electron-phonon interaction, it would seem that such large vibrations could occur without ionic diffusion. Perhaps the flat elastic potential is a result of the near degeneracy in the cation and anion d and p levels. This would make charge transfer between them a low energy process, resulting in a high lattice polarizability, or equivalently easily deformable ions. Such a picture is usually invoked in discussing fast ion conductors. So it could be that most of the systems that are potentially interesting superconductors of this type will also coincidentally be superionic conductors.

While a continued search for copper oxides containing mixed Cu^{+2}/Cu^{+3} formal valence states will be extended, perhaps a single example of another compound that should reproduce all the above features will suffice for illustrative purposes. Consider an oxide containing Ni^{+3}/Ni^{+4}. The formal valence state is higher than copper, so the Ni d states should be pulled down in energy, hopefully enough to again produce strong mixing with the oxygen. The electron configuration is d^7, which in a distorted cubic or square planar environment will be spin 1/2. A tetrahedral Ni enviornment in the low crystal field limit produce a spin 3/2 state, but in the large crystal field limit a spin 1/2 state would again be produced. If the compound is metallic, the band states at the Fermi level would be anti-bonding, thus further increasing the mixing and producing large electron-phonon coupling. Such Ni oxides would therefore be almost exact analogues of the high T_c copper oxides. Experimentally, the preparation of nickel oxides

with an average formal valence greater than +3 may require high
oxygen pressures or low temperature techniques.

Other anions

Other anions that are likely candidates include nitrogen,
sulfur, and chlorine. Fluorine is unlikely to produce considerable
mixing, because the p levels are a few volts below those of
oxygen, and therefore fluorides are unlikely candidates. Further,
fluorides tend to be rather ionic and very likely to be insulators
(without a literature search, I can only think of a few conducting
fluorides, Ag_2F and graphite intercalated with fluorine or
fluorine containing anions). Mixing anions may produce small
shifts in the cation and other anion levels and may be a suitable
way to fine tune the properties. But the anion p levels are
sufficiently different in energy that it is unlikely that both
anions would mix with the cation levels to the same extent; one
would always dominate.

Nitrides and sulfides that meet the conditions necessary for
extensive mixing of the states at the Fermi level may be prepared
if the cations are chosen from elements to the left of copper,
since the anion p states are higher in energy than those of oxygen
by about 1.5eV. If it is important to be near a metal-insulator
transition, then choosing a compound based on a 3d transition
element will enhance that probability. Empirically, such
transitions, as a function of composition or temperature, occur
most frequently in that part of the periodic table. 4d and 5d
compounds tend to be more metallic than the equivalent 3d
compounds, since the radial extent of the d wavefunctions is
larger than for 3d cations. However, there is no hard and fast
rule that will exclude other possibilities; witness the metal-
insulator transition in $Ba(Pb/Bi)O_3$ as a function of composition.
Depending upon the oxidation state, transition elements from
nickel to vanadium should be considered. If the d occupancy
becomes small (say less than 4), the states at the Fermi level
will be predominantly non-bonding and will tend to have somewhat
less mixing than the unoccupied anti-bonding states at higher
energy. This will make the largest difference to mechanisms based
on phonon or exciton interactions, but may not be important to the
RVB state at all.

Chlorides may also be good candidates, if metallic phases can
be prepared. However, the vast majority of chlorides are
insulators. The p states of chlorine are close to those of oxygen,
perhaps 0.5eV higher. Most of the same considerations applied
above to the case of nitrides would apply to the chlorides as
well.

Summary

The unusual features of the new high temperature
superconductors have been outlined. It is suggested that further
synthetic studies, using some general guidelines coupled with a
few physical property measurements, of new oxides as well as
nitrides, sulfides, hydrides, and perhaps even chlorides may well
lead to more of these fascinating novel superconductors.

ACKNOWLEDGMENTS

This work was supported in part by the National Science Foundation through the Materials Science Center at Cornell University.

LITERATURE CITED

1.   J. G. Bednorz and K. A. Muller   Z. Phys. 1986, B64, 189
2.   R. J. Cava, R. B. van Dover, B. Batlogg, and E. A.
     Rietman   Phys. Rev. Letts. 1987, 58, 408
3.   M. K. Wu, J. R. Ashburn, C. J. Torng, P. H. Hor, R. L.
     Meng, L. Gao, Z. J. Huang, Y. Q. Wang, C. W. Chu   Phys. Rev.
     Letts. 1987, 58, 908
4.   R. J. Cava, B. Batlogg, R. B. van Dover, D. W. Murphy, S. A.
     Sunshine, T. Siegrist, J. P. Remeika, E. A. Reitman, S.
     Zahurak, and G. P. Espinosa   Phys. Rev. Letts. 1987, 58,
     1676
5.   M. A. Beno, L. Soderholm, D. W. Capone, D. G. Hinks, J. D.
     Jorgensen, I. K. Schuller, C. U. Segre, K. Z. Lang, J. D.
     Grace   Appl. Phys. Letts. 1987, (to be published)
6.   E. M. Engler, V. Y. Lee, A. I. Nazzal, R. B. Beyers, G. Lim,
     P. M. Grant, S. S. P. Parkin, M. L. Ramirez, J. E. Vasquez,
     and R. J. Savoy J. Amer. Chem. Soc. (to be published)
7.   D. B. Mitzi, A. F. Marshall, J. Z. Sum, D. J. Webb, M. R.
     Beasley, T. H. Geballe, and A. Kapitulnik   Phys. Rev. B (to
     be published)   .
8.   L. F. Mattheiss and D. R. Hamann Phys. Rev. 1983, B28, 4227
9.   J. M. Tarascon, L. H. Greene, W. R. McKinnon, G. W. Hull, and
     T. H. Geballe Science 1987, 235, 1373
10. L. F. Matheiss Phys. Rev. Letts. 1987, 58, 1028
11. J. Yu, A. J. Freeman, and J. H. Xu Phys. Rev. Letts. 1987,
     58, 1035
12. M. H. Whangbo, M. Evian, M. A. Beno, and J. M. Williams
     Inorg. Chem. 1987 (to be published)
13. L. F. Schneemeyer, J. V. Waszczak, S. M. Zahurak, R. B. van
     Dover, and T. Siegrist Mat. Res. Bull. 1987 (to be
     published)
14. A. Santoro, S. Miraglia, F. Beech, S. A. Sunshine, D. W.
     Murphy, L. F. Schneemeyer, and J. V. Waszczak   Mat. Res.
     Bull. 1987 (to be published)
15. T. M. Rice Z. fur Physik B 1987 (to be published)
16. P. Preovsek, T. M. Rice, and F. C. Zhang J. Physics C 1987,
     20L
17. P. W. Anderson Science 1987, 235, 1196
18. T. Oguchi, H. Nishimori, and Y. Taguchi J. Phys. Soc. Japan
     1986, 55, 323
19. V. J. Emery Phys. Rev. 1987 (to be published)
20. Z. Kresin 1987 (to be published)
21. C. M. Varma, S. Schmidt-Rink, and E. Abrahams Solid State
     Commun. 1987 (to be published)

RECEIVED July 6, 1987

Chapter 6

Chemistry of High-Temperature Superconductivity in $YBa_2Cu_3O_7$

Michael L. Norton

Department of Chemistry, University of Georgia, Athens, GA 30602

Chemical substitution has been used as a probe of the origin of the high transition temperature in $YBa_2Cu_3O_7$. The properties of this material are best understood when the material is considered to be a semiconductor doped to metallicity. When oxygen doped, the material becomes a 1 dimensional excitonic superconductor. The intrinsic semiconducting energy gap then provides the mechanism for the extremely high critical temperatures observed in this class of materials.

Using the knowledge that one of the strongest Jahn-Teller ions, Cu(+2), should have the largest electron-phonon interactions and thus should lead to high temperature superconductivity, Bednorz and Muller (1) discovered that a K_2NiF_4 structured La-Ba-Cu-O compound superconducts near 30K.

Although the phonon mechanism has not been proven to be the operative one, this exciting discovery does fit well within our current understanding of phonon mediated superconductivity, in which electronic motions and lattice deformations are strongly correlated. This mechanism is easily defensible up to transition temperatures (Tc's) of 40K (2), and perhaps to slightly higher temperatures. The observation of Tc's near 90K discovered by Wu, Chu, and co-workers demands, in the opinion of many, another explanation (3).

One possible mechanism which must be considered is the excitonic mechanism. We suggested that mixed valence, layered perovskite related structured materials would be best suited as systems to support this mechanism due to the flexibility of the structure and chemistry of such systems (4). The major requirements of such a model are the existence of a local polarizable medium (a semiconductor or an insulator), in extremely close proximity to a conducting (metallic) moiety such as a layer (5), or, in the case of the title compound, a conducting chain.

Our laboratories have conducted chemical substitution experiments to discover the crystallochemical correlation of composition and properties of superconductors of the $YBa_2Cu_3O_7$

family. Substitutions have been performed on the Y, Ba, Cu and O sites. The Ba and Cu substitutions are most relevant to the determination of the mechanism operative in this system and will be discussed in this paper. Additionally, optical characterization of the semiconducting (reduced) form of this material has been conducted.

Experimental

Synthesis. In a typical experiment, a total of 2 grams of the component oxides (Y_2O_3, ZnO, and CuO) and carbonates ($BaCO_3$ and K_2CO_3) (reagent grade or better, and used as received with no attempt at purification) were mixed and ground in an agate mortar and pestle, cold pressed for 5 minutes at 10,000 psi in a pellet press into 0.2g pellets, without the use of chemical binders. The pellets were fired using alundum substrates, in quartz envelopes, under a constant flow of oxygen. The usual firing schedule involves a 12 hour firing at 950C, followed by slow cooling to 500C, where cooling is arrested to allow oxygen uptake for 2 hours. The samples are then slow cooled to room temperature. Samples were then stored in air tight bottles (but without protective ambient atmosphere) until studied.

Structural Characterization. X-ray diffraction patterns of powder samples were made by using a Phillips generator with a copper target and nickel filter, and a scanning powder diffractometer.

Compositional Characterization. The composition of the bulk samples was not determined since the components to these refractories were not considered sufficiently volatile to distort the nominal stoichiometry. Micro-analysis of the materials was performed in an effort to determine the homogeneity of the product crystals. These EDAX (Energy Dispersive X-ray Analysis) studies were performed by use of a Phillips model 505 scanning electron microscope. Microscopic investigation of the macrostructure of the ceramic were also performed using this instrument.

Optical Characterization. The relative reflectivity measurements were taken at 300K by use of a diffuse reflectance attachment on a FT-IR instrument with the sample in the form of a sintered pellet. In this configuration, it was determined that the reflectance in the investigated energy range is dominated by the specular component of the sample's reflectivity. Well ground powder samples were also investigated to insure the specular nature of the reflectance spectrum observed with the pellet samples, and to help differentiate particle size effects from intrinsic absorption effects. No particle size artifacts were identified.

Characterization of Superconducting Properties. The real part of the complex AC magnetic susceptibility has been used as a tool for the detection of the superconducting transition temperature. The apparatus used was similar to that described in reference (6), with modifications made for use of an electronic thermometer with analog output and a nitrogen gas flow system for cooling the sample/coil system. Samples were mounted directly on the junction of a copper/constantan thermocouple. The transition temperature is taken to be the point of intersection of the baseline susceptibility of the sample with a line drawn through the susceptibility curve in the diamagnetic region. This roughly identifies the point of maximum inflexion in the curve, and indicates the point of onset of bulk superconductivity. Transition temperatures determined by this technique are usually lower than bulk conductivity onsets as the contribution of minority or filiamentary phases is minimized, rather than maximized by the magnetic measurements.

Results and Discussion

Potassium Substitution. The results of these experiments show that the incorporation of potassium into the phase has a major effect on the superconducting properties. EDAX measurements show that the limit of solid solubility is 2.6 mole percent. The superconducting measurements found that bulk high temperature superconductivity (above 77K) was quenched in these multiphase samples when annealed in oxygen, yet superconductivity with a magnetically determined onset temperature of 96K was found upon annealing the same samples in air. The transition was not complete by 77K, yielding a minimum width for the transition of 19K, much larger than for a 123 standard (3 to 4K) with a Tc onset of 93K. Crystallochemically, it is to be expected that potassium can readily be substituted for barium on the basis of their similiarity of size. The major impact of such substitution is expected to be, assuming constant oxygen content, a rise in the oxidation state of copper, the only variable valence element in the compound. In the case of a stoichiometry of Y1:Ba2:Cu3:07, one expects to find two of the coppers in the +2 state for one copper in the +3 state. Recent studies of the oxygen stoichiometry (7) yield values of 6.85 for the oxygen content. This data has been interpreted in terms of +2 coppers on the square pyramidal site, and a 70%:30% Cu+3:Cu+2 ratio on the square planar site. In this case, we expect that K substitution leads to an increase in +3 copper content in the copper chain (square planar) site. This will then decrease the carrier concentration since Cu+2 can be assumed to be the carrier donor, and the transition temperature should be found to drop. The finding that superconductivity is returned upon annealing in air is consistent with a decrease in the stability of the +3 state in this medium of lower oxidizing potential. In most of the superconductors of this family it has been found that maximizing the +3 content has led to the highest transition temperatures. These experiments show that there is an optimal +3 content which must be set for each material by choice of equilibrium oxidizing atmosphere composition. The increase in transition temperature observed for the K doped material relative to the usual 123 compound may be due to achievement of an optimal +3 content in a part of the sample. The breadth of the transition may be due either to use of a nonoptimal oxidizing atmosphere, or to a decrease in the chain length caused by an oxygen vacancy content greater than 0.15. The majority of this doped phase material does not display optimal superconducting properties.

<u>Zinc Substitution</u>. Although the solid solution limit for the substitution of zinc for copper has not been determined, it has been found that even at low concentrations (3% i.e. Cu:Zn 2.9:0.1) high temperature superconductivity (Tc>77K) is quenched, although the electrical conductivity remains quite high. Crystallochemically, it is to be expected that Zn+2 can readily substitute for Cu+2, although some local lattice contraction is to be expected. Currently, it is considered that the square pyramidal sites make a negligible contribution to the superconductivity on the grounds that the substitution of large moment rare earths in their proximity has no effect upon Tc other than that which can be ascribed to size (carrier density) effects (8). These zinc substitution effects support this point of view. Zinc substitutions will act as insulators in the two dimensional network due to the closed shell nature of Zn+2. Even if we assume that there is mixed valence on the plane, the introduction of 5% resistive elements would not be expected to significantly influence the percolation conductivity in this layer, since the percolation threshold is ten times this amount, or 50% in two dimensions (9). In the one dimensional chains, substitution of Zn+2 for the oxygen vacancy induced Cu+2 would result in a loss of charge carriers (carrier density) as well as a loss of delocalization in the chain. The former effect would lead to the lowering of Tc in an excitonic superconductor.

<u>Oxygen Nonstoichiometry</u>. Samples quenched from oxygen at 750C and air at 950C have been studied. A value near 6.5 for the oxygen stoichiometry is expected for both of these samples (10). The samples are found to be semiconducting by measurement of their resistance vs temperatures curves. The reflectance spectrum of the air quenched sample has been taken to directly determine the value of the optical band gap in these materials. The reflectance spectrum in the infrared is presented in Figure 1. As in photoacoustic spectroscopy, a value of the electronic energy gap can be found by determining the point of maximum absorption (minimal reflectivity) at the base of the absorption edge. This value is 640 cm-1 or 0.08 eV. We attribute this transition to excitation of a square pyramidal Cu+2 valence electron onto a Cu+3 localized square pyramidal site. An oxygen stoichiometry value greater than 6.5 is required by this interpretation, since the copper valence is exactly 2 at stoichiometry 6.5. Although other explanations, such as the onset of plasma (Drude) excitations, are possible, this valence band - conduction band transition interpretation is consistent with the continuous, high absorptivity observed at photon energies above the gap energy. Either of these electronic excitations can form the basis of excitonic type theories.

There are many aspects of these materials which any suggested mechanism must be consistent with, including, but not limited to: 1. The observed absence of an oxygen isotope effect upon the transition temperature (11). This observation may be taken as a clear indication that a phonon mechanism is not operative

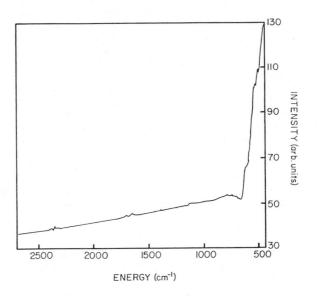

Figure 1.   Infrared reflectance spectrum of reduced $YBa_2Cu_3O_7$ displaying onset of electronic absorption with a maximum near 640 cm.$^{-1}$

in this system. This is especially convincing in that the oxygen vibrations should be primarily involved in the Jahn-Teller, or electron lattice interaction. 2. The bimodal nature of the transition temperature as oxygen is extracted from the lattice (Tarascon, J.M., Materials Research Society Symposium on High Temperature Superconductors, 1987 Spring Meeting). This has already been used as evidence that the linear Cu-O-Cu chains are required for the observation of high Tc. 3. The extremely high Tc. As suggested above, the phonon mechanism has been shown to be inadequate for the production of superconductivity at 93K. 4. The increase in Meissner effect seen when the stoichiometry is changed to $YBa_3Cu_4O_x$ (Bednorz, J., Materials Research Society Symposium on High Temperature Superconductors, 1987 Spring Meeting). This structure is thought to be one in which the Y site is partially occupied by Ba. This would be expected to increase the local lattice parameter, and, most importantly, increase the a and b lattice parameter. This would have the effect of stretching the chain, and increasing the planarity of the oxygen coordination at the base of the square pyramid. In the event that this substitution relaxes the oxygen buckling in the square pyramidal plane without increasing the Cu-O distance, the difference in energies of a d9 electron on the two copper lattice sites would become smaller, with the Cu+2 character of Cu2 (square pyramidal) experiencing little change, while Cu1's (square planar) Cu+3 stability is lowered due to loss of Cu-O chain interaction. This would increase the transition probability, or the coupling of the chain electrons with the electronic excitations of Cu2. A more obvious effect would appear to be the change in the number of electrons donated from the Y+3 layer to the Ba-Cu-O layer assembly. This could be expected to yield, as in the K substitution case, a larger Cu+3 concentration. However, self compensation, in the form of additional oxygen vacancy formation, could operate to maintain an optimal +3:+2 ratio. 5. The observation of raised Tc when the stoichiometry is changed to $YBa_4Cu_5O_x$ (Cohen, M.L., Materials Research Society Symposium on High Temperature Superconductors, 1987 Spring Meeting). Although the structure is not known at this time for this composition, it is probable that there is some substitution of Ba on the Y site at this composition. An even more intriguing possibility is that extra layers containing the one dimensional chains are inserted into the structure. This would yield a higher volume fraction of the material as superconducting, causing, through a proximity effect, a rise in the transition temperature. The proximity effect has not been demonstrated in an excitonic superconductor yet, but few excitonic theories reject the possibility of such an effect. 6. The small bandgap observed in the semiconducting form of the material. Most metals cannot be produced in a semiconducting form. This is one of the interesting aspects of metal oxides near the metal-insulator (semiconductor) transition. Although the bandgap cannot be observed in the material in its metallic state due to its high plasma reflectivity, the forbidden gap in electronic states remains. Since this is a significant aspect of the electronic structure of the materials, and it no doubt is a common feature to all of the high Tc materials since it is

Cu-O based, any theory should note the impact of such a gap on the superconducting mechanism. The exciton model (12) requires the existence of such an electronic excitation. 7. The observation of low temperature superconductivity in $LaBa_2Cu_3O_7$ (13). This material appears to form with the cubic perovskite structure, although the ordering of the oxygen vacancies has not been determined. The observation of true superconductivity at 48K in this material may indicate the presence of two dimensional planes of Cu-O in the structure, as seen in the original La-Ba-Cu-O system. Clearly high Tc is correlated with structural diversity not available in cubic materials. The exciton mechanism requires such structural diversity in order for part of the structure to act as a metal, and part of the structure to act as a local electronic resonator, in close proximity to the conducting portion of the material. The best realization of this mechanism should be in systems in which these moieties can be integrated on the unit cell level, as appears to be true in this current family of superconductors. 8. These materials appear to have a quite low conduction electron density. This is entirely consonant with the exciton mechanism which actually requires low carrier densities, on the order of 10^{19-20} (12). 9. The observation of a BCS type superconducting energy gap of approximately the right magnitude for BCS superconductivity in tunnelling experiments (Geballe, T., Materials Research Society Symposium on High Temperature Superconductors, 1987 Spring Meeting) could be interpreted to lend support to a phonon mechanism. However, BCS type gaps are predicted for the exciton mechanism also. At this point, the exciton mechanism is compatible with the data known about the new family of superconductors. It is important, then, that attempts to modify the materials be based upon optimization of this mechanism, rather than toward optimization of a BCS type material.

The implications of this work are two fold, in that (1) the excitonic mechanism directly leads to certain restraints on the structures and properties of the superconducting materials, and (2) new materials have been generated with electronic properties, as well as physical macrostructures quite different from those of $YBa_2Cu_3O_7$ prepared under similar conditions.

Previously it was expected that cubic materials would generate the highest transition temperatures. The current highest Tc materials are far from isotropic. Although it may be possible to produce materials with a three dimensional structure, yet composed of metallic and semiconducting moieties, the inhomogeneity of the interface of these two electronic phases would certainly result in the degradation of their coupling. The optimum cases collapse to a one dimensional model and the two dimensional model. The title system represents a hybrid of these two possibilities. The interchain coupling mediated by the layer moiety provides a weak coupling, relaxing the mono-dimensionality of the sytem. It would appear that optimization of the current Cu containing systems will consist of improving the properties of each of the moieties. The chain connectivity is broken by oxygen vacancies in the known systems. Since the exciton mechanism does not require that the metallic segment be one dimensional, an increase to two dimensionality may improve the properties. This may even be the basis for superconductivity at 155K observed recently (14). Under this model, the chain oxygen atoms would be replaced with two fluorine atoms, providing a two dimensional metallic moiety. Another component that could perhaps be optimized is the semiconducting gap. Decreasing the gap would increase the polarizability, but there will be a limit to this effect as the polarizations may freeze in, leading to an actual loss in polarizability. A similar situation in the BCS superconductors results in the martensitic phase transformation. A search for other natural superlattice materials which can be made as conducting analogs is in progress, since quite complex layer sequences, within a large lattice parameter unit cell may be required to observe extremely high Tc's.

On a more technical level, the observation of high Tc's in saturated potassium containing systems is important. Most ceramics require a sintering aid for their production. This aid should not inordinately degrade the intrinsic properties of the material, although, often, tradeoffs have to be made. In the case of K substitution, large grain ceramics have been produced under the same conditions usually leading to one micron particle size ceramics. Although the width of the transition has been degraded, the material remains a high Tc material.

Conclusion

Chemical substitution experiments have been performed which support the suggestion that the one dimensional chains are integral to the high Tc in the title compound, and indicate that the superconducting properties are sensitive to the mixed valence state of the chain. A consideration of the function and contribution of the two dimensional planar portion of the structure has led to the conclusion that it is not the conduction properties of this moiety that are significant, but the electronic properties. A band gap like structure attributed to electronic excitations of this portion of the structure can provide the mechanism for high temperature one dimensional superconductivity through the exciton mechanism. This mechanism has been shown to be consistent with our current knowledge of this family of high temperature superconductors. Since the upper limit for Tc utilizing the exciton mechanism is near 1700K (12), there is no theoretical impediment to production of materials and devices which can be operated at temperatures 100's of degrees below their transition temperatures at room temperature. Only an understanding of the chemistry of high temperature superconductors can lead us directly to production of these materials.

Acknowledgments

I acknowledge very valuable discussions with R.L. Greene and J.A. de Haseth. I also thank JAH for use of his facilities, and Ray Robertson for taking the IR spectra. This work is supported in part by a grant from the National Science Foundation, DMR-8600224.

Literature Cited

1. Bednorz, J.G.; Muller, K.A.Z. Phys. B: Condens. Matter 1986, 64, 189-93.
2. Shuvayev, V.P.; Sazhin, B.I. Poly. Sci. U.S.S.R. 1982, 24, 229-36.
3. Anderson, P.W.; Abrahams, E. Nature 1987, 327, 363.
4. Norton, M.L.; Wolfe, L.G. Solid State Ionics 1986, 22, 75-83.
5. Allender, D.; Bray, J.; Bardeen, J.; Phys. Rev. B. 1973, 7, 1020-9.
6. Norton, M.L. J. Phys. E: Sci. Instrum. 1986, 19, 268-70.
7. David, W.I.F.; Harrison, W.T.A.; Gunn, J.M.F.; Moze, O.; Soper, A.K.; Day, P.; Jorgensen, J.D.; Hinks, D.G.; Beno, M.A.; Soderholm, I.; Capone, C.W.; Schuller, I.K.; Segre, C.U.; Zhang, K.; Grace, J.D. Nature 1987, 327, 310-2.
8. Hor, P.H.; Meng, R.L.; Wang, Y.Q.; Gao, L.; Huang, Z.J.; Bechtold, J.; Forster, K.; Chu, C.W. Phys. Rev. Lett. 1987, 58(18), 1891-4.
9. Adkins, C.J.J. Phys. C: Solid State Phys. 19798, 12, 3389-93.
10. Strobel, P.; Capponi, J.J.; Chaillout, C.; Marezio, M.; Tholence, J.L. Nature 1987, 327, 306-8.
11. Batlogg, B.; Cava, R.J.; Jayaraman, A.; van Dover, R.B.; Kourouklis, G.A.; Sunshine, S.; Murphy, D.W.; Rupp, L.W.; Chen, H.S.; White, A.; Short, K.T.; Mujsce, A.M.; Reitman, E.A. Phys. Rev. Lett. 1987, 58(22), 2333-6.
12. Collins, T.C. Ferroelectrics 1987, 73, 469-80.
13. Murphy, D.W.; Sunshine, S.; van Dover, R.B.; Cava, R.J.; Batlogg, B.; Zhurak, S.M.; Schneemeyer, L.F. Phys. Rev. Lett. 1987, 58(18), 1888-90.
14. Ovinshinsky, O.R.; Young, R.T.; Allred, D.D.; DeMaggio, G.; Van der Leeden, G.A. Phys. Rev. Lett. 1987, 58(24), 2579-81.

RECEIVED July 6, 1987

Chapter 7

Chemical Problems Associated with the Preparation and Characterization of Superconducting Oxides Containing Copper

S. Davison, K. Smith, Y-C. Zhang, J-H. Liu, R. Kershaw, K. Dwight, P. H. Rieger, and A. Wold

Chemistry Department, Brown University, Providence, RI 02912

The superconducting oxides $La_{1.8}Sr_{.2}CuO_4$ and $Ba_2YCu_3O_7$ were prepared by decomposition of mixed metal nitrates. Thermogravimetric analysis and electron spin resonance measurements indicated the presence of Cu(I) and Cu(III) in the $La_{1.8}Sr_{.2}CuO_4$ phase. The compound $Ba_2YCu_3O_7$ prepared from the nitrates and subjected to an oxygen anneal at 425°C gave a sharp superconducting transition at 92 K. The phase was stoichiometric but readily decomposed when kept in contact with moist air.

In the last six months the compounds $La_{1.8}Sr_{.2}CuO_4$ and $Ba_2YCu_3O_7$ have received much attention in the literature because of their high temperature superconducting transitions. $La_{1.8}Sr_{.2}CuO_4$ was reported to have a superconducting transition at 36 K [1-3]. (See also Bednorz, J.G.; Takashige, M.; Muller, K.A.; to be published). A single crystal study has indicated that the compound crystallizes in the space group of I4/mmm [4]. The coordination geometry around the copper atoms is a tetragonally elongated octahedron with four short copper-oxygen bonds d[Cu-O(1)] = 1.90 Å and two long bonds d[Cu-O(2)] = 2.41 Å. The high temperature superconducting transition has been attributed to the existence of mixed oxidation states of copper in the structure [1, 3], but the valence and valence distribution have not been determined. The determination of copper valences in $La_{1.8}Sr_{.2}CuO_4$ could be a key for the understanding of the superconducting mechanism of the compound. The first part of this paper will deal with how the oxidation state of copper in $La_{1.8}Sr_{.2}CuO_4$ can be determined.

The compound $Ba_2YCu_3O_7$ shows an even higher superconducting transition (∿ 93 K) and crystallizes as a defect perovskite. The structure of $Ba_2YCu_3O_7$ has been determined by neutron diffraction analysis. The space group is Pmmm with a = 3.8198, b = 3.8849 and c = 11.6762 Å. Barium and yttrium are ordered on the A site to give a tripled cell along c and the oxygens occupy 7/9 of the anion sites. One third of the copper is in 4-fold coordination and 2/3 are five-fold coordinated (Gallagher, P.K.; O'Bryan, H.M.; Sunshine, S.A.; Murphy, D.W.; to be published).

0097–6156/87/0351–0065$06.00/0

Very little has appeared in the literature concerning the optimum conditions for the synthesis of $Ba_2YCu_3O_7$ and only the above study has attempted to determine the degree and nature of the oxidation state of copper. The preparative conditions necessary to give pure $Ba_2YCu_3O_7$ and the chemical instability of the pure phase towards moist air will be discussed in the second part of this paper.

La_2CuO_4 and $La_{1.8}Sr_{.2}CuO_4$

Preparation of La_2CuO_4 and $La_{1.8}Sr_{.2}CuO_4$. Samples of La_2CuO_4 and $La_{1.8}Sr_{.2}CuO_4$ were prepared by codecomposition of the corresponding nitrates. The starting compounds were high purity copper metal (Matthey S.50250), La_2O_3 (Lindsay #528 99.999%) and $SrCO_3$ (Matthey 2H118, 99.999%). The copper was prereduced in 85%Ar/15%H_2 at 450°C for 6 hours. The La_2O_3 was heated at 800°C for 8 hours to drive off adsorbed CO_2 and water. The molecular weight of the $SrCO_3$ was analyzed as 147.6 by thermogravimetric analysis. A mixture of 254.2 mg copper metal, 1172.9 mg La_2O_3 and 118.1 mg $SrCO_3$ was dissolved in 6 ml of concentrated nitric acid to convert all of the initial compounds to nitrates. The solution was dried at 150°C for 12 hours and then predecomposed at 500°C for 4 hours. The sample was ground and heated at 970°C for 120 hours. During the heating the sample was taken out and ground 4 times. Finally, the sample was quenched to room temperature by removing it from the furnace at the elevated temperature. The sample of pure La_2CuO_4 was prepared by the same procedure.

Characterization of Products. X-ray powder diffraction patterns of the samples were obtained using a Philips diffractometer and monochromated high intensity CuKα_1 radiation (λ = 1.5405 Å). The diffraction patterns were taken in the range 12° < 2θ < 75° with a scan rate of 1° 2θ/min and a chart speed of 30 in/hr.

Temperature programmed reduction of prepared samples was carried out using a Cahn System 113 thermal balance. The samples were purged in a stream of 85%Ar/15%H_2 for 2 hours. Then the temperature was increased to 990°C at a rate of 50°C/hr. The flow rate of the gas mixture was 60 ml/min.

Magnetic susceptibility measurements were carried out using a Faraday Balance from 77 to 300 K with a field strength of 10.4 kOe. Honda-Owen (field dependency) measurements were carried out at both 77 and 296 K.

Electron spin resonance spectra of $La_{1.8}Sr_{.2}CuO_4$ and La_2CuO_4 were recorded using a Bruker ER220D Spectrometer at room temperature. The frequency was ν_0 = 9.464 GHz, the microwave power was 200 mW, and the field modulation amplitude was 1 Gauss.

Results and Discussion. The x-ray powder diffraction patterns of La_2CuO_4 and $La_{1.8}Sr_{.2}CuO_4$ are given in Figure 1(a) and (b). La_2CuO_4 shows a single phase which can be indexed on the basis of a distorted K_2NiF_4 type structure. The data obtained is consistent with those of orthorhombic La_2CuO_4 reported by J. M. Longo (5). $La_{1.8}Sr_{.2}CuO_4$ was also prepared as a single phase and could be indexed on the basis of an undistorted tetragonal K_2NiF_4 type structure. The tetragonal

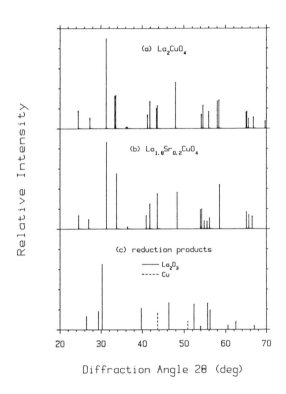

Figure 1. X-ray diffraction patterns of (a) La_2CuO_4, (b) $La_{1.8}Sr_{.2}CuO_4$, (c) reduction products of $La_{1.8}Sr_{.2}CuO_4$.

phase is the superconducting phase with a transition reported to be about 36 K (1-3).

The average oxidation state of copper in the $La_{1.8}Sr_{.2}CuO_4$ was determined from the TPR results shown in Figure 2. X-ray diffraction confirmed that $La_{1.8}Sr_{.2}CuO_4$ decomposed completely during the reduction, the detectable reduction products being La_2O_3 and metallic copper (Figure 1(c)). The reduction proceeded according to the following equation:

$$10La_{1.8}Sr_{.2}CuO_{(2.90+0.5x)} + 5xH_2 \rightarrow 9La_2O_3 + 2SrO + 10Cu + 5xH_2O$$

where x is the average oxidation state of the copper ions. From the ratio of final and initial weights the average valence of the copper ions was determined as 2.00($\pm$.04). There are several possibilities consistent with these results. The copper could remain all Cu^{2+}, disproportionate to Cu^+ and Cu^{3+}, or be a mixture of all three valencies. In order to help determine the most probable oxidation state, an electron spin resonance study was carried out.

In the La_2CuO_4 structure, the C ion has an elongated octahedral coordination with respect to oxygen (3, 5). The lengths of the four short planar CuO bonds are 1.90 Å, and the two long Cu-O bonds along the +z and -z direction are 2.43 Å. Hence, the copper is actually in a square planar coordination with respect to oxygen. Wang et al. have reported that $La_{1.85}Sr_{.15}CuO_4$ has a tetragonal K_2NiF_4 structure with space group I4/mmm. The four short [Cu-O(1)] bonds and the two long [Cu-O(2)] bonds are 1.90 Å and 2.41 Å, respectively (4). They found that copper ions were also in a tetragonally distorted octahedral site. The copper ions in $La_{1.85}Sr_{.15}CuO_4$ are also primarily square planar coordinated with respect to oxygen. Since there is an apparent structural similarity of Cu in the two compounds, Cu^{2+} in La_2CuO_4 can be used as a standard (6) for the ESR study of $La_{1.8}Sr_{.2}CuO_4$.

The electron spin resonance spectrum of La_2CuO_4 (ν_0 = 9.464GHz, room temperature) is shown in Figure 3(a). The spectrum can be interpreted with

$$g_{||} = 2.310\pm0.002, \; g_{||} = 2.062\pm0.002,$$

$$A_\perp = (132\pm2)\times10^{-4}cm^{-1}, \text{ and } A_\perp = (21\pm)\times10^{-4}cm^{-1}$$

Such parameters are typical of Cu(II) in a square planar or tetragonally distorted octahedral site. ESR spectra of $La_{1.8}Sr_{.2}CuO_4$ show a feature with the same g-value as the perpendicular features seen in the La_2CuO_4 spectrum. However, the size of this resonance decreases with sample purification and appears to be due to a trace of the La_2CuO_4 phase. The spectrum of one such sample, shown in Figure 3(b) (same experimental conditions as the spectrum of Figure 3(a)) has an intensity about 1% that of pure La_2CuO_4 This fact means that only about 1% of the copper atoms in the $La_{1.8}Sr_{.2}CuO_4$ heated at 970°C for 120 hours are in the Cu(II) state. Considering that copper in $La_{1.8}Sr_{.2}CuO_4$ has an average valence of 2.0, it is proposed that the copper has disproportionated into Cu(I) and Cu(III). Cu(II) is generally the most stable copper ion; therefore, the disproportionation is a unique characteristic of

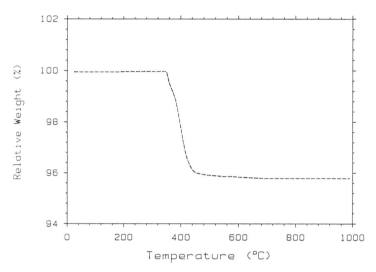

Figure 2.   Temperature programmed reduction profile of La$_{1.8}$Sr$_{.2}$CuO$_4$ in 85%Ar/15%H$_2$.

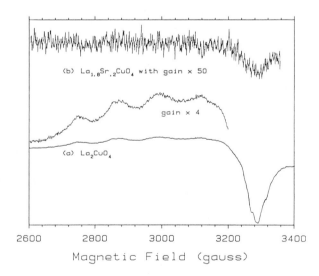

Figure 3.   X-band ESR spectra of (a) La$_2$CuO$_4$ and (b) La$_{1.8}$Sr$_{.2}$CuO$_4$.(Microwave power, 200 mW, field modulation amplitude, 1 Gauss.)

$La_{1.8}Sr_{.2}CuO_4$. This unique property is probably related to the observed superconductivity at high temperature.

The magnetic susceptibility of $La_{1.8}Sr_{.2}CuO_4$ is plotted against temperature in Figure 4. The absence of any temperature dependence demonstrates Pauli paramagnetism over the temperature range from 77 to 300 K. Cu(II) $3d^9$ electrons are usually localized and characterized by Curie-Weiss behavior; such results were obtained by Ganguly and Rao for La_2CuO_4 (7). Since the Pauli-paramagnetic behavior of $La_{1.8}Sr_{.2}CuO_4$ is consistent with delocalized electrons, this would also indicate a high probability for the existence of Cu(I)-Cu(III) formed as a result of disproportionation of Cu(II).

$Ba_2YCu_3O_7$

Attempted Preparation of $Ba_2YCu_3O_7$ from $BaCO_3$, Y_2O_3 and CuO. Samples were prepared by the usual ceramic techniques using predried $BaCO_3$ (400°C), Y_2O_3 heated at 800°C for 6 hours and analyzed CuO. The mixture of the carbonate and oxides was heated at 900°C for 24 hours.

Preparation of $Ba_2YCu_3O_7$ from the Nitrates. The nitrate samples were prepared from stoichiometric quantities of $BaCO_3$, Y_2O_3 and Cu metal. The appropriate weights were put into a porcelain crucible and dissolved by the addition of concentrated nitric acid (20 ml/g Cu). The excess acid was boiled off and the product was heated at 130°C for 6 hours and then at 400°C for 3 hours in order to decompose most of the copper and yttrium nitrates. The product was then removed, ground and transferred to a platinum or stabilized zirconia crucible. The powder was heated to 800°C for 12 hours and cold-pressed into pellets at 6 Kbar. The pellets were heated to 950°C for 12 hours and then annealed in oxygen at 425°C for 6 hours. They were finally cooled to room temperature at 50°C/hr. Additional powder samples, used for TGA studies, were prepared by partial decomposition of the nitrate mixture at 400°C followed by a second heating at 900°C for 24 hours.

Characterization of Products: Chemical Analysis. Determination of the stoichiometry of $Ba_2YCu_3O_7$ was performed using a Cahn System 113 thermal balance. An atmosphere of of 85%Ar/15%H_2 was used for reduction in these studies. The gas was predried by passing it through a P_2O_5 column. The sample was heated at 50°C/hr until no further weight change was noted.

Characterization of Products: Electrical Measurements. The van der Pauw method was used to measure the electrical resistivities. Contacts were made by the ultrasonic soldering of indium directly onto the sintered discs of $Ba_2YCu_3O_7$, and their ohmic behavior was established by measuring their current-voltage characteristics. The sample and a thermocouple were mounted in a small cavity inside a massive copper body which was first cooled to 77 K and then allowed to warm very slowly.

Results and Discussion. $Ba_2YCu_3O_7$ was prepared both by the reaction of $BaCO_3$, Y_2O_3 and CuO according to published procedures and by the

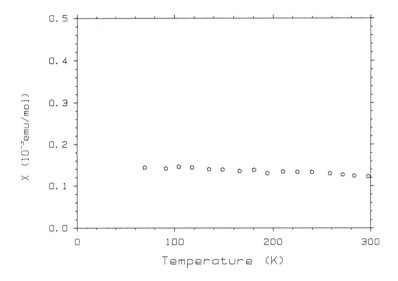

Figure 4. Temperature dependence of the magnetic susceptibility of $La_{1.8}Sr_{.2}CuO_4$.

decomposition of the mixed metal nitrates. The changes in weight are plotted as functions of temperature for both reaction mixtures in Figure 5. Copper nitrate and yttrium nitrate are completely converted to oxides by 400°C, so that the TGA results shown in Figure 5 represent the ease of decomposition of barium nitrate compared to barium carbonate. It can readily be seen that barium nitrate is decomposed by 650°C whereas the carbonate does not decompose substantially until 900°C. It should therefore be possible to prepare $Ba_2YCu_3O_7$ more readily via the decomposition of the mixed metal nitrates. From Figure 6 it is seen that the phase $Ba_2YCu_3O_7$ is not completely formed even after a 24 hour heat treatment of barium carbonate, yttrium oxide and copper(II) oxide at 900°C. However, from Figure 7 the product is readily formed from the nitrates after being heated at 900°C for 24 hours.

An important part of the preparative procedure is annealing the phase formed at 900-950°C in a flowing oxygen atmosphere at 425°C. This step assures a product which closely approaches the stoichiometric composition of $Ba_2YCu_3O_7$. Samples which were annealed at 425°C in oxygen and allowed to cool slowly at 50°C/hr were

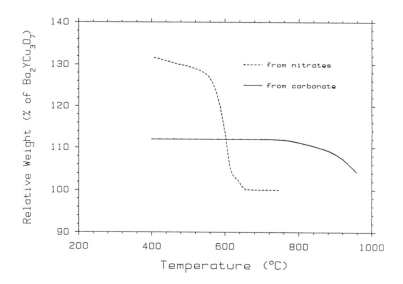

Figure 5. Decomposition in air of a mixture of $BaCO_3$, Y_2O_3 and CuO compared with that of a mixture of the nitrates.

analyzed by reduction in an 85%Ar/15%H_2 atmosphere. The results are plotted in Figure 8. As determined from the measured weight loss, the formula may be represented as $Ba_2YCu_3O_{7.01(2)}$.

The electrical resistivity and magnetic susceptibility of these samples were measured as functions of temperature and the results are plotted in Figures 9 and 10. As can be seen from Figure 9, the resistivity data defines an extremely sharp transition at a temperature of 92 K. The superconducting nature of the compound below this temperature is verified by the Meissner effect evident in the susceptibility data shown in Figure 10.

The pure samples of $Ba_2YCu_3O_7$ prepared from the nitrates were placed on a watch glass which was transferred to a desiccator containing water in place of the desiccant. After 48 hours, the sample was removed and subjected to visual and x-ray analysis. The black product contained numerous aggregated white particles. The x-ray analysis of the product is shown in Figure 11 and indicates the formation of $BaCO_3$. These results would indicate that the product is attacked by moist air and the instability of $Ba_2YCu_3O_7$ will present problems for any practical device application.

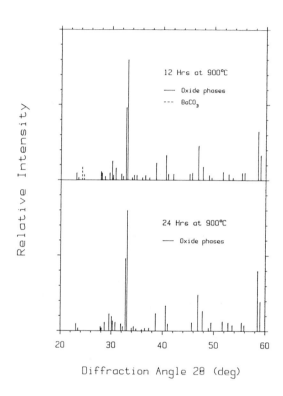

Figure 6. X-ray diffraction patterns of a mixture of BaCO₃, Y₂O₃ and CuO heated at 900°C showing the presence of unreacted BaCO₃ after 12 hrs and a multiphase product after 24 hrs.

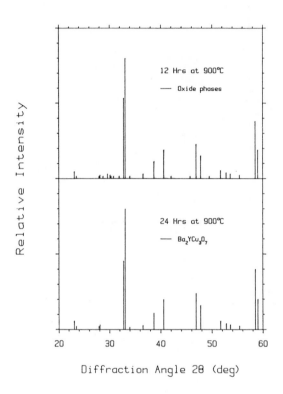

Figure 7. X-ray diffraction patterns of a mixture of the nitrates heated at 900°C for 12 and 24 hrs showing the final product to be single-phase $Ba_2YCu_3O_7$.

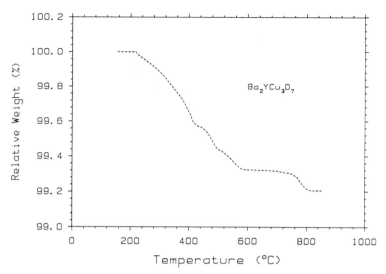

Figure 8. Analysis of $Ba_2YCu_3O_7$ by temperature-programmed reduction in 85%Ar/15%H_2. The observed weight loss corresponds to the formula $Ba_2YCu_3O_{7.01(2)}$.

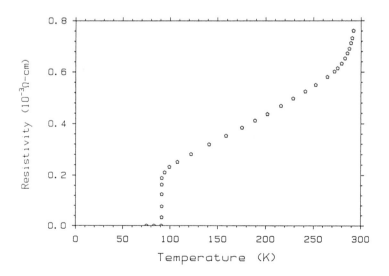

Figure 9. Resistivity of $Ba_2YCu_3O_7$ as a function of temperature.

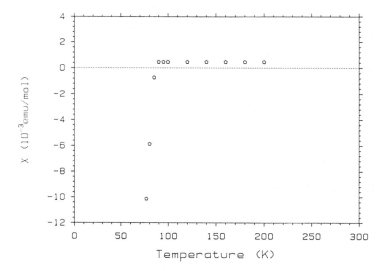

Figure 10. Magnetic susceptibility of $Ba_2YCu_3O_7$ as a function of temperature.

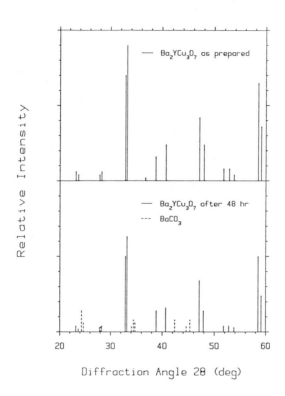

Figure 11. X-ray diffraction patterns of Ba$_2$YCu$_3$O$_7$ as prepared for the nitrates and after 48 hrs of exposure to moist air, showing partial decomposition of the sample.

Acknowledgments

This research was partially supported by the Office of Naval Research and by NSF Grant. No. DMR-820-3667. The authors also express their appreciation for the use of Brown University's Materials Research Laboratory which is funded by the National Science Foundation.

Literature Cited

1. Rao, C.N.R.; Ganguly, P. Current Science 1987, 56(2), 47.
2. Bednorz, J.G.; Muller, K.A.; Takashige, M. Science 1987, 236, 73.
3. Cava, R.J.; van Dover, R.B.; Batlogg, B.; Rietman, E.A., Phys. Rev. Lett. 1987, 58(4), 408.
4. Wang, H.H.; Geiser, M.; Thorn, R.J.; Carlson, K.D.; Beno, M.A.; Monaghan, M.R.; Allen, T.J.; Proksch, R.B.; Stupka, D.L.; Kwok, W.K.; Crabtree, G.W.; William, J.M. Inorg. Chem. 1987, 26, 1190.
5. Longo, J.M.; Raccah, P.M. J. Solid State Chem. 1973, 6, 526.
6. Jorgensen, J.D.; Schüttler, H-B.; Hinks, D.G.; Capone II, D.W.; Zhang, K.; Brodsky, M.B.; Scalapino, D.J. Phys. Rev. Letts. 1987, 58(10), 1024.
7. Ganguly, P.; Rao, C.N.R. J. Solid State Chem. 1984, 53, 193.

RECEIVED July 6, 1987

Chapter 8

Single-Crystal Growth of the High-Temperature Superconductor YBa$_2$Cu$_3$O$_x$

F. Holtzberg, D. L. Kaiser, B. A. Scott, T. R. McGuire, T. N. Jackson, A. Kleinsasser, and S. Tozer

Thomas J. Watson Research Center, IBM, Yorktown Heights, NY 10598

We report a set of conditions for crystal growth of the high-temperature superconductor YBa$_2$Cu$_3$O$_x$. The as-grown single crystals have critical temperatures up to 85 K. Preliminary studies show that the transition temperature can be increased by thermal annealing in oxygen, as in ceramic samples. The crystals are of suitable dimensions for definitive magnetic, optical and transport measurements.

After the major discovery of the high temperature superconductor La$_{2-x}$Ba$_x$CuO$_4$. by Bednorz and Muller (1) and the subsequent identification (2) and isolation (3-5) of the higher T$_c$ YBa$_2$Cu$_3$O$_x$ (YBC) phase, there followed a tidal wave of publications. The proliferation of papers attests to the seeming ease of fabrication of ceramic samples which have superconducting properties. It is evident on the other hand that there is a rather limited literature on the growth of single crystals of YBC. Indeed to our knowledge, there are few publications which give serious consideration to the methods and conditions for single crystal growth of YBC having dimensions, purity and homogeneity suitable for determination of the intrinsic properties of this superconductor. The following communication addresses conditions for crystal growth and establishes some relevant aspects of the phase relationships in the system containing the desired YBC compound.

Initial crystal growth experiments using the more common low temperature fluxes such as B$_2$O$_3$, KF and PbO were unsuccessful. Subsequent DTA measurements suggested that YBC decomposes peritectically at about 1020°C. To overcome this difficulty, we searched for a liquidus field for crystal growth below the decomposition temperature. We concentrated our efforts on the pseudoternary YBC-BaCuO$_2$-CuO system.

Experimental Procedures

Selection of Crucible Material. Initial experiments (6) were carried out in platinum crucibles. Crystals up to 1×2×0.01 mm were found primarily in cavities within the

0097–6156/87/0351–0079$06.00/0
© 1987 American Chemical Society

solidified mass. The dimensions of the crystals appeared to be independent of the amount of charge. Yield of YBC crystals was considerably less than 1%.

Several problems were encountered with crystal growth from the platinum crucibles. The liquid wet the Pt surface and could not be completely contained. In addition, the liquid reacted extensively with the crucible, forming mixed platinates. Consequently, the charges were depleted of yttrium and barium oxides, lowering the YBC crystal yield. A search was therefore made for other crucible materials.

Samples of the constituent compounds YBC, $BaCuO_2$, and CuO and also composition 'a' in Fig. 1 were heated in air at 950°C for 16 hrs on Pt, Au and Cu, and single crystals of sapphire, Y-stabilized zirconia (12% Y), magnesia, magnesium aluminate ($MgAl_2O_4$) and quartz. All of the potential crucible materials showed little or no visible reaction with the three compounds. Extensive melting of composition 'a' occurred on all support materials with the exception of Pt. Single crystal magnesia (which is difficult to obtain) and gold showed the least apparent reactivity with the melt.

Determination of Melting Behavior. On the basis of the above results, gold crucibles were used in a study of melting and crystal growth in the YBC-$BaCuO_2$-CuO pseudoternary system, Fig. 1. Compositions along the pseudobinaries $BaCuO_2$-CuO and YBC-CuO and within the pseudoternary at molar Ba/Y ratios of 9/1, 4/1 and 7/3 and CuO mole fractions from 0.5 to 0.8 were studied. The sample with Ba/Y = 9/1 and CuO mole fraction = 0.7 (composition 'a' in Fig. 1) showed the greatest degree of melting. Samples were prepared from mixtures of YBC, $BaCuO_2$ and CuO. Use of these compounds, rather than their oxide or carbonate precursors, greatly improved solid state reactivity, allowing complete reaction in a much shorter period of time. YBC and $BaCuO_2$ were synthesized by heating pressed pellet mixtures of $BaCO_3$, CuO and Y_2O_3 at 940-950°C in flowing O_2(g) for 16 hrs on a platinum sheet. The pellets were then ground, remixed, pressed and the heating procedure repeated. In each case the samples were fired on a bed of powder of the same mixture. Following the final reaction period, they were cooled at 50°C/hr to 400°C. The resulting compounds YBC and $BaCuO_2$ were examined by x-ray powder diffraction, and found to give single phase patterns in good agreement with those published in the literature (3,7).

To determine a suitable growth temperature for composition 'a', samples were heated to temperatures between 950 and 1000°C and cooled at 25°C/hr. Crystal growth was most favorable at 970°C for this composition.

Crystal Growth. For the crystal growth runs, a pellet of composition 'a' weighing 2 gms was placed in a 0.35 cm³ Au crucible which was supported on a Au sheet on an alumina plate. The sample was heated at 200°C/hr to 970°C in air, held at 970°C for 1.5 hrs and cooled at 25°C/hr to 400°C.

The liquid moved along the inner and outer surfaces and therefore could not be contained in the crucible. Free-standing, well-formed YBC crystals were found in the cavity between the crucible bottom and the support plate where the liquid moved away from the crystals, as can be seen in the micrograph, Fig. 2. YBC crystals were rectangular prisms of two types. Thin platelets (up to 1×2×0.01 mm) were observed similar to those grown in Pt crucibles. The other type had sharp edges of nearly equal lengths, up to 0.5×0.5×0.2 mm. These crystals were fairly easy to separate from the Au. Smaller crystals (up to 0.2×0.2×0.1 mm) of YBC also grew on the inner walls of the crucible and in the solidified mass. The latter had residual melt on the surfaces or were embedded in the solid, making it difficult to isolate them for measurement.

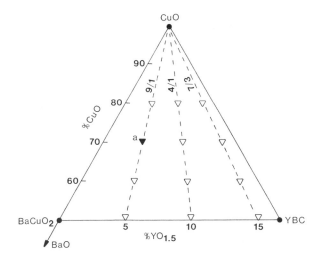

Figure 1. Pseudoternary system YBC-BaCuO$_2$-CuO showing compositions along constant Ba/Y lines used in melting studies in Au crucibles at 950°C in air. Composition 'a' was used in all crystal growth runs. Compositions are given in terms of mole percent of the binary oxides BaO, CuO and YO$_{1.5}$.

Figure 2. Scanning electron micrograph of YBC crystals grown in a Au crucible. Note the extremely sharp edges and flat facets. (Micrograph courtesy of J. G. Clabes and D. A. Smith)

Results

X-Ray and Compositional Analyses. Single crystal x-ray precession photographs of YBC crystals from the Pt and Au crucible runs confirmed their perovskite-related structure. Crystal composition was determined by electron microprobe analysis. The Y/Ba/Cu ratios were 1/2/3 in each crystal within the accuracy of the microprobe measurements. However, the crystals grown in Au crucibles contained approximately 0.7 mole% Au.

Magnetic Measurements. To determine superconducting transition temperatures, magnetic measurements were made using a SHE 905 magnetometer equipped with a 40 KG superconducting solenoid. A typical magnetization vs. temperature curve for a crystal from a Au crucible run is shown in Fig. 3. The diamagnetic shielding and Meissner effects were observed on heating and cooling respectively in a low field (14 Oe). Reasonable agreement was obtained between calculated and measured susceptibilities after correction for demagnetization. The low Meissner effect of 20% is typical of these oxide superconductors (8). We estimate the critical temperature for this crystal to be 80 K or higher. In general, our crystals have T_c values in the range of 55 to 85 K. As in ceramic samples (4,9,10), the critical temperatures are strongly dependent on oxygen concentration and have been increased by annealing in oxygen. We are currently searching for optimal annealing conditions.

Resistivity Measurements. Four-probe resistivity measurements were also used to determine critical temperatures. The resistance vs. temperature profile for the crystal shown in Fig. 4 gives a critical temperature of about 85 K. Electrical contact was made to the crystal by wire-bonding aluminum to the surface; however, this method is not always successful. Alternative techniques have given ohmic contacts with low resistance which have eliminated the self-heating problem associated with the aforementioned method. These results will be the subject of another publication.

Conclusions

In summary, we have developed a technique for growing single crystals of YBC reproducibly under carefully controlled conditions. The as-grown crystals are superconducting with transition temperatures in the range 55 to 85 K; post-annealing in oxygen increases the T_c. Detailed transport, magnetic and optical properties of the YBC single crystals are currently under investigation.

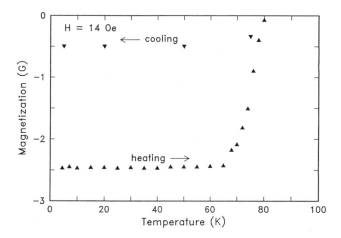

Figure 3. Temperature-dependent magnetization data at low field for a crystal from a Au crucible run. Diamagnetic shielding and the Meissner effect occur on heating and cooling respectively.

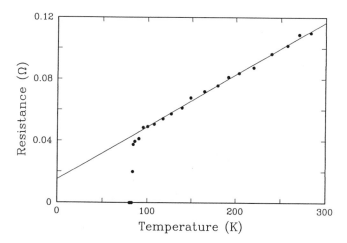

Figure 4. Temperature-dependent resistance for a crystal from a Pt crucible run.

Acknowledgments

We thank S. LaPlaca for single crystal x-ray diffraction measurements, H. Lilienthal for magnetization measurements and K. Kelleher for microprobe measurements.

Literature Cited

1. Bednorz, J. G.; Muller, K. A. Z. *Phys.* 1986, **B64**, 189.

2. Wu, M. K.; Ashburn, J. R.; Torng, C. J.; Hor, P. H.; Meng, R. L.; Gao, L.; Huang, Z. J.; Wang, Y. Q.; Chu, C. W. *Phys. Rev. Lett.* 1987, **58**, 908.

3. Cava, R. J.; Batlogg, B.; van Dover, R. B.; Murphy, D. W.; Sunshine, S.; Siegrist, T.; Remeika, J. P.; Rietman, E. A.; Zahurak, S.; Espinosa, G. P. *Phys. Rev. Lett.* 1987, **58**, 1676.

4. Grant, P. M.; Beyers, R. B.; Engler, E. M.; Lim, G.; Parkin, S. S. P.; Ramirez, M. L.; Lee, V. Y.; Nazzal, A.; Vazquez, J. E.; Savoy, R. J. *Phys. Rev. B* 1987, **35**, 7242.

5. LePage, Y.; McKinnon, W. R.; Tarascon, J. M.; Greene, L. H.; Hull, G. W.; Hwang, D. M. *Phys. Rev B* 1987, **35**, 7245.

6. Kaiser, D. L.; Holtzberg, F.; Scott, B. A.; McGuire, T. R. submitted to *Appl. Phys. Lett.*

7. Migeon, N.; Jeannot, F.; Zanne, M.; Aubry, J. *Rev. Chim. Miner.* 1976, **13**, 440.

8. Dinger, T. R.; Worthington, T. K.; Gallagher, W. J.; Sandstrom, R. L. submitted to *Phys. Rev. Lett.*

9. Qadri, S. B.; Toth, L. E.; Osofsky, M.; Lawrence, S.; Gubser, D. U.; Wolf, S. A. *Phys. Rev. B* 1987, **35**, 7235.

10. Strobel, P.; Capponi, J. J.; Chaillout, C.; Marezio, M.; Tholence, J. L. *Nature*, 1987, **237**, 306.

RECEIVED July 6, 1987

Chapter 9

Broad Search for Higher Critical Temperature in Copper Oxides

Effects of Higher Reaction Temperatures

J. B. Torrance, E. M. Engler, V. Y. Lee, A. I. Nazzal, Y. Tokura, M. L. Ramirez, J. E. Vazquez, R. D. Jacowitz, and P. M. Grant

Almaden Research Center, IBM, San Jose, CA 95120

A wide variety of the ternary (L-M-Cu) and binary (L-Cu and M-Cu) oxides involving rare earths (L = Y, La, Lu, and Sc) and alkaline earths (M = Sr, Ba) with copper were prepared. The vast majority of the 300 samples prepared were not metallic. Several contained known superconducting phases. In the Y-Ba-Cu-O system, many samples prepared at 1050 and 1200 C show some partial melting and deviate strongly from the predictions of the phase diagram determined at 950 C. In a few samples, anomalous resistance decreases were observed at temperatures as high as 260K. In each case, however, these anomalies could be directly traced to problems with strongly temperature dependent contact resistance and phase problems due to high degree of sample inhomogeneity. Multiphase samples prepared at the higher temperatures (i.e. 1050 and 1200 C) appear most prone to such resistance artifacts.

Since the initial discovery of higher temperature superconductivity in perovskite copper oxide compounds (1), the race for further improving T_c has gone on unabated. The superconducting behavior of $La_{2-x} Ba_x Cu O_{4-y}$ and $Y Ba_2 Cu_3 O_{9-y}$ and their isostructural derivatives has been well documented by numerous research groups (2). Superconductivity in these compounds is reproducible, stable and confirmed by a variety of physical measurements in many laboratories. The typical value of T_c for $Y Ba_2 Cu_3 O_{9-y}$ lies between 95 and 100K. In contrast, a small, but growing number of isolated reports have appeared (3-9), where much higher superconducting transitions are claimed in short-lived and irreproducible samples. Considering the unprecedented breakthroughs of the last six months, such fleeting behavior might be signalling another round of increases in T_c.

With the hope that there exist copper oxide compounds with higher T_c, we have undertaken an extensive matrix experiment, in which composition and processing conditions were varied for a wide variety of rare earth-alkaline earth-copper-oxide combinations. Most of the higher T_c, reports have involved Y-Ba-Cu-O compositions, and so this system received the major portion of our attention. We also examined other isoelectronic elements such as Sc, La, and Lu substituting for Y and, to a lesser extent, Sr for Ba. Binary rare earth-

0097–6156/87/0351–0085$06.00/0
© 1987 American Chemical Society

copper and alkaline earth-copper and ternary rare earth-alkaline earth-copper combinations were prepared.

Sample Preparation

The starting compositions that were prepared are outlined in Figure 1. Figures 2 and 3 provide the pseudo-ternary phase diagram for La-Ba-Cu and Y-Ba-Cu. For La-Ba-Cu, we concentrated on the unexplored Cu-rich region (Figure 2), neglecting other areas where there has been extensive work. For Y-Ba-Cu, most of the compositions were chosen outside of the shaded area of Figure 3, since in that area the known $Y Ba_2 Cu_3 O_x$ phase is expected to be formed (10-12). We searched in the "$K_2 Ni F_4$" region (where several high temperature anomalies have been reported (3-6)) as well as the unexplored Cu-rich compositions. All compositions were initially calcined in flowing O_2 for 12 hours at 700 C, cooled, reground, and then pressed into pellets for a final firing under O_2. The main process parameters examined were temperature and quench rate. Each composition was heated at 875 and 1050 C for 16 hours in flowing oxygen. Most of the Y-Ba-Cu compositions were also fired at 1200 C for several hours. Quench rate effects were studied by either removing the sample rapidly from the oven or by allowing the oven to slowly cool to room temperature over about 5 hours. Since some reports of higher T_c claim the behavior is time dependent, disappearing after a few hours to a few days, our electrical measurements were carried out typically within 2-12 hours after removal from the oven. In our experiments, reaction temperature was found to be a more important variable than the cooling rate and most of our effort was concentrated on the effects of the former.

We use phase sensitive detection as the basis of our four-probe measurements. The voltage source is the reference channel output of a Princeton Applied Research model 124A lock-in amplifier set to 1 volt rms at a frequency of 100 Hz. A typical value of Rs is 100 Kohm or 1M Ohm, yielding a nominal constant current through the sample of 10 or 1 microamperes, respectively, under conditions of low contact impedance. The voltmeter is the signal channel of the lock-in, a PAR model 116 differential amplifier with 100M Ohm input impedance. The phase angle between reference and signal channel is adjusted for maximum voltage signal. Normally the current and voltage are then in-phase if the contacts have no capacitive component. Our contacts are comprised of 10 mil Au wire imbedded into silver paste stripes painted on the sample. The current contacts completely cover the sample ends, while the voltage contacts are applied around the body of the material.

Results and Discussion

The preliminary results are summarized in Figures 2 and 3 for the La-Ba-Cu and Y-Ba-Cu systems, respectively. Semiconducting samples are marked with a solid triangle, metals with a circle, and superconductors with a square. Samples which melted are shown as a solid dot. The midpoint temperature of any resistance anomaly is indicated in the circle or square. From the point of view of discovering new superconducting systems, the results were disappointing and indicative of the difficulty of making such a discovery. We found no metallic samples in our limited search in the Lu-Ba-Cu, Sc-Ba-Cu, Sr-Cu, or Ba-Cu systems (under the conditions employed). The resistance anomaly at 32K in $La_2 Cu_5$ is probably associated with the superconductivity discovered at 40K in samples of nominal $La_2 Cu O_4$ (13-15). The other anomalies observed (Figure 2) may be associated with superconductivity reported in La_3

(a)

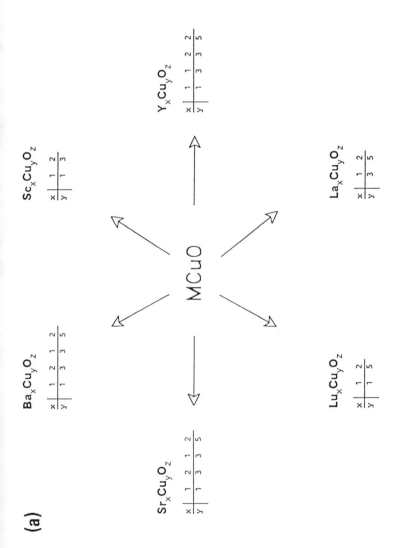

Figure 1a. Schematic diagram of compositions of pseudobinary copper oxides prepared using Ba, Sr, Lu, La, Y, and Sc.

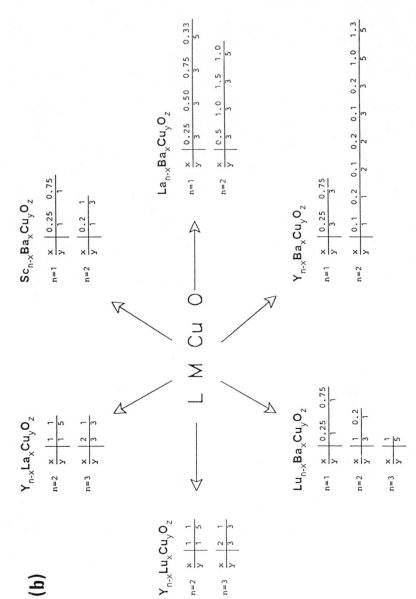

Figure 1b. Schematic diagram of pseudoternary copper oxides prepared in this work.

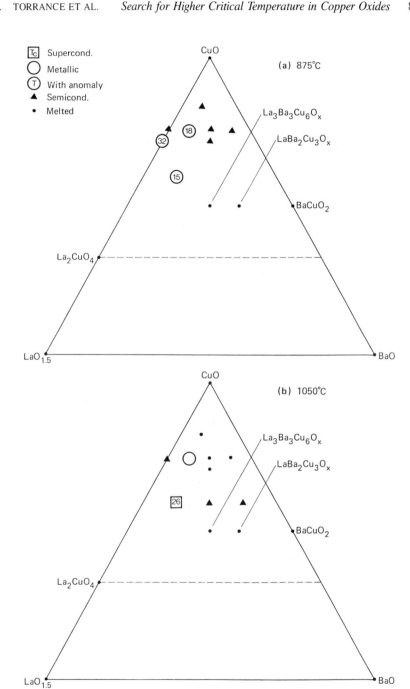

Figure 2. Compositions of La–Ba–Cu–O system and summary of results for samples prepared at (a) 875 °C and (b) 1050 °C.

Ba_3 Cu_6 O_x, (16) or La Ba_2 Cu_3 O_x (17-18). In the Y-Ba-Cu system (Figure 3), the anomalies observed near 90K are probably associated with the super-conducting phase Y Ba_2 Cu_3 O_x (Tc = 95K) (2).

The unexpected results of this study have come from the samples prepared at higher than normal temperatures (i.e. 1050 and 1200 C vs 950 C). The first result concerns the phase diagram for Y-Ba-Cu determined (10,11) at 950 C, some of the tie lines of which are shown in Figure 3a,b,c. Samples reacted at 875 (Fig.3a) behave as expected: that is, samples with compositions within the shaded area should contain some of the phase Y Ba_2 Cu_3 O_x and hence become superconducting near 90K, as observed. Compositions outside this shaded area were found to be semiconducting.

When these same compositions are prepared by reacting the material at 1050 and 1200 C, different results are obtained, as readily seen as shown in Figs. 3b and c, where a number of samples with compositions outside the shaded area exhibit superconductivity. Samples prepared at 1050 C tend to have superconducting transitions near 45K, while those prepared at 1200 C have T_c near 90K. Preliminary powder X-ray diffraction results indicated that $Y_{1.8}$ $Ba_{0.2}$ Cu_2 O_x prepared at 1050 C contains primarily Y_2 Cu_2 O_5 and Y Ba_2 Cu_3 O_x, with very little of the green Y_2 Ba Cu O_5. Large amounts of this latter phase would have been expected on the basis of the 950 C phase dia-gram. These data suggest that a quite different phase diagram which is ap-propriate for the 1050/1200 C samples. In fact, at these higher temperatures it appears that the tie line extends NOT from Y_2 Ba Cu O_5 to CuO, but from Y_2 Cu_2 O_5 to Y Ba_2 Cu_3 O_x. Partial melting of some of these samples indicates that this behavior may arise from complications of liquidus effects. More ex-tensive measurements are necessary to confirm this conjecture.

For all the highly conducting samples made in this study, care was taken to measure the conductivity quickly after removal from the oven, and to look carefully between 300K and 90K. Generally, no indication of any resistance anomaly was observed, consistent with all previous work in our laboratory. On several samples prepared at 1050 and 1200 C, however, anomalies in the measured voltage were observed at high temperatures. In Fig. 4, an example of these data is shown for a sample with a composition of $Y_{1.5}$ $Ba_{.5}$ Cu O_x.

In this case, the effect observed is produced by temperature dependent current contacts and NOT by superconductivity. Here the room temperature current contact resistance was about 4K ohm. This resistance increased to around 10^9 ohms at 77 K, thus, over most of the temperature range of the measurement, the voltage source was not behaving as an ideal current source, resulting in an apparent drop in resistance as the temperature was lowered, producing the 220 K "onset" shown in Fig. 4.

In one of our samples of composition $Y_{1.2}$ $Ba_{0.8}$Cu O_x, the voltage contact resistance varied over several megaohms between room temperature and 77 K. Under these conditions, the 100 megaohm input impedance of the signal channel no longer represented an ideal voltmeter and significant fluctuating current was drawn through the voltage contacts resulting in an apparent resistive anomaly in the 200-300 K range. This anomaly disappeared under temperature cycling as the voltage contact impedance decreased.

These results demonstrate that high temperature resistive anomalies can be produced by temperature dependent contact resistance. Moreover, our ex-perience has been that the contacts can become quite capacitive as the tem-perature is lowered, shifting the voltage output into quadrature, again producing an apparent drop in resistance.

Given the fact that resistive anomalies can arise from the above variety of measurement artifacts, we suggest that workers observing indications of

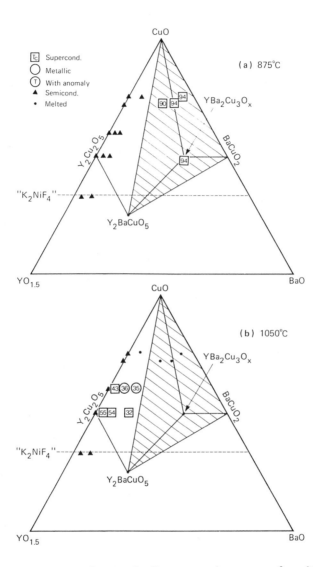

Figure 3. Compositions of Y–Ba–Cu–O system and summary of results for samples prepared at (a) 875 °C and (b) 1050 °C. *Continued on next page.*

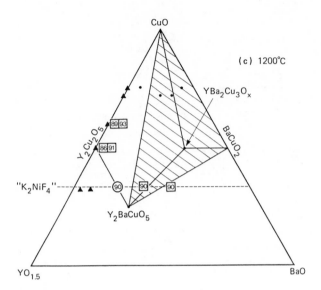

Figure 3. *Continued.* Compositions of Y–Ba–Cu–O system and summary of results for samples prepared at (c) 1200 °C.

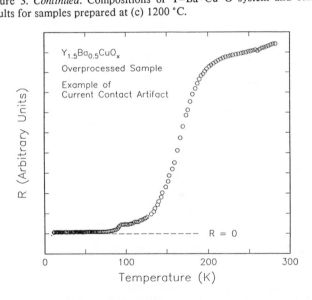

Figure 4. Resistance of sample prepared at 1200 °C exhibiting anomalous drop at 170 K. It is shown that the behavior observed in this sample is caused by an artifact rather than superconductivity.

superconductivity above 100 K during four-probe measurements follow the following guidelines in reporting their results:

- Measure temperature dependence of contacts by two-probe pairwise measurements and report numerical values.
- Permute current and voltage contacts to check for topological artifacts (19-20).
- Monitor quadrature signal throughout temperature scan for capacitive effects and report results.
- If anomalies are observed, try several different sample currents and signal frequencies.
- Look at second harmonic component via oscilloscope for evidence of contact/sample nonlinearities.
- Report fully on all observations listed above.

Conclusion

A broad search of 57 copper oxide compositions prepared at three different temperatures has indicated the difficulty of discovering new superconducting compounds. The dependence of the results on the preparation temperature suggests that the effective phase diagram may be quite different at different temperatures. Finally, it was found that samples prepared with multiple phases at too high temperatures can show resistivity anomalies at temperatures as high as 260K, which are not an indication of superconductivity. Rather, they are artifacts caused by the extreme microscopic inhomogenity of such samples and the difficulties of making good contacts to them.

Acknowledgment

We gratefully acknowledge helpful discussions with R. Beyers and K.G. Frase.

Literature Cited

1. Bednorz, J. G. and Muller, K. A. Z. Phys. B 1986 64, 189
2. For a general review see:Engler, E. M.; CHEMTECH , 1987, to be published
3. Chu, C. W. Proc. Natl. Acad. Sci., to be published
4. Chen, J. T.; Wenger, L. E.; McEwan, C. J.; and Logothetis, E. M.. Phys. Rev. Lett. 1987, 58, 1972
5. Hayri, E. A.; Ramanujachary, K. V.; Li, S.; Greenblatt, M.; Simizu, S.; and Friedberg, S. A. ibid, submitted.
6. Politis, C.; Geerk, J.; Dietrich, M.; Obst, B.; and Luo, H. L. Z. Phys. B, 1987, 66, 279.
7. Geballe, T. H.; Kapitulnik, A.; Beasley, M. R.; Hammond, R. H.; Webb, D. J.; Mitzi, D. B.; Sun, J. Z.; Kent, A. D.; Hsu, J. W.; Arnason, S.; Gusman, M.I.; Hildenbrand, D. L.; Johnson, S. M.; Quinlan, M. A.; and Rorocliffe, D. J. Mater. Res. Soc. Bull., to be published.
8. Ovshinsky, S. R.; Young, R. T.; Allred, D. D.; DeMaggio, G.; and Van der Leeden, G. A. Phys. Rev. Lett. to be published; Zettl, A., Phys. Lett. to be published.
9. Press reports from USA, India, USSR and Japan have claimed 240 K and even room temperature superconductivity in undisclosed compositions.
10. Frase, K. G.; Diniger, E. G.; and Clarke, D. R. Commun. Amer. Ceramic Soc. 1987.to be published.

11. Hwu, S. J.; Song, S.N.; Ketterson, J. B.; Mason, T. O.; and Poeppelmeier, K. R. ibid 1987,to be published.
12. Che, G. C.; Liang, J. K.; Chen, W.; Yang, Q. A.; Chen, G. H. and Ni, Y. M. J. Less-Common Metals, to be published.
13. Grant, P. M.; Parkin, S. S. P.; Lee, V. Y.; Engler, E. M.; Ramirez, M. L.; Vazquez, J. E.; Lim, G.; Jacowitz, R. D. and Greene, R. L. Phys. Rev. Lett. 1987 58 , 2482
14. Beille, N. J.; Cabanal, R.; Chaillout, C.; Chevallier, B.; Demazeau, G.; Deslandes, F.; Etourneau, J. Lejay, P.; Michel, C.; Provost, J.; Raveau, B.; Sulpice, A.; Tholence, J.L. and Thornier, R. Physique Mat. Condensie 1987, to be published.
15. Sekizawa, K.; Takano, Y.; Takigami, H.; Tasaki, S.; and Inaba, T. Jpn. J. Appl.Phys. 1987. 26 , L840.
16. Mitzi, D. B.; Marshall, A. F.; Sun, J. Z.; Webb, D. J.; Beasley, M. R.; Geballe, T. H.; and Kapitulnik, A. to be published.
17. Hor, P. H.; Meng, R. L.; Wang, Y. Q.; Gao, L.; Huang, Z. J.; Bechtold, J.; Forster, K.; and Chu, C. W. Phys. Rev. Lett. 1987 58, 1891.
18. Murphy, D. W.; Sunshine, S.; van Dover, R. B.; Cava, R. J.; Batlogg, B.; Zahurak, S. M.; and Schneemeyer, L. F. ibid 1888.
19. Shafer, D. E.; Wudl, F; Thomas, G. A.; Ferraris, J. P.; and Cowan, D. O. Solid State Commun. 1974 14, 347.
20. Bickford, L. R.; and Kanazawa, K. K. J. Phys. Chem. Solids. 1976 37, 839.

RECEIVED July 6, 1987

Chapter 10

Effect of Nominal Composition on Superconductivity in La_2CuO_{4-y}

S. M. Fine[1], M. Greenblatt[1], S. Simizu[2], and S. A. Friedberg[2]

[1]Department of Chemistry, Rutgers—The State University of New Jersey, New Brunswick, NJ 08903
[2]Department of Physics, Carnegie-Mellon University, Pittsburgh, PA 15213

Samples of La_2CuO_4 prepared with a nominal stoichio-metry slightly deficient in lanthanum (i.e. $La_{1.9}CuO_4$) showed a greater diamagnetic susceptibility and a sharper resistive transition than samples prepared from stoichiometric amounts of starting materials. The effect of adding K_2CO_3 to the reaction mixture is also discussed.

Recently Bednorz and Muller (1) reported possible superconductivity at 30 K in the metallic-oxide La-Ba-Cu-O system. Subsequent work by Tanaka and co-workers (2) showed that the high temperature supercon-ductivity is due to the solid solution $La_{2-x}Ba_xCuO_{4-y}$ which forms in the K_2NiF_4-type structure. Making solid solutions with strontium instead of barium, Cava et. al. (3). were able to show bulk super-conductivity at 38 K in $La_{2-x}Sr_xCuO_{4-y}$. While solid solutions of the type $La_{2-x}M_xCuO_{4-y}$ (M = Sr, Ba) have been known for some time, (4-5) magnetic and electron transport properties of these compounds had not previously been measured at low temperatures (i.e. <120 K). More recently, (6-7) several groups have observed superconductivity in the unsubstituted phase La_2CuO_{4-y} at temperatures just below the metal to insulator phase transition. However, in this case the Meissner effect measurements were consistent with only a small fraction of the sample being in the superconducting state. While the magnetic and electron transport properties of La_2CuO_{4-y} had been measured some time ago, (8) superconductivity was not previously observed. Thus, the presence or absence of superconductivity in samples of La_2CuO_{4-y} is a sensitive function of the method of sample preparation.

The observation of superconductivity in nominally pure La_2CuO_{4-y} is of great interest, particularly considering the recent reports of an antiferromagnetic ordering in La_2CuO_{4-y} near 220 K. (9) Matheiss (10) has suggested that near-perfect Fermi-surface nesting makes orthorhombic La_2CuO_{4-y} (space group Cmca) susceptible to a Peierls distortion. Kasowski et. al. (11) have suggested that orthorhombic

La_2CuO_{4-y} is metallic, and that the sharp increase in the resistivity
of La_2CuO_{4-y} at low temperatures is due to a phase transition to a
lower symmetry space group. Thus, the filamentary superconductivity
in nominally pure La_2CuO_{4-y} can be attributed to regions of the
sample in which the metal to insulator phase transition has been
suppressed.

In the present study we discuss the nature of the
superconductivity in La_2CuO_{4-y}. More specifically, it was found that
purposefully preparing La_2CuO_4 with a nominal stoichiometry
deficient in lanthanum significantly sharpened the superconducting
resistive transition and increased the magnitude of the diamagnetic
susceptibility. The effect of adding K_2CO_3 to the starting mixture
is also discussed.

Samples were prepared by the solid state reaction of appropriate
amounts of La_2O_3 and CuO. In some cases K_2CO_3 was added to the
mixture of starting materials. Since basic oxides are known to
stabilize transition metals in higher oxidation states, we wanted
to: 1) see if we could synthesize solid solutions of the type $La_{2-x}K_xCuO_{4-y}$. 2) see if adding K_2CO_3 to the starting mixture would
effect the superconductivity in La_2CuO_{4-y}. Intimate mixtures of the
reactants were fired at 900° for 16 h, pressed into pellets, then
sintered at 1050°C for 16 h. The samples were then annealed for 10
h at 700°C in a flowing oxygen atmosphere. Resistivity measurements
were done using the four probe van der Pauw technique with a
measuring current of between 0.2 and 1 mA. Electrical contacts to
the sample were made with ultrasonically soldered indium. The
temperature was measured using a carbon-glass thermometer which had
been calibrated against a silicon diode. The X-ray diffraction data
was collected on Scintag PAD V diffractometer using CuKa radiation
and a germanium single crystal detector. The AC susceptibilities of
flat disc shaped pellets were measured with an oscillatory field of
ca. 30 Oe at a frequency of 80 Hz. The oxides La_2O_3 and CuO were
mixed in appropriate amounts to form samples with nominal
stoichiometries La_2CuO_{4-y} and $La_{1.9}CuO_{4-y}$ respectively. Each mixture
of starting materials was then divided into three portions; 5% K_2CO_3
and 10% K_2CO_3 were added to the second and third portion of each
mixture respectively. A sample of nominal composition $La_{1.9}K_{0.1}CuO_{4-y}$
was also prepared from appropriate amounts of La_2O_3, CuO, and K_2CO_3.

All compositions were either single phase or nearly single phase
as determined by X-ray powder diffraction. In the least pure sample
there was only one diffraction peak with 3% of the maximum intensity
which could not be indexed in the orthorhombic La_2CuO_{4-y} structure
(Cmca). None of the samples to which K_2CO_3 had been added showed
diffraction peaks which could be attributed to either K_2O or K_2CO_3,
suggesting that either K_2O sublimed out of the sample, or was
present in a poorly crystalline form. The lattice parameters
calculated for all seven compositions are the same within two
standard deviations, and are consistent with the published values
for the lattice parameters for La_2CuO_{4-y}. Thus, the implicatons is
that in all cases the bulk of the sample is stoichiometric,
La_2CuO_{4-y}, and that a solid solution of the type $La_{2-x}K_xCuO_{4-y}$ did not
form.

The resistivity vs. temperature data for three of the
compositions are shown in Fig. 1. As can be seen, the sample with

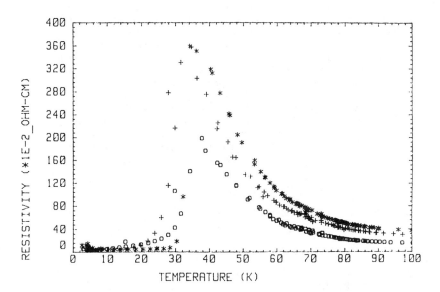

Figure 1. Temperature dependence of the resistivity for three nominal compositions, La_2CuO_{4-y} (open circles), $La_{1.9}CuO_{4-y}$ (asterisks), and La_2CuO_{4-y} + 5% K_2CO_3 (plus signs).

nominal compositon $La_{1.9}CuO_{4-y}$ shows a large negative temperature
coefficient of resistivity just above T_c and a sharp transition into
the superconducting state. The transition into the superconducting
state is significantly sharper in $La_{1.9}CuO_{4-y}$ than in either La_2CuO_{4-y}
or La_2CuO_{4-y} to which 5% K_2CO_3 had been added. The temperature
dependence of the resistivity for the sample of nominal composition
$La_{1.9}K_{0.1}CuO_{4-y}$ is similar to that of $La_{1.9}CuO_{4-y}$. This is consistent
with our X-ray diffraction data which indicates that in both
$La_{1.9}CuO_{4-y}$ and $La_{1.9}K_{0.1}CuO_{4-y}$ the bulk of the sample is stoichiometric
La_2CuO_{4-y} and that a solid solution of the type $La_{2-x}K_xCuO_{4-y}$ did not
form. The effect of adding 5% K_2CO_3 to the sample with nominal
composition La_2CuO_{4-y} is less dramatic. The resistivity of the
sample to which 5% K_2CO_3 was added showed a larger resistivity just
above T_c and a slightly sharper transition into the superconducting
state than the sample to which no K_2CO_3 had been added. The
resistivity data for all seven compositions are shown in Table I.

Table I.   Superconducting Critical Temperatures for the Samples
            Prepared in This Study as Determined by Resistivity vs.
            Temperature Measurements. The $T_{minimum}$ is the Temperature
            at Which the Minimum Resistivity is Measured and R
            is the Resistivity Just Above T_{onset}

Nominal Composition	T_{onset}(K)	$T_{midpoint}$(K)	$T_{minimum}$(K)	R (ohm-cm)
La_2CuO_4	37	32	4	2.0
La_2CuO_4 + 5% K_2CO_3	35	29	8	3.6
La_2CuO_4 + 10% K_2CO_3	30	22	3	5.9
$La_{1.9}CuO_4$	34	32	22	3.5
$La_{1.9}CuO_4$ + 5% K_2CO_3	33	28	9	7.8
$La_{1.9}CuO_4$ + 10% K_2CO_3	33	28	8	4.3
$La_{1.9}K_{0.1}CuO_4$	35	32	17	21.4

The AC susceptibility data for three of the compositions are
shown in Fig. 2. The samples with nominal compositions $La_{1.9}K_{0.1}CuO_{4-y}$
and $La_{1.9}CuO_{4-y}$ respectively exhibit a slope discontinuity at ca.
33 K which appears to indicate the onset of superconductivity. The
diamagnetic susceptibility increases only slowly as the temperature
is lowered, and at 4.2 K reaches a value of $3x10^{-5}$ cm^3/g for
$La_{1.9}CuO_{4-y}$ and $5x10^{-5}$ cm^3/g for $La_{1.9}K_{0.1}CuO_{4-y}$. These values
correspond to between 0.3% and 0.4% of the diamagnetism expected for
a bulk superconductor. The imaginary ac susceptibility component
(χ'') is shown as a function of temperature in Fig. 3. As we found
in $Ba_2LnCu_3O_7$ (ln = Y, Ho, Er), (12) χ'' increases below T_c. However
in neither $La_{1.9}CuO_{4-y}$ nor $La_{1.9}K_{0.1}CuO_{4-y}$ did we find a maximum in χ''
above 4.2 K. Slow increase in χ' and a non-zero χ'' seem to arise
from a finite resistance, perhaps at grain boundaries, against a
macroscopic shielding current induced by the oscillatory field. The
resistance evidently decreases as the temperature is lowered.
Although the resistivity of La_2CuO_{4-y} also exhibits an abrupt drop

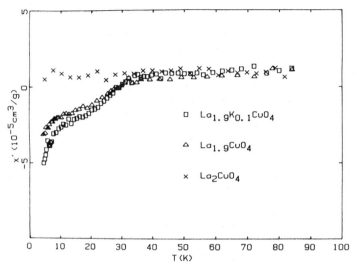

Figure 2. Temperature dependence of the AC susceptibility for La$_{1.9}$K$_{0.1}$CuO$_{4-y}$, La$_{1.9}$CuO$_{4-y}$, and La$_2$CuO$_{4-y}$.

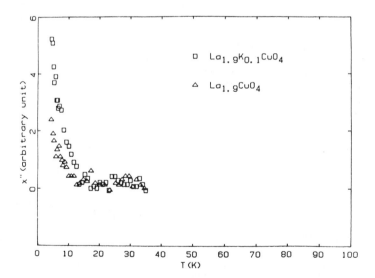

Figure 3. Temperature dependence of the imaginary part of the susceptibility (in arbitrary units) of La$_{1.9}$K$_{0.1}$CuO$_{4-y}$ and La$_{1.9}$CuO$_{4-y}$.

below 35 K, the susceptibility remains constant and positive down to 4.2 K. Thus, the fraction of this sample which is in the superconducting state is either very small or zero.

To summarize, the effect of preparing lanthanum cuprate with a slightly lanthanum deficient nominal stoichiometry (i.e. $La_{1.9}CuO_4$) is to: 1) increase the resistivity just above T_c, 2) sharpen the resistive transition into the superconducting state, 3) increase the temperature at which the minimum resistivity is measured and 4) increase the magnitude of the diamagnetic susceptibility. The reason for these effects is unclear, but is probably related to lanthanum deficient regions within the sample in which the mixed valency of copper (Cu(II), Cu(III)) is stabilized and the metal to insulator phase transition has been suppressed. Raveau and co-workers (7) found that the samples of La_2CuO_4 which were annealed under high oxygen pressure showed a much larger diamagnetic susceptibility than samples which were simply fired in air. High oxygen pressure also should stabilize the mixed valency of copper (Cu(II), Cu(III)). The importance of mixed valent copper for super-conductivity in solid solutions of the type $La_{2-x}M_xCuO_{4-y}$ (m = Sr, Ba) has been suggested by many groups. (2,3,13) Samples of $La_{2-x}M_xCuO_{4-y}$ which are annealed in oxygen show sharp transitions into the superconducting state, while those annealed in vacuum are no longer superconducting.(14)

Acknowledgments

We acknowledge helpful conversations with W. H. McCarroll and W. McLean. This work was supported by a National Science Foundation - Solid State Chemistry Program Grant (DMR84-04003) and a National Science Foundation - Materials Research Instrument Grant (DMR84-08266). The magnetic susceptibility studies were done at Carnegie-Mellon University and were supported by a National Science Foundation Grant (DMR84-05833).

Literature Cited

1. Bednorz, J.G.; Muller, K.A. Phys. 1986, B 64, 189.
2. Takagi, H.; Uchida, S.; Kitazawa, K.; Tanaka, S. to be published in Jpn. J. Appl. Phys.
3. Cava, R.J.; van Dover, R.B.; Batlogg, B.; Rietman, E.A. Phys. Rev. Lett. 1987, 58, 408.
4. Nguyen, N.; Choisnet, J.; Herview, M.; Raveau, B. J. Solid State Chem. 1981, 39, 120; Nguyen, N.; Studer, F.; Raveau, B. J. Phys. Chem. Solids 1983, 44, 389.
5. Shaplygin, I.S.; Kakhan, B.G.; Lazarev, V.B. J. Inorg. Chem. 1979, 24, 620.
6. Grant, P.M.; Parkin, S.S.P.; Greene, R.L.; Lee, V.Y.; Engler, E.M.; Ramirez, M.L.; Vazquez, J.E.; Lim, G.; Jacowitz, R.D. submitted to Phys. Rev. Lett.
7. Beille, J.; Cabanel, R.; Chaillout, C.; Chevallier, B.; Demazeau, G.; Deslandes, F.; Etourneau, J.; Lejay, P.; Michel, C.; Provost, J.; Raveau, B.; Suplice, A.; Tholence, J.; Tournier, R. submitted to Compts. Rendus Acad. Sc.
8. Ganguly, P.; Rao, C.N.R. Mat. Res. Bull. 1973, 8, 405.
9. Vaknin, D.; Sinha, S.K.; Moncton, D.E.; Johnston, D.C.; Newsam, J.; Safinya, C.R.; King, Jr., H.E. submitted to Phys. Rev. Lett.

10. Mattheiss, L.F. <u>Phys. Rev. Lett.</u> 1987, <u>58</u>, 1028.
11. Kasowski, R.V.; Hsu, W.Y.; Herman, F. to be published.
12. Hayri, E.; Ramanujachary, K.V.; Li, S.; Greenblatt, M.;
     Friedberg, S.A.; Simizu, S. submitted to <u>Solid State Commun.</u>
13. for example see: Politis, C.; Geerk, J.; Dietrich, M.; Obst,
     B. <u>Z. Phys. B</u> 1987, <u>66</u>, 141.
14. Takagi, H.; Uchida, S.; Kitazawa, K.; Tanaka, S. <u>Japan. J.
     Appl. Phys.</u> 1987, <u>26</u>, L218.

RECEIVED July 6, 1987

Chapter 11

Interplay of Synthesis, Structure, Microstructure, and Superconducting Properties of $YBa_2Cu_3O_7$

G. F. Holland[1,2], R. L. Hoskins[1,2], M. A. Dixon[1,2], P. D. VerNooy[1,2],
H.-C. zur Loye[1,2], G. Brimhall[3], D. Sullivan[4], R. Cormia[4], H. W. Zandbergen[2],
R. Gronsky[2], and Angelica M. Stacy[1,2]

[1]Department of Chemistry, University of California—Berkeley, Berkeley, CA 94720
[2]Lawrence Berkeley Laboratory, Berkeley, CA 94720
[3]Department of Geology, University of California—Berkeley, Berkeley, CA 94720
[4]Surface Science Laboratories, Charleston Road, Mountain View, CA 94043

Magnetic susceptibility, high resolution transmission electron microscopy and X-ray photoelectron spectroscopy (XPS) have been used to examine the physical properties, microstructure, and surface chemical composition of the high temperature superconductor $YBa_2Cu_3O_7$. Meissner effect and persistent current measurements indicate a lower limit of 35% superconducting material and the presence of flux trapping in sample nominally 90% pure by powder X-ray diffraction. Atomic resolution electron microscopy reveals substantial amounts of intergrowths and structural defects. Unusually high amounts of surface carbon and oxygen have been observed by XPS; these are ascribed to adsorbed CO_2 and H_2O. These microscopic inhomogeneities will limit the technological applications of $YBa_2Cu_3O_7$, in particular the critical current J_c. In order to improve sample homogeneity, we also report results from synthetic efforts aimed at preparation of more uniform $YBa_2Cu_3O_7$ material.

The recently discovered high temperature superconductor $YBa_2Cu_3O_7$ (1-5) has been studied extensively in an effort to understand how the structure and properties relate to the superconducting behavior. Considerable enthusiasm has been generated for the potential technological applications of these materials. However, it is evident that the microstructure of these materials, which depends on the method of preparation, affects the superconducting properties. Prior to application, chemical processing of these complex oxides needs to be modified in order to optimize the superconducting properties. In particular, sample homogeneity is of great importance in maximizing the critical current J_c. In this report, we present a study of the physical and structural properties of $YBa_2Cu_3O_7$, including an analysis of the microstructure. In order to determine whether the superconducting grains in a sintered piece are isolated from each other by resistive material at the grain boundaries, persistent currents in a ring of the superconductor were measured using a SQUID magnetometer. High resolution transmission electron microscopy, together with X-ray photoelectron spectroscopy, have been used to probe

0097–6156/87/0351–0102$06.00/0
© 1987 American Chemical Society

the grain boundaries. These results are correlated with bulk structural studies using powder X-ray diffraction. The microstructural analysis of this oxide shows much more heterogeneity than is evident on a bulk scale. In order to improve microscopic homogeneity, one needs to develop improved synthetic approaches. To that end, we also present our efforts involving hydrothermal processing and molten salt fluxes for the preparation of $YBa_2Cu_3O_7$.

EXPERIMENTAL

All materials used in this study were obtained from commercial suppliers and were of reagent quality or better. Magnetic susceptibility measurements were performed on a SHE SQUID magnetometer in an applied field of 12 Gauss; the field was calibrated using a superconducting Sn sphere. X-ray powder diffraction work was done on a Picker θ-θ diffractometer equipped with a graphite monochromator using Cu $K\alpha_1$ radiation and a scan rate of 1 deg 2θ/min.

Samples for electron microscopy were prepared by crushing the samples under liquid nitrogen in an agate mortar, suspending in ethanol, dispersing the powder ultrasonically, and collecting on a carbon-coated Triafol film bonded to a 200 mesh copper grid. High resolution phase contrast imaging was performed in the JEOL JEM ARM-1000 atomic resolution microscope at the National Center for Electron Microscopy (NCEM). All images were recorded at 1000 kV accelerating potential, using through-focus series settings that bracketed the Scherzer condition (6). Microchemical analysis was performed in the JEOL JEM 200CX analytical electron microscope at NCEM, equipped with Kevex System 8000 software and an ultrathin window X-ray detector sensitive to oxygen. Both elemental maps and point counting were used for analysis.

X-ray photoelectron spectroscopic (XPS) studies were conducted using a Surface Science Laboratories X-ray photoelectron spectrometer. Wavelength-dispersive electron microprobe results were obtained by Mr. John Donovan at the Department of Geology microanalytical facility at UC Berkeley.

<u>Preparation of $YBa_2Cu_3O_7$ via Nitrate Route</u>. Stoichiometric amounts of Y_2O_3, CuO, and $BaCO_3$ were weighed out and then dissolved in concentrated nitric acid (100 ml/5 g sample). The solution was evaporated to dryness, care being taken to avoid decomposition of the nitrates formed. This nitrate mixture was then decomposed at 500°C for 4 h. The resultant grey material was ground mechanically to assure homogeneity, and refired at 750°C for 12 h. The grinding and heating steps were repeated at 800, 900 and 950°C. A final annealing step consisted of heating the sample to 950°C in a pure O_2 flow, followed by cooling to room temperature over 8 h. The X-ray patterns of the samples showed all the lines assigned to the superconducting phase (4,5), together with a small amount of CuO.

In order to examine this route in more detail, the initial nitrate mixture was decomposed at varying temperatures for varying lengths of time. The products formed after each decomposition were analyzed by powder X-ray diffraction.

<u>Attempted Syntheses of $YBa_2Cu_3O_7$ in Molten Salt Fluxes.</u> A large variety of molten salt fluxes ($CuCl_2$, $PbCl_2$, $PbCl_2$/KCl, $PbCl_2$/Ba(OH)$_2$, $PbCl_2$/BaCl_2$, $PbCl_2$/KNO_3$, PbO, KCl/KNO_3$, $BaCl_2$/Ba(OH)$_2$) were explored

in an attempt to synthesize the $YBa_2Cu_3O_7$ superconducting phase. A typical experiment involved combining stoichiometric amounts of Y_2O_3, BaO_2 and CuO, with an excess of flux (5-50x) in an alumina crucible. The reaction mixture was then heated to 500-950°C for periods ranging from 1-4 days, followed by cooling to room temperature. The product obtained was examined by powder X-ray diffraction and/or single crystal X-ray techniques.

Attempted Recrystallization of $YBa_2Cu_3O_7$ via Hydrothermal Routes. Growth of single crystal $YBa_2Cu_3O_7$ was attempted following Hirano' and Takahashi's hydrothermal synthesis of various barium-bismuth-lead oxides (7). Powders of the pre-formed superconductor were placed into small gold or platinum capsules with enough KCl to make a 4.5 M solution at room temperature and pressure (although the salt will not completely dissolve until higher temperatures are reached). Approximately 0.4 ml of solvent was then added, and the capsules are welded shut. Distilled water or 2% H_2O_2 were used in early runs. In an attempt to improve the stability of $YBa_2Cu_3O_7$ in aqueous media, later runs were performed under high pH conditions by the addition of KOH or $Ba(OH)_2$. A side-by-side comparison of the use of peroxide solutions to the use of distilled water showed little difference. Quantities of the superconductor were varied from 2.9 to 40 weight percent. Temperatures were varied from 250 - 550°C, and the pressure for all runs was approximately 1 kilobar.

Persistent Current Measurements. A ring of superconducting material was prepared by drilling a hole through the center of a pellet of $YBa_2Cu_3O_7$. This ring was placed into a SQUID magnetometer with the plane of the ring both perpendicular and parallel to the applied field. In each case, the ring was cooled from 100 K to various temperatures below T_c in an applied field of 42 Gauss. The applied field was turned off and a large positive magnetic moment was detected due to trapped flux.

RESULTS

Preparation via the nitrate route yields materials whose X-ray patterns can be readily indexed based on the published single crystal structure for $YBa_2Cu_3O_7$ (4); CuO is frequently observed as a minor impurity in the powder pattern. Given the approximately 5% detection limit in powder X-ray diffraction, crystalline impurities account for less than 10% of the sample. Assuming the absence of any amorphous impurity phases, this places an upper limit on the sample purity at 90% crystalline $YBa_2Cu_3O_7$. Magnetic susceptibility per gram versus temperature plots, Figure 1, indicate a sharp diamagnetic transition beginning at T_c = 91 K, with a width (10% to 90%) of 10 K. The observed Meissner effect of only 35% may be low due to flux that is not expelled but is trapped by defects.

    The presence of a persistent current is one unique characteristic of superconductors. This persistence is dependent on the sample being chemically pure and free of structural defects. For a ring of pure superconducting substance oriented with the plane of the ring perpendicular to an applied magnetic field, H_{appl}, one should observe a large positive magnetic moment when the sample is cooled in this field to some temperature below T_c, and the field turned off, due to flux which is trapped in the center of the ring. No such magnetic moment should be observed when the ring plane is oriented parallel to H_{appl}. However, we observe large positive moments for $YBa_2Cu_3O_7$ in both orientations.

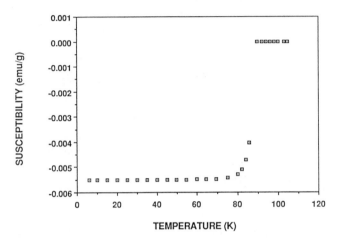

Figure 1. Magnetic susceptibility vs. temperature plot for $YBa_2Cu_3O_7$, measured in a field of 12 Gauss.

Incongruities between X-ray diffraction data and the magnetic susceptibility and persistent current measurements necessitated an examination of the microstructure of the $YBa_2Cu_3O_7$. High resolution electron microscopy revealed a very complex structure. Although quite pure by powder X-ray diffraction, images were dominated by considerable intergrowths of impurity phases like $BaCuO_2$ and Y_2BaCuO_5. Imaging of ordered $YBa_2Cu_3O_7$, Figure 2, may be compared with Figure 3 in which regions of inhomogeneous intergrowths are labelled by an arrow. In addition, the regions of the pure $YBa_2Cu_3O_7$ showed a considerable degree of structural flaws such as stacking faults as shown in Figure 4.

X-ray photoelectron spectral data are useful for study of the surface composition of these materials. The spectra show the presence of substantial amounts of impurities on the surface of powdered $YBa_2Cu_3O_7$. Specifically, we observe high percentages of carbon and proportionately less oxygen and metals than expected (see Table 1).

TABLE I. XPS Results for $YBa_2Cu_3O_7$

element	at. % calc.	obs.	peak	eV
Y	7.7	4.8	$3d_{5/2,3/2}$	157.1, 159.1
Ba	15.4	8.3	$3d_{5/2,3/2}$	779.3, 794.6
Cu	23.1	5.5	$2p_{3/2,1/2}$	932.9, 940.6
O	53.9	40	1s	529.8, (531.1)
C	0	41	1s	284.6

Furthermore, analysis of the metals content at the surface (top 100 Å) shows lower amounts of Cu at the surface compared to the theoretical ratio of Y:Ba:Cu of 1:2:3. Wavelength dispersive electron microprobe is an extremely sensitive probe of the bulk, rather than surface composition. Data acquired by this technique gave a metals stoichiometry much closer to the expected $YBa_2Cu_3O_7$, but now somewhat high in barium and copper, $Y_{1.01(6)}Ba_{2.34(6)}Cu_{3.14(9)}$.

The heterogeneities observed in these microscopic studies gave cause to re-examine our synthetic approach. The conditions of nitrate decomposition proved to be quite important. Thermal decomposition at temperatures less than 750°C produced predominantly $BaCuO_2$, the yttrium remaining as the oxide. Extended (>12h) reaction times, however, lead to formation of Y_2BaCuO_5. We found no evidence of $YBa_2Cu_3O_7$ forming below 750°C based on powder X-ray diffraction of intermediates. The $BaCuO_2$ and Y_2BaCuO_5 phases formed at lower temperatures persisted at the higher temperatures required for formation of $YBa_2Cu_3O_7$, but after several heat-and-grind cycles they could no longer be detected by powder X-ray diffraction. Higher reaction temperatures for the nitrates were found to avoid the formation of these troublesome intermediates, as long as the temperature does not exceed 970 C. $YBa_2Cu_3O_7$ itself is unstable above 975°C, decomposing to the green Y_2BaCuO_5 phase. Reactions of the nitrates between 750°C and 950 C quickly lead to formation of the superconducting phase. Flash decomposition of the dried nitrate, accomplished by inserting the crucible in an 800°C furnace for 4 h, followed by an air quench to room temperature, shows predominantly the $YBa_2Cu_3O_7$ X-ray pattern.

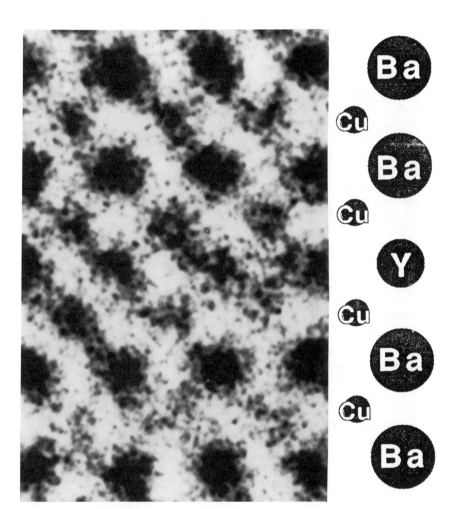

Figure 2. High resolution electron micrograph of $YBa_2Cu_3O_7$ viewed along [100] or [010] projection.

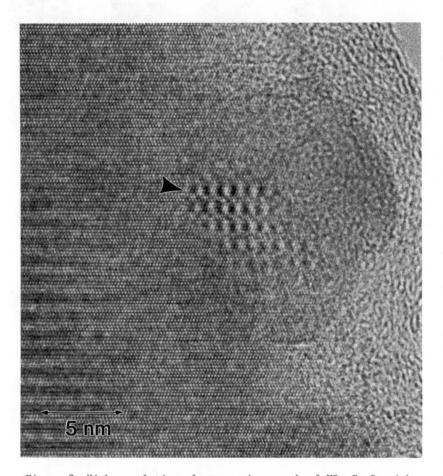

Figure 3. High-resolution electron micrograph of $YBa_2Cu_3O_7$ with local heterogeneities marked by an arrow. Grain surface is located in upper right corner.

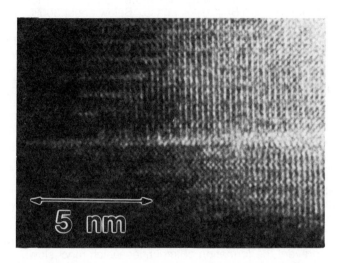

Figure 4. High-resolution electron micrograph of $YBa_2Cu_3O_7$ with fault plane extending parallel to crystallographic c-axis.

Efforts to avoid formation of stable impurity phases lead to attempts to synthesize $YBa_2Cu_3O_7$ by different preparative schemes. Reaction of Y_2O_3, BaO_2 and CuO in molten salt fluxes, over temperatures ranging from 400°C to 950°C, frequently lead to the isolation of highly crystalline binary or ternary oxides. However, these syntheses resulted in the $YBa_2Cu_3O_7$ compound in only one case, in which an excess of $Ba(OH)_2$ was used as both flux and Ba source. While most crystals isolated were identified as CuO or various mixed metal oxides, a number of unidentified single crystals await further examination.

We also explored hydrothermal routes as an alternate means for recrystallization of the $YBa_2Cu_3O_7$ phase. Below about 300°C, very little of the oxide dissolves. At elevated temperatures, clear, high quality crystals of yttrium or barium oxide are isolated. At 450°C and 32 weight percent superconductor, we found large single crystal needles of CuO. Adding a 7 weight percent excess of BaO_2 or CuO produced only a bluish, noncrystalline mud. We also attempted to increase the stability of the superconductor by increasing the pH with various amounts of $Ba(OH)_2$ and KOH, but with no success to date. Reaction at temperatures ranging from 350- 550°C, in solutions of varying ionic strength, and over an extended range in pH has not yet yielded a Y-Ba-Cu-O compound of the appropriate stoichiometry.

DISCUSSION

Preparation of the $YBa_2Cu_3O_7$ via decomposition of the precipitated nitrates yields material that is of high purity based on powder X-ray diffraction. It appears that this nitrate route gives a more homogeneous product in a shorter time than does simple reaction of the metal oxides, provided the nitrates are decomposed directly at temperatures greater than 750°C to prevent the formation of impurities which react slowly to form the superconducting phase. The nitrates are a more intimately mixed, homogeneous starting material, decreasing the importance of repeated heat-and-grind cycles found necessary in the straight oxide route. In either case, however, material judged pure by powder diffraction standards proves to be quite imperfect when examined on a finer scale.

Magnetic and electronic properties of the $YBa_2Cu_3O_7$ phase are very sensitive to sample purity. The sharp superconducting transition, taken from magnetic susceptibility measurements, Figure 1, suggests pure, well ordered material. Using the experimentally determined density to convert the gram susceptibility to the volume susceptibility, the Meissner effect for our samples is typically 35%. Demagnetization effects, i.e. corrections for the shape of the sample, reduce the maximum possible Meissner effect to some value less than 100% for any sample shape other than a long thin needle. For samples in the shape of cylinders as used here, the Meissner effect for a pure sample which expells all the flux should be near 65%. Therefore, the low value of 35% indicates that either portions of the sample are not superconducting or flux is trapped by impurities and defects. This is consistent with the persistent currents measurements.

A classic experiment in the field of superconductivity is the trapping of magnetic flux in a ring of superconducting material. If one takes a ring of superconducting material above T_c, passes a magnetic field through it such that the plane of the ring is perpendicular to the field, and then lowers the temperature below T_c, the magnetic flux will be expelled from the superconducting material. Some flux will be going through the hole of the ring, and some around the ring. If the magnetic field is then turned off, the flux lines through this hole will be trapped. This creates a supercurrent which will flow through the ring to maintain the ring flux. This supercurrent gives rise to a large positive magnetic moment which can be measured. If the same experiment is done with the plane of the ring parallel to the applied magnetic field, no flux will be trapped, no supercurrent will be produced, and no positive magnetic moment will be observed.

Cooling a drilled pellet of $YBa_2Cu_3O_7$ through its critical temperature and in the presence of an applied 42 Gauss magnetic field produced the expected large magnetic moment in the sample when the ring plane was perpendicular to the field. However, we also detected a large magnetic moment when the ring plane was oriented parallel to the applied field. Observation of a similar magnetic moment in both orientations is due to trapping of magnetic flux at impurities and defects. This suggests the presence of many small isolated current loops throughout the sample.

The fine detail obtained in high resolution electron microscope images of $YBa_2Cu_3O_7$ indicated a wide variety of structural and compositional deviations from the idealized case. Atomic resolution imaging of ordered $YBa_2Cu_3O_7$, Figure 2, shows nearly perfect modelling of the experimentally observed image. This contrasts with images from another section having fault planes parallel to the crystallographic c-axis, Figure 4. Further evidence of local heterogeneities, labelled with an arrow in the high resolution image of Figure 3, were frequently observed near the surface of the grains. On a much larger scale, X-ray mapping revealed that segregation of the elements appears to be a more universal problem. The microscopic structural defects and compositional variations will have a large, adverse effect on the electronic and magnetic properties of the superconductors, for example in limiting the critical current and leading to the locally trapped flux observed in persistent current measurements described above.

Deviations from the idealized stoichiometry are also manifested in the XPS results. The unusually high content of carbon and oxygen at the surface can be understood in light of recent temperature programmed desorption (TPD) studies of Keller and co-workers (8). These TPD experiments evidence evolution of CO_2 and H_2O when the sample was heated in a He flow. A second feature seen in the XPS involves the copper deficiency at the surface. One interpretation of the results is that there is a microscopic phase separation of components near the surface, with CuO_x diffusing away from the surface, into the bulk. Electron microprobe results, however, are less sensitive to surface effects and show a stoichiometry much closer to the ideal $YBa_2Cu_3O_7$. Hence it seems that these high surface area powders are especially susceptible to degradation through atmospheric water and CO_2 and surface phase segregation.

The high resolution electron microscope images of $YBa_2Cu_3O_7$ confirm the presence of non-superconducting phases and structural defects as suggested by Meissner effect calculations and persistent current measurements. In addition, further evidence for contamination at the surface and at grain boundaries has come from XPS results. These inhomogeneities and structural discontinuities severely limit the current carrying effectiveness of this material. The small current loops implied by the persistent current measurements are a manifestation of this limitation. The presence of small current loops precludes the formation of defect-free extended current loops, which translates to a low critical current, J_c. Any attempt to process these materials must address these shortcomings.

Continued studies of different preparative routes to $YBa_2Cu_3O_7$ are warranted judging from the inadequacies pointed out in the straight oxide and nitrate decomposition routes. Two alternate schemes to the preparation of $YBa_2Cu_3O_7$ involve molten salts and hydrothermal processing. Although we have tried a wide variety of melts as well as thermal and electrochemical approaches, we have prepared the superconducting phase only in the basic $Ba(OH)_2$ melt. In general it appears that the more basic melts offer the most promise. Given the wide variety of melts still untested, and recent encouraging results in $Ba(OH)_2$, we remain optimistic about this approach.

Hydrothermal preparations have not yet produced the superconducting phase. Although there are many things still to be tried, it is already apparent that the superconductor is not stable in water at high temperatures, and that the component oxides crystallize more readily than does the superconductor. This is consistent with the observation that $YBa_2Cu_3O_7$ does not form below 750°C when processing the metal nitrates. If the conditions do exist that will allow the recrystallization of the quaternary oxide, they are likely to be very exacting, and thus difficult to find without mapping all the solubility curves of the system.

SUMMARY

Thermal decomposition of the appropriate metal nitrates produces $YBa_2Cu_3O_7$ that is of high purity based on powder X-ray diffraction. The metal nitrates must be processed at temperatures greater than 750°C because the formation of $YBa_2Cu_3O_7$ is not observed at temperatures lower than 750°C, but below 970°C at which temperature the $YBa_2Cu_3O_7$ decomposes. This material shows a high superconducting T_c (91 K) and a narrow transition width (10%-90% over 10K) in accord with well-ordered, pure superconductor. The Meissner effect puts put a lower limit of 35% on superconductor content and indicates that not all the flux is expelled. The persistent currents measured when a ring of superconducting material was oriented parallel to H_{appl} also indicates that there is some mechanism of flux trapping, for example by voids or structural defects. These features are not observed on the macroscopic scale, but their presence is confirmed by atomic resolution electron microscopy. AREM images show a considerable degree of sample heterogeniety from both phase intergrowths and structural defects. Surface composition is also seen to deviate from the ideal; XPS results indicate contamination of the surface by adsorbed gases. There also appears to be some degree of phase segregation at the surface, there being a deficiency in copper. Finally, efforts to synthesize $YBa_2Cu_3O_7$ free from chemical contaminants and structural defects has proceeded via molten salt and hydrothermal routes. Results thus far are encouraging; it appears that greater emphasis needs to be given to extremely basic media.

ACKNOWLEDGMENTS

G.F.H. thanks T. Nguyen for help in organizing the synthetic results. A.M.S. thanks Prof. J. Clarke (Department of Physics, UC Berkeley) for his advice on the measurements of persistent currents. This research was supported by a Presidential Young Investigator Award from the National Science Foundation (Grant CHE83-51881) to A.M.S. with matching funds from E.I. DuPont de Nemours and Company and from Dow Chemical Company, and by the Director, Office of Basic Energy Sciences, Materials Science Division and U.S. Department of Energy under contract No. DE-AC03-76SF0098.

REFERENCES

1. Wu, M.K.; Ashburn, J.R.; Torng, C.J.; Hor, P.H.; Meng, R.L.; Gao, L.; Huang, Z.J.; Wang, Y.Q.; Chu, C.W. Phys. Rev. Lett. **1987**, 58, 908.
2. Hor, P.H.; Gao, L.; Meng, R.L.; Huang, Z.J.; Wang, Y.Q.; Forster, K.; Vassilious, J.; Chu, C.W. Phys. Rev. Lett. **1987**, 58, 911.
3. Tarascon, J.M.; Greene, L.H.; McKinnon, W.R.; Hull, G.W. Phys. Rev. B **1987**, 35, 7115.
4. Siegrist, T.; Sunshine, S.; Murphy, D. W.; Cava, R. J.; Zahurak, S. M. Phys. Rev. B **1987**, 33, 7137.
5. Stacy, A.M.; Badding, J. V.; Geselbracht, M. J.; Ham, W. K.; Holland, G. F.; Hoskins, R. L.; Keller, S. W.; Millikan, C. F.; zur Loye, H.-C. J. Am. Chem. Soc. **1987**, 109, 2528.
6. Scherzer, O. J.Appl. Phys. **1949**, 20, 20.
7. Hirano, S.; Takahashi, S. J. Cryst. Growth **1986**, 79, 219-222.
8. Keller, S.W.; Leary, K.J.; Stacy, A.M.; Michaels, J.N.; Matt. Lett., in press.

RECEIVED July 6, 1987

Chapter 12

Superconductivity in $YBa_2Cu_3O_x$ for x Greater Than 7.0

Steven W. Keller[1,2], Kevin J. Leary[2,3], Tanya A. Faltens[1,2], James N. Michaels[2,3], and Angelica M. Stacy[1,2]

[1]Department of Chemistry, University of California—Berkeley, Berkeley, CA 94720
[2]Materials and Chemical Sciences Division, Lawrence Berkeley Laboratory, Berkeley, CA 94720
[3]Department of Chemical Engineering, University of California—Berkeley, Berkeley, CA 94720

The effect of oxygen stoichiometry on the magnetic properties of $YBa_2Cu_3O_x$ has been studied for $x \geq 7.0$. The oxygen content of the samples was determined using temperature-programmed desorption (TPD) and temperature-programmed reduction (TPR). For x = 7.1, the material exhibited a sharp diamagnetic transition with an onset temperature of 90 K and a transition width of 10 K (10% - 90%). The magnitude of the Meissner effect for the $YBa_2Cu_3O_{7.1}$ sample was 28 % at 5 K. When x was increased to 7.2, the diamagnetic transition became broader and the magnitude of the Meissner effect decreased, but the onset temperature remained constant.

The recent discovery of superconductivity above 90 K in samples of yttrium-barium-copper oxides has generated much enthusiasm among scientists and engineers in a variety of disciplines (1). The 90 K superconductor has been identified as $YBa_2Cu_3O_x$ where x = 6.9 - 7.0 (2-5). Several studies on this material have focussed on the effect of oxygen stoichiometry on the superconducting properties; in particular, for ranges of x between 6.5 and 7.0 (6-8). In this study, we show that samples with x between 7.0 and 7.5 can be prepared. The value of x was measured using temperature-programmed desorption (TPD) and temperature-programmed reduction (TPR). We found that as x was increased above 7.0, the onset temperature remained constant, but the width of the superconducting transition increased and the magnitude of the Meissner effect decreased.

EXPERIMENTAL

MATERIALS. A $YBa_2Cu_3O_x$ sample was prepared by reacting stoichiometric ratios of Y_2O_3 (Alpha Products), $BaCO_3$ (Fisher Scientific), and CuO (Fisher Scientific). An intimate mixture of finely divided particles was obtained by dissolving the reactants in concentrated nitric acid and evaporating the solution to dryness at 120 C. The nitrates were decomposed at 500 C for 4 h in air, ground, and reacted further at 750 C for a few days in flowing O_2 with several intermediate grindings. Finally, the

0097–6156/87/0351–0114$06.00/0

sample was fired in flowing O_2 for 12 h at 950 C, annealed in oxygen for 12 h at 500 C, and slow cooled to room temperature. Powder X-ray diffraction of the material showed a single phase, and the oxygen content was determined as shown below to be x = 7.1.

Three other samples of varying oxygen content were prepared using the YBa$_2$Cu$_3$O$_{7.1}$ sample as a starting material. In each case, a portion of the YBa$_2$Cu$_3$O$_{7.1}$ sample was first reduced by heating in helium at 830 C (1100 K) for 5 min. Then by annealing the reduced material in oxygen under different conditions, the oxygen content was varied. The three different annealing conditions used were: 300 C for 14 h, 400 C for 30 min, and 400 C for 2 h.

DETERMINATION OF OXYGEN CONTENT. The oxygen content of the samples was determined using temperature programmed desorption (TPD) and temperature programmed reduction (TPR). By measuring the amount of oxygen which desorbed when the sample was heated in helium and the amount of water which desorbed when the sample was reduced subsequently in hydrogen, the oxygen content of each of the samples was determined.

The apparatus used for these experiments has been described previously (10). In a typical TPD experiment, 25 mg of sample were placed in a quartz microreactor which was mounted inside a furnace. Following evacuation for 1 h at room temperature, helium was flowed over the sample at a rate of 100 cc/min (STP) and the temperature was raised at 0.5 K/s. During heating, the desorption products were swept from the reactor by the helium stream and monitored downstream with a UTI Model 100 C quadrupole mass spectrometer. Upon completion of each TPD experiment, the mass spectrometer was calibrated for oxygen as described below, and then a TPR experiment was performed using a hydrogen flow rate of 200 cc/min (STP). After each TPR experiment, the mass spectrometer calibration was repeated.

To calibrate the mass spectrometer for oxygen, known quantities of oxygen were pulsed into a 100 cc/min (STP) helium stream and the oxygen concentration in the pulse was monitored with the mass spectrometer. The integrated area of the oxygen signal as a function of time for each pulse varied linearly with the amount of oxygen pulsed. Therefore a calibration factor was obtained by dividing the amount of oxygen in a pulse by the integrated area. This factor was multiplied by the area under the oxygen desorption curve in each TPD spectrum to determine the amount of oxygen which desorbed. The accuracy of the calibration was estimated to be better than $\pm$ 2 %.

The mass spectrometer was calibrated for water by performing TPR experiments with known quantities of CuO. In each case, a hydrogen flow rate of 200 cc/min was used. During TPR, the CuO was completey reduced to copper metal. The area under the TPR spectrum was found to vary linearly with the amount of CuO, allowing a calibration factor for water to be determined. Following each TPR of the CuO samples, the mass spectrometer was calibrated for oxygen in the manner described above. From these results, the relative sensitivity of the mass spectrometer to water compared to oxygen was determined. Since we found that the relative sensitivity remained constant, it was only necessary to calibrate the mass spectrometer for oxygen after each experiment. From the CuO experiments, the accuracy of this calibration was determined to be $\pm$3 %.

RESULTS

Figure 1 shows the TPD spectrum obtained on a sample of $YBa_2Cu_3O_{7.1}$ sample which had been exposed to the air for one week. During exposure to the air, these materials incorporate a significant amount of water, CO, and CO_2 (9). As shown in Figure 1, these impurities desorb over the entire temperature range. Figure 1 also shows that oxygen desorbs during heating in helium in a large peak centered at approximately 815 K, and in a second much smaller peak at 1035 K. By integrating the area under the oxygen desorption spectrum, we found that $0.6 \pm .01$ moles of oxygen atoms (0.3 moles of O_2) were removed from the sample per mole of starting material.

Upon completion of the TPD experiment and calibration of the mass spectrometer, the sample was heated in hydrogen to 1200 K. Under these conditions, the sample was reduced to Y_2O_3, BaO, and Cu (2). The TPR spectrum is shown in Figure 2. The amount of water produced was determined by integrating the TPR spectrum. This amount corresponded to the removal of $3.0 \pm .09$ moles of oxygen atoms per mole of starting material. Adding this to the 0.6 moles removed by TPD and the 3.5 moles of oxygen in the reduction products, we obtain a stoichiometry of the starting material of $YBa_2Cu_3O_{7.1}$. The uncertainty in the value of x is ± 0.1.

The oxygen contents of the other three samples were determined in the same manner. The sample which was prepared by heating $YBa_2Cu_3O_{7.1}$ in helium at 1100 K for 5 min and annealing in oxygen at 300 C for 14 h was found to have an oxygen stoichiometry of x = 7.0. The samples which were annealed at 400 C for 30 min and 2 h had oxygen contents of x = 7.2 and 7.5, respectively. We should emphasize that the exact times and temperatures needed to vary the oxygen content are sample dependent.

In each case, the differences in oxygen content were reflected in the differences in the amount of oxygen which desorbed during the TPD experiment. The TPR spectra of all the samples looked very similar to the one shown in Figure 2, although the peak positions and the relative heights of the peaks varied slightly from sample to sample. Thus, heating in helium to 1100 K reduced the samples to $YBa_2Cu_3O_{6.5}$. From these results, we conclude that for single phase $YBa_2Cu_3O_x$ samples with $x \geq 6.5$, the oxygen stoichiometry can be determined by TPD alone. Further reduction of the samples in hydrogen is unnecessary.

One advantage of using TPD to determine the oxygen content of the samples is that TPD is also useful for characterizing these materials. The effect of oxygen content on the oxygen desorption spectrum of $YBa_2Cu_3O_x$ is shown in Figure 3 for values of x between 7.0 and 7.5. The spectrum for x = 7.0 contains a large desorption peak near 860 K and a much smaller peak near 1015 K. This spectrum looks similar to that shown in Figure 1 for x = 7.1, and appears to be characteristic of a material with a sharp diamagnetic transition. As x increases to 7.2, a third small peak begins to grow in as a shoulder on the low temperature peak between 900 and 950 K. This shoulder is never observed on samples with sharp diamagnetic transitions. As x increases still futher to 7.5, the shoulder does not increase in size, but the size of the high temperature peak between 1000 and 1050 K increases dramatically.

The magnetic susceptibilities of the starting material $YBa_2Cu_3O_{7.1}$ and a sample which was determined to have the stoichiometry $YBa_2Cu_3O_{7.2}$ were measured as a function of temperature, and the results are shown in Figure 4. The magnetic measurements were made using a SHE SQUID magnetometer in a field of 12 Gauss; the Meissner effect was determined by

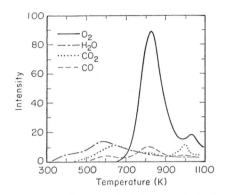

Figure 1:   TPD spectrum of YBa$_2$Cu$_3$O$_{7.1}$ for a helium flow rate of 100 cc/min (STP) and a heating rate of 0.5 K/s.

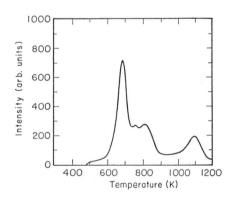

Figure 2:   TPR spectrum of YBa$_2$Cu$_3$O$_{7.1}$ for a hydrogen flow rate of 200 cc/min (STP) and a heating rate of 0.5 K/s.

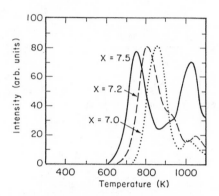

Figure 3:   Effect of oxygen content on the oxygen desorption spectrum
of $YBa_2Cu_3O_x$ for values of x between 7.0 and 7.5. In each
case the material is reduced to $YBa_2Cu_3O_{6.5}$.

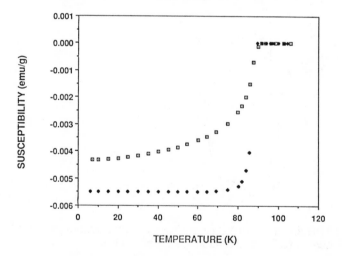

Figure 4:   Magnetic susceptibility vs. temperature curves for (♦)
$YBa_2Cu_3O_{7.1}$ and (□) $YBa_2Cu_3O_{7.2}$.

measuring the magnetization as the samples were cooled from 110 K. For both samples, the onset of diamagnetism occured at 90 K. The $YBa_2Cu_3O_{7.1}$ sample had a sharp transition and a transition width of 10 K as shown in Figure 4. A 28 % Meissner effect was calculated for this sample using the measured density of 3.92 g/cc. For samples with x > 7.1, the transitions were always broader, but the Meissner effect varied depending on the length of time the samples were exposed to air. The magnetic data for the $YBa_2Cu_3O_{7.2}$ sample after exposure to the air for one month is shown in Figure 4. The results in Figure 4 suggest that the magnitude of the Meissner effect at 5 K decreases when x is increased from 7.1 to 7.2. Further experiments are in progress to quantify these results.

DISCUSSION

Most of the studies on the effect of oxygen stoichiometry on the superconducting properties of $YBa_2Cu_3O_x$ have concentrated on ranges of x between 6.0 and 7.0. It has been shown that the reduction of $YBa_2Cu_3O_7$ to $YBa_2Cu_3O_{6.5}$ corresponds to the removal of half of the oxygen atoms in the chains between the planes of CuO (11). However, single crystal X-ray analysis has shown that the oxygen content can be reduced further to $YBa_2Cu_3O_6$ (13). Our results suggest that the first 0.5 moles of oxygen atoms are removed rapidly but that further reduction requires longer times and/or higher temperatures. We postulate that the change in the kinetics of oxygen removal may be associated with the orthorhombic to tetragonal phase transition; the mobility of oxygen in the orthorhombic phase is substantially higher than in the tetragonal modification.

Because these materials are reduced reproducibly to $YBa_2Cu_3O_{6.5}$ by heating in helium to 1100 K, the oxygen content can be determined by TPD alone. One advantage of using TPD to characterize $YBa_2Cu_3O_x$ samples is that it is very sensitive to minor differences between samples. For example, differences between the $YBa_2Cu_3O_{7.1}$ and $YBa_2Cu_3O_{7.2}$ samples were clearly evident in the oxygen desorption spectra shown in Figure 3. Another advantage is that impurities such as absorbed water and carbon oxides can be detected easily as shown in Figure 1. These impurities can not be detected directly with a technique like thermal gravimetric analysis (TGA). Therefore if TGA is used to measure oxygen addition and removal from these materials, significant errors can be made unless care is taken to first remove these impurities (9). These impurities also may have an effect on the electronic and magnetic properties of these materials. It has been reported that water has a detrimental effect on the superconductive properties of lanthanum-strontium- copper oxides (11). The effects of these impurities on $YBa_2Cu_3O_x$ samples is under further investigation.

In summary, we have shown that samples of $YBa_2Cu_3O_x$ can be prepared with x > 7.0. For x = 7.1, the material exhibits a sharp diamagnetic transition with an onset temperature of 90 K and a transition width of 10 K (10% - 90%). When x is increased to 7.2, the transition is broadened substantially, and the magnitude of the Meissner effect at 5 K is decreased.

ACKNOWLEDGMENTS

The authors would like to thank William K. Ham and Hans-Conrad zur Loye for help with sample preparation. This research was supported by a Presidential Young Investigator Award from the National Science Foundation to AMS with matching funds from E.I. DuPont de Nemours and

Company and from Dow Chemical Company, and by the Director, Office of Basic Energy Sciences, Materials Science Division and U.S. Department of Energy under contract No. DE-AC03-76SF0098.

LITERATURE CITED

1. Wu, M.K.; Ashburn, J.R.; Torng, C.J.; Hor, P.H.; Meng, R.L.; Gao, L.; Huang, Z.J.; Wang, Y.Q.; Chu, C.W. Phys. Rev. Lett., 1987, 58, 908.
2. Cava, R.J.; Batlogg, B.; VanDover, R.B.; Murphy, D.W.; Sunshine, S.;Siegrist, T.; Remeika, J.P.; Rietman, E.A.; Zahurak, S.; Espinosa, G.P. Phys. Rev. Lett., 1987,   58, 1676.
3. Beyers, R.; Lim, G.; Engler, E.M.; Savoy, R.J.; Shaw, T.M.; Dinger, T.R.; Gallagher, W.J.; Sandstrom, R.L. Appl. Phys. Lett., submitted for publication.
4. Le Page, Y.; McKinnon, W.R.; Tarascon, J.M.; Greene, L.H.; Phys. Rev. Lett., submitted for publication.
5. Okamura, F.P.; Sueno, S.; Nakai, I.; Ono, A.; submitted for publication.
6. Swinnea, J.S.; Steinfink, H.; preprint.
7. Morris, D.E.; Scheven, U.M.; Bourne, L.C.; Cohen, M.L.; Crommie, M.F.; Zettl, A.; Proceedings from the Materials Research Conference. Symposium S. Anaheim, CA. April, 1987, in press.
8. Tarascon, J.M., Proceedings from the International Conference on Novel Mechanisms of Superconductivity, Berkeley, CA, June, 1987, to be published.
9. Keller, S.W.; Leary, K.J.; Stacy, A.M.; Michaels, J.N.; Matt. Lett., in press.
10. Leary, K.J.; Michaels, J.N.; Stacy, A.M.; J. Catal., 1986, 101, 301.
11. Beno, M.A.; Soderholm, L.; Capone, D.W. II; Hinks, D.G.; Jorgensen, J.D., Schuller, I.K.; Segre, C.U.; Zhang, K.; Grace, J.D. submitted to Appl. Phys. Lett. ,1987.
12. Kisho, K.; Sugii, N.; Kitazawa, K.; Fueki, K.; Jpn. J. Appl. Phys., 1987, 26, L466.
13. Steinfink, H; Swinnea, J.S.; Sui, Z.T.; Hsu, H.M.; and Goodenough, J.B.; J. Amer. Chem. Soc.; in press.

RECEIVED July 6, 1987

STRUCTURE-PROPERTY RELATIONSHIPS

Chapter 13

Mixed-Valence Copper Oxides Related to Perovskite

Structure and Superconductivity

B. Raveau, C. Michel, and M. Hervieu

Laboratoire de Cristallographie et Sciences des Matériaux, ISMRa,
Bd du Maréchal Juin, Université de Caen, 14032 Caen Cedex, France

Until 1986, superconductivity was limited to the low-temperature range, and particularly to the use of liquid helium. The record of critical temperature ($T_c \simeq 23.3K$) was held by Nb_3Ge, prepared in the form of thin layers. Moreover, from the point of view of many physicists, it was rather clear that the increase of T_c of a dozen degrees or even a few degrees was unlikely in spite of the expectations made by Little about organic superconductors. The discovery of the mixed-valence copper oxides, characterized by a high T_c value and a high critical field opens a new door as well in fundamental physics as in technology.

The Superconducting Phase in the Ba-La-Cu-O System was not the Expected One

The system Ba-La-Cu-O chosen by Bednorz and Muller (1) for their first investigation of superconductivity was connected to the existence of metallic properties of the oxide $BaLa_4Cu_5O_{13+\delta}$ recently isolated in Caen (2). However the method used for the synthesis did not lead to the expected oxide but to a mixture of phases. Fortunately, one of these latter phases was found to be at the origin of superconductivity. In fact the concerned oxide is a K_2NiF_4-type oxide, $La_{2-x}Ba_xCuO_{4-y}$, isolated and studied for its electron transport properties several years before in Caen (3-4). It was then shown by several authors (5-9) that the strontium isostructural oxide $La_{2-x}Sr_xCuO_{4-y}$ (10-11) and the calcium compound (3-4) were also superconductors with T_C values ranging form 20K to 40K.

From these first observations it appeared that strong electron-phonon couplings might be at the origin of those superconducting properties, which could be related to the existence of the mixed valence of copper $Cu(II)-Cu(III)$ and to its Jahn Teller

0097-6156/87/0351-0122$06.00/0
© 1987 American Chemical Society

effect. Moreover the recent model of labbé and Bock (12) suggests
that the bidimensional character of the structure plays an important role in the superconductivity of those oxides. The two conditions - mixed valence and anisotropy - were those which we used
several years ago, for the research of copper oxides with exotic
metallic properties.

Generation of Bidimensional Metallic Conductors by Creation of
Ordered Oxygen Vacancies in the Perovskite Framework

The tervalent state of copper is difficult to stabilize at normal
pressure. When we started this study, the only known oxide was
$LaCu^{III}O_3$, which was synthesized under high pressure (65 kbars)
(13). In order to favour the partial oxidation of Cu(II) into
Cu(III) under normal oxygen pressure, we decided to insert a
basic A ion such as an alkaline earth cation (Ba, Sr, Ca) into
the MO_n framework ; exotic upper oxidation states like Bi(V) or
Pb(IV) are indeed stabilized under normal oxygen pressure by such
cations as shown for instance for the superconducting perovskite
$Ba(Pb_{1-x}Bi_x)O_3$ (14). Anisotropy can be introduced by building a
mixed framework in which copper exhibits several coordinations -
octahedral, tetrahedral, pyramidal or square planar -. In this
respect, the ordered elimination of oxygen atoms or of rows of
oxygen atoms in the perovskite host lattice is susceptible to
lead to very anisotropic structures.

Two types of oxygen deficient perovskites, characterized by
a dimensional character have been isolated. The first one,
$La_3Ba_3Cu_6O_{14}$ (15) results from the ordered elimination of rows of
oxygen atoms parallel to the <110> direction of the cubic
perovskite. It results that the host lattice of the oxide (Figure
1) is built up from corner-sharing CuO_6 octahedra, CuO_5 pyramids
and CuO_4 square planar groups. It is worth pointing out that the
limit described here is not really reached, so that the actual
formulation $La_3Ba_3Cu_6O_{14+\delta}$, corresponds to the intercalation of
oxygen between the layers according to the applied oxygen pressure : δ = 0.05 for the oxide synthesized in air at 1050°C and
quenched at room temperature, and δ = 0.43 for the compound annealed under oxygen (1 atm) at 450°C. It must also be noticed
that the conductivity increases with the amount of oxygen intercalated (16), going from a semiconducting behaviour for δ = 0 to
a semimetallic behaviour for δ = 0.43. However, no sign of superconductivity was detected for this oxide in spite of its tendancy
for a bidimensional structure. The second oxygen deficient perovskite is represented by the orthorhombic oxide $YBa_2Cu_3O_{8-x}$ which
was recently isolated by two groups (17-18) almost
simultaneously. The structure of this compound (19-20) is closely
related to the one of $La_3Ba_3Cu_6O_{14}$. It can be described (figure
2) as formed of triple layers $[Cu_3O_7]_\infty$ of corner-sharing CuO_5
pyramids and CuO_4 square planar groups. Each triple layer is
built up from two $[CuO_{2.5}]_\infty$ pyramidal layers connected through
one $[CuO_2]_\infty$ layer formed of CuO_4 square planar groups. In this
oxide, barium and yttrium ions are ordered in such a way that two

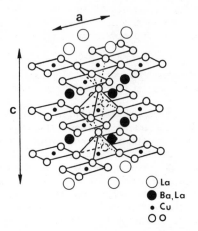

Figure 1. Structure of the oxide $Ba_3La_3Cu_6O_{14}$.

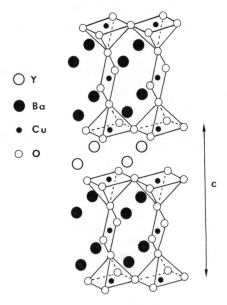

Figure 2. Structure of the oxide $Ba_2YCu_3O_7$.

barium planes alternate with one yttrium plane along c : the
barium ions are located in the triple layers. This structure can
be described from that of the perovskite by an ordered elimination of rows of oxygens parallel to the $[010]_p$ direction i.e. at
the same level along c as the yttrium ions and as the copper ions
of the $[CuO_2]_\omega$ layer. This oxygen deficient ordered perovskite
exhibits a metallic conductivity and a zero resistance below 91K
(figure 3) in agreement with the signs of superconductivity observed for the first time in a mixture of oxides with composition
$Y_{1.2}Ba_{0.8}CuO_{4-y}$ by Chu et al. (21). Recently we have also isolated a similar superconducting oxide in the case of lanthanum
$LaBa_2Cu_3O_8$ (22) characterized by a lower critical temperature T_c
= 75K. Nevertheless it must be pointed out that for this latter
oxide, the examination of the X-ray powder diffraction pattern
suggests a tetragonal symmetry. A careful study by electron microscopy will be necessary to understand this difference with
$YBa_2Cu_3O_{8-y}$ and to correlate it with superconductivity.

In order to introduce anisotropy in the electron transport
properties of those oxides another direction was investigated.
The idea was to realize the intergrowth of "insulating" sodium
chloride-type layers with conducting oxygen-deficient perovskite
type layers. Two series of oxides were then synthesized and characterized the oxide $La_{2-x}A_xCuO_{4-x/2+\delta}$ (figure 4a) and the oxides
$La_{2-x}Sr_{1+x}Cu_2O_{6-x/2+\delta}$ (figure 4b). Those compounds can be described as the members n = 1, and n = 2 respectively of the intergrowths $AO (ACuO_{3-y})_n$ where AO corresponds to the stoichiometric
sodium chloride type layer and $(ACuO_{3-y})_n$ corresponds to the
oxygen deficient multiple perovskite layer. One of those oxides,
$La_{2-x}A_xCuO_{4-x/2+\delta}$, represents the first generation of high T_c
superconductors with critical temperatures ranging from 30K to
40K as shown from figure 5 . Among these latter oxides those of
strontium exhibit the highest T_c. It must be pointed out that
their chemistry is rather complex : they exhibit three different
domains whose homogeneity range depends upon the method of
preparation, especially upon the temperature, and upon oxygen
pressure, i.e. upon oxygen stoichiometry δ. For instance, for the
oxides heated in air at 1050°C one observe the following domains:

- An orthohombic domain (O) for $0 \le x \le 0.1$, which corresponds to
  a monoclinic distortion of the perovskite cubic cell leading to
  the following relations :

$$a_0 \simeq a_p \sqrt{2} \simeq a_{La2CuO4} \; ; \; b_0 \simeq a_p \sqrt{2} \simeq b_{La2CuO4} \; ; \; c \simeq c_{La2CuO4}$$

- A tetragonal domain (T) for $0.1 < x < 1$, in which the
  Cu(III)/Cu(II) ratio is maximum at about x = 0.30 with the following parametric relationships :

$$a \simeq a_p \; ; \; c \simeq c_{La2CuO4}$$

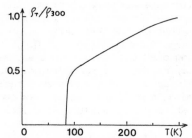

Figure 3. Resistivity ratio (ρ_T/ρ_{300K}) versus temperature of the oxide $Ba_2YCu_3O_7$.

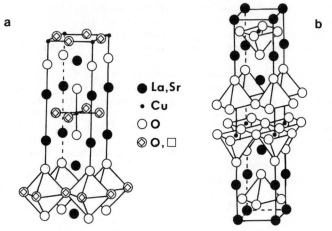

Figure 4. Structure of the oxides $La_{2-x}A_xCuO_{4-x/2+\delta}$ (a) and $La_{2-x}Sr_{1+x}Cu_2O_{6-x/2+\delta}$ (b).

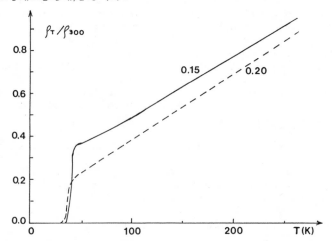

Figure 5. Resistivity ratio (ρ_T/ρ_{300K}) versus temperature of the oxides $La_{2-x}A_xCuO_{4-x/2+\delta}$.

- A tetragonal domain with superstructures (T.S), in which 90 % of the crystals investigated by electron diffraction show superstructures with respect to the tetragonal cell :

$$a \simeq n \, a_p \quad ; \quad c \simeq c_{La2CuO4}$$

with n = 4, 5, 6, 4.5, 5.3, 5.6, etc...

Among these three domains only the second one was found to present superconductivity for x ranging from about 0.12 to 0.20 (5-6). It is worth pointing out that no sign of superconductivity was observed in the orthorhombic domains, and that La_2CuO_4 was described as a semi-conductor. Recently, we have shown in collaboration with the Universities of Bordeaux and Grenoble, that for x = 0, superconductivity could be obtained : La_2CuO_4 prepared under high oxygen pressure, or "non-stoichiometric" La_2CuO_4, which can be formulated $La_{1.9}CuO_4$ prepared under normal pressure were prepared for the first time as superconductors with a T_c close to from 37K (Figure 6). Moreover, the magnetic measurements carried out on those samples, show 50 % to 90 % diamagnetism, in agreement with a bulk superconductivity. There results can be interpreted from a chemical point of view by considering that the preparation of La_2CuO_4 at normal oxygen pressure, does not allow the mixed valence of copper to be reached, but that the crystals contain Schottky defects on all their sites according to the formulation $La_{2-2\epsilon}Cu^{II}{}_{1-\epsilon}O_{4-4\epsilon}$. The application of a high oxygen pressure allows the anionic vacancies to be filled leading to the mixed valence oxide $La_{2-2\epsilon}Cu^{II,III}{}_{1-\epsilon}O_4$. In the same way, the necessity to use an excess of copper oxide for the synthesis of "La_2CuO_4" at normal oxygen pressure can be interpreted by the fact that the formation of anionic vacancies is favoured in a first step leading to the intermediate species $La_{1.9}Cu^{II}O_{3.85}$ which tends to capt oxygen at normal pressure leading to the formulation $La_{1.9}Cu^{II,III}O_4$.

$YBa_2Cu_3O_{7-\epsilon}$: a Complex Crystal Chemistry

The recent investigations of the phase transformations carried out by Roth et al. (23) show clearly that the orthorhombic superconducting oxide $YBa_2Cu_3O_{7-\epsilon}$ transforms into a tetragonal form above 750°C, and loses oxygen above 930°C. It is also now well established that superconductivity decreases as the oxygen content decreases. Recently it was shown that it was possible to stabilize the tetragonal form by quenching at room temperature (24). Thus it appears from X-ray diffraction (19, 24) and neutron diffraction studies (20, 25) that two limits exist : the orthorhombic superconductor $YBa_2Cu_3O_7$ and the tetragonal oxide $YBa_2Cu_3O_6$. However, from the chemical analysis and structural studies that the upper limit is rarely reached and that for the different samples the oxygen content may vary between these two limits. Consequently, physical measurements as well as structural observations which are carried out without taking into account the chemical composition of the bulk will be difficult to use for a serious interpretation in the near future. We summarize here the recent observations we have made by high resolution electron microscopy (26-29) on two compounds : the superconducting oxide $YBa_2Cu_3O_{6.85}$ characterized by an orthorhombic symmetry and the tetragonal phase $YBa_2Cu_3O_{6.5}$ obtained by quenching the sample from 950°C at room temperature which does not exhibit any super-conductivity.

Orthorhombic $YBa_2Cu_3O_{6.85}$ (T_C = 91K). In spite of its great purity, deduced from the X-ray and neutron diffraction patterns, the orthorhombic superconductor $YBa_2Cu_3O_{6.85}$ is charac-terized by a complex crystal chemistry, when examined by high resolution electron microscopy.

The first class of "inhomogeneities" is due to the fact that the CuO_4 square planar groups belonging to the $[CuO_2]_\infty$ slabs can take a different orientation from one crystal to the other or form one region to the other. Consequently, three phenomena can be distingued : twin domains, orientated domains and disorienta-ted domains. Most of the crystals are twinned due to the transi-tion orthorhombic-tetragonal (figure 7). Those twinning domains which extend over distances ranging from 500 Å to 1000 Å, can be described from the consideration of the vacancy ordering along [010].

The change in the direction of the oxygen vacancy ordering implies a perpendicular orientation of the CuO_4 groups from one domain to the adjacent one. Possible models of the junction bet-ween the domains, with or without mirror plane (110) are drawn in figure 8 ; only one layer out of three polyhedral layers has been presented in order to simplify the representation ; we have indeed choosen the $[CuO_2]_\infty$ layer built up from CuO_4 groups. The first model (Figure 8a), with a mirror plane, is based on the presence of additonal CuO_6 octahedra and CuO_5 pyramids, in the $[CuO_2]_\infty$ layer, at the boundary. In the second model, the junction could only be ensured through pyramids models CuO_5, as shown in figure 8b . In must be pointed out those these two models imply the presence of additonnal oxygen at the boundary but no drastic displacement in the cation framework. A third model, where the junction is ensured through tetrahedra can also be imagined but the existence of such CuO_4 tetrahedra which involves a displace-ment of the Cu atoms appears to be really less favourable (figure 8c). The observation of the crystals along [010] and [100] show a second type of domains called orientated domains (figure 9a) : area labelled exhibits the typical contrast of [100] image for t = 30 Å and Df = -430 Å whereas area labelled 2 is characteristic of a [010] image, under the same conditions. Such domains are interpreted by a change in the orientations of the $[CuO_2]_\infty$ layers, from one triple polyhedral layer to the adjacent one, forming "orientated" slices perpendicular to the c axis ; the idealized model of the "orientated" domains is proposed in figure 9b . It is worth pointing out that, contrary to the twin-ning domains, those domains are characterized by a junction in-

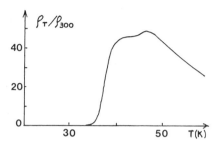

Figure 6. Resistivity ratio (ρ_T/ρ_{300K}) versus temperature of the superconductor La_2CuO_4.

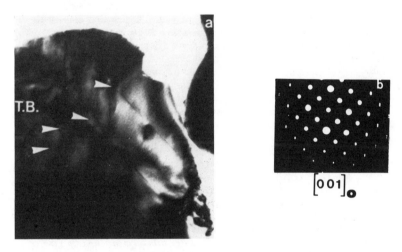

Figure 7. a) Image of a crystal exhibiting twinning domains. The twin boundaries (T.B) are white arrowed. b) electron diffraction pattern characterized by a splitting of the spots hh0.

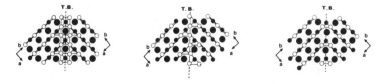

Figure 8. Idealized models of junction between twinning domains.

volving a juxtaposition of two different parameters "a" and "b" respectively at the boundary. It results that the domain interfaces are particularly disturbed. Another typical structural feature corresponding to misorientation of crystal areas, must be outlined (figure 10). Such domains which exhibit a variable β angle between their c axis, are due to the difficulty to adapt the "a" parameter of one domains to the "b" parameter of the adjacent one, especially when these domains are very narrow.

Numerous defects are distributed overall the matrix, forming domains whose size varies from about 30 Å to some hundred Å. The types of defects have been frequently observed, corresponding to an excess of oxygen, an oxygen substoichiometry and a cationic disorder respectively. An example of oxygen stoichiometry is presented by the HREM micrograph of figure 11 , where a variation of the contrast is clearly observed. Such a variation, is interpreted, by additionnal oxygen atoms in the $[CuO_2]_\infty$ layer in agreement with the calculated images (figure 11). It is obvious that such a contrast variation gives evidence of additionnal oxygens but does not allow an exact occupancy ratio of the corresponding sites to be determined ; thus the stoichiometry of these domains can be characterized by a formula $YBa_2Cu_3O_{8-\delta}$. The structure is then described as a succession of one octahedron CuO_6 and two pyramids CuO_5 along the c axis, the idealized model shown in figure 12 was previously described for the hypothetical limit $YBa_2Cu_3O_8$ (3). It is worth noting that, owing to their size, such domains can be only observed, and interpreted, on the thin edges (< 200 Å) of the crystals. Regions of oxygen substoichiometry are also often observed as shown from the observation of the crystals along [100] (figure 13). One can indeed observe a variation of the spacing of the rows of spots which move apart, implying a bending of the adjacent double row and an extra row of small white spots. Such a feature can be explained by the arising of a double row of edge-sharing CuO_4 groups similar to those observed in the $SrCuO_2$ (30) framework (figure 14). The third important type of defect deals with variations in the Y, Ba cations ordering (figure 15a) ; the c/3 shifting of the fringes is observed in the top of the micrograph (arrowed). This defect is easily explained by a reversal of one barium and one yttrium layer form one part of the defect to the other ; while the second barium layer remains unchanged through the defect, as shown from idealized model (figure 15b). On that example, the defect extends over twelve triple layers $YBa_2Cu_3O_7$ with a clear contrast and the junction of the reverse layers is regularly translated, perpendicularly to the c axis.

Semi-conducting $YBa_2Cu_3O_{6.50}$. After heating the mixture of oxides Y_2O_3, CuO and carbonate $BaCO_3$ at 950°C, and quenching them at room temperature, the oxide $YBa_2Cu_3O_{6.50}$ is obtained. The X-ray diffraction pattern of this compound can be indexed in a tetragonal cell with a = 3.860 Å, c = 11.803 Å. The examination of the crystals of this composition by electron diffraction and microscopy shows that the twinning domains have disappeared in

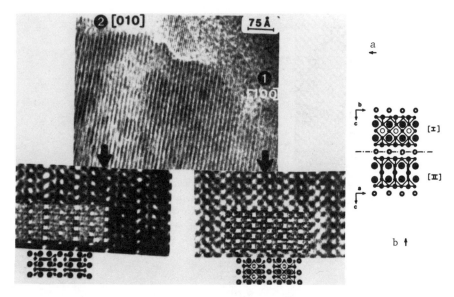

Figure 9. a) High-resolution image of a crystal exhibiting orientated domains. The simulated images are compared to the enlargement of the experimental images. b) Idealized model of the junction.

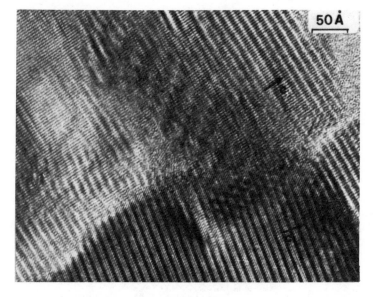

Figure 10. High-resolution image of misorientated domains. The overlapping of such areas leads to Moiré patterns.

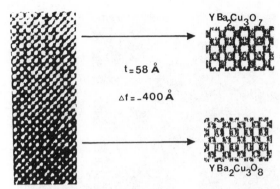

Figure 11. High resolution image and calculated images of an area corresponding to an oxygen overstoichiometry.

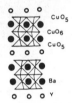

Figure 12. Idealized model of $YBa_2Cu_3O_8$.

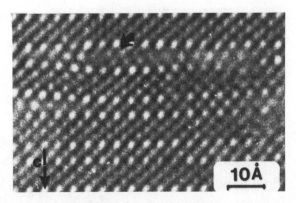

Figure 13. High resolution image of a defect corresponding to the existence of a double row of edge-sharing CuO_4 groups similar to those observed in $SrCuO_2$.

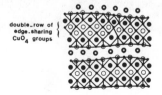

Figure 14. Idealized model of the defect.

agreement with the tetragonal symmetry. However it is again worth pointing out that the crystals are characterized by a great inhomogeneity. We can indeed notice that most of the diffraction patterns exhibit diffusion streaks, signature of a disordered state of the material. More or less extended zones of disorder can be observed by HREM (figure 16), whereas a great number of crystals are wrapped in an amorphous layer (figure 17).

As a conclusion, these results show clearly that the crystal chemistry of those oxides is very complex and it could be possible that the presence of domains and defects would influence the superconducting properties of these oxides.

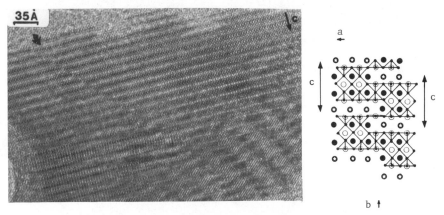

Figure 15. High resolution image (a) and idealized models (b) of a defect corresponding to a change in the Y-Ba ordering.

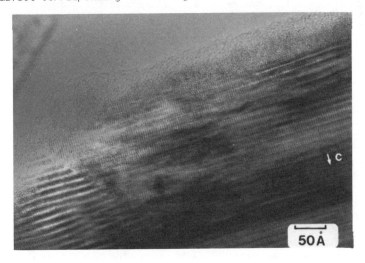

Figure 16. Image of a disordered crystal of $YBa_2Cu_3O_{6.5}$.

Figure 17. High resolution image of a crystal of $YBa_2Cu_3O_{6.5}$ showing the amorphous layer (some tens Å) which systematically appears in that sample.

Literature cited

1. Bednorz, J.G. ; Müller, K.A. Z. Phys. B. 1987, 64, 189.
2. Michel, C. ; Er-rakho, L. ; Raveau, B. Mater. Res. Bull. 1985, 20, 667.
3. Er-rakho, L. Thesis, Université de Caen, 1981.
4. Michel, C. ; Raveau, B. Rev. Chim. Miner., 1984, 21, 407.
5. Tarascon, J.M. ; Greene, L.H. ; McKinnon, W.R. ; Hull, G.W.; Geballe, T.H., Science 1987, 235, 1373.
6. Cava, R.J. ; Van Dover, R.B. ; Batlogg, B. ; Rietman, E. Phys. Rev. Lett., 1987, 58, 408.
7. Bednorz, J.G. ; Müller, K.A. ; Takashige, M. Science 1987,
8. Kishio, K. ; Kitazawa, K. ; Kanbe, S. ; Yasuda, I. ; Sugii, N. ; Takagi, H. ; Uchida, S. ; Fucki, K. ; Tanaka, S. Chim. Lett., 1987.
9. Rao, N.C.R. ; Ganguly, P. Current Science, 1987 56, 47.
10. Nguyen, N. ; Choisnet, J. ; Hervieu, M. ; Raveau, B. J. Solid State Chem., 1981, 39, 120.
11. Nguyen, N. ; Studer, F. ; Raveau, B. J. Phys. Chem. Solids 1983, 44, 389.
12. Labbé, J. ; Bok, J. Europhysics Lett., 1987, to be published
13. Demazeau, G. ; Parent, C. ; Pouchard, M. ; Hagenmuller, P. Mat. Res. Bull., 1972, 7, 913.
14. Sleight, A.W. ; Gilson, J.L. ; Bierstedt, P.E. Solid State Comm. 1975, 17, 27.
15. Er-rakho, L., ; Miche, C. ; Provost, J. ; Raveau, B. J. Solid State Chem. 1981, 37, 151.
16. Provost, J. ; Studer, F. ; Michel, C. ; Raveau, B. Synthetic Metals 1981, 4, 147 ; 4, 157.
17. Cava, R.J. ; Batlogg, B. ; Van Dover, R.B. ; Murphy, D.W. ; Sunshine, S. ; Siegrist, T. ; Remeika, J.P. ; Rietman, E.A.; Zahurak, S. ; Espinosa, G.P. Phys. Rev. Lett. 1987, 58, 1676

18.   Michel, C. ; Deslandes, F. ; Provost, J. ; Lejay, P. ; Tour-
      nier, R. ; Hervieu, M. ; Raveau, B. C.R.Acad. Sci. 1987, 304
      II, 1059.
19.   Lepage, Y. ; McKinnon, W.R. ; Tarascon, J.M. ; Greene, L.H.;
      Hull, G.W. ; Hwang, D.M. Phys. Rev. B, 1987, 35, 7245.
20.   Caponi, J.J. ; Chaillout, C., Hewat, A.W., ; Lejay, P. ;
      Marezio, M. ; Nguyen, N. ; Raveau, B. ; Soubeyroux, J.L. ;
      Tholence, J.L. ; Tournier, R. Europhysics Lett. 1987,12,1301.
21.   Wu, M.K. ; Ashburn, J.R. ; Torng, C.J. ; Hor, C.J. ; Meng,
      R.L. ; Gao, L. ; Huang, Z.J. ; Wang, Y.Q. ; Chu, C.W. Phys.
      Rev. Lett. 1987, 58, 908.
22.   Michel, C. ; Deslandes, F. ; Provost, J. ; Lejay, P. ; Tour-
      nier, R. ; Hervieu, M. ; Raveau, B. C.R.Acad. Sci. 1987,
      304 II, 1169.
23.   Roth, G. ; Ewert, D. ; Heger, G. ; Hervieu, M. ; Michel,
      C.; Raveau, B. ; d'Yvoire, F. ; Revcolevski, A. Z. Phys. B,
      1987, to be published.
24.   Roth, G. ; Renker, D. ; Heger, G. ; Hervieu, M. ; Domenges,
      B. ; Raveau, B. Z. Phys., 1987, submitted.
25.   Bordet, P. ; Chaillout, C. ; Capproni, J.J. ; Chevanas, J. ;
      Marezio, M. Nature, 1987, submitted.
26.   Hervieu, M. ; Domengès, B. ;Michel, C. ; Heger, G. ; Provost
      J. ; Raveau, B. Phys. Rev. B, 1987, to be published.
27.   Hervieu, M. ; Domengès, B. ; Michel, C. ; Provost, J. ;
      Raveau, B. J. Solid State Chhem., submitted.
28.   Hervieu, M. ; Domengès, B. ; Michel, C. ; Raveau, B. Euro-
      physics Lett., submitted.
29.   Domengès, B. ; Hervieu, M. ; Michel, C. ; Raveau, B. Euro-
      physics, submitted.
30.   Müller-Bushbaum, H.K. ; Wollschläger, W. Z. anorg. allg.
      Chem., 1975, 414,56.

RECEIVED July 7, 1987

Chapter 14

Variation in the Structural, Magnetic, and Superconducting Properties of $YBa_2Cu_3O_{7-x}$ with Oxygen Content

D. C. Johnston[1], A. J. Jacobson[1], J. M. Newsam[1], J. T. Lewandowski[1], D. P. Goshorn[1], D. Xie[2], and W. B. Yelon[2]

[1]Exxon Research and Engineering Company, Route 22 East, Annandale, NJ 08801
[2]Missouri University Research Reactor, Columbia, MO 65211

Several samples of $YBa_2Cu_3O_{7-x}$ with controlled stoichiometries, $0.04 \leq x \leq 1.00$, have been prepared and characterized by thermogravimetric analysis, differential scanning calorimetry, magnetic susceptibility and Meissner effect measurements. Structural details for representative samples were also determined by full profile refinements of powder X-ray and neutron diffraction data. Quantitative relationships between the magnetic and superconducting properties and the composition and structure are presented. The relative contributions of the Cu-0 chains and layers to the superconducting transition temperature are discussed.

The discovery of high T_c superconductivity in the La-(Ba,Sr)-Cu-O systems at 30-50 K ([1-7]) was quickly followed by identification of the phase responsible for the high T_c to be $La_{2-x}(Ba,Sr)_xCuO_4$ with either the K_2NiF_4 structure or an orthorhombically distorted version thereof ([2,4,7,8]). An even higher T_c of ~100K was found in the Y-Ba-Cu-O system ([9]), for $YBa_2Cu_3O_7$, with an orthorhombic structure related to the perovskite and K_2NiF_4 structures ([10-15]). Qualitatively, it is known that the superconducting properties of these compounds are sensitive to the oxygen stoichiometry (see e.g [1,6,7,16-23]), but the quantitative relationships are not known for either system.

Herein, we report the results of coordinated thermogravimetric, differential scanning calorimetry, neutron and X-ray diffraction structural analyses, and Meissner effect and magnetic susceptibility studies on a series of $YBa_2Cu_3O_{7-x}$ samples prepared with different oxygen contents ($0 < x \leq 1$). From these measurements, we have quantitatively correlated the crystallographic, magnetic and superconducting properties of the $YBa_2Cu_3O_{7-x}$ system with the oxygen stoichiometry. Such correlations provide part of the foundation for identifying the mechanism for the high T_c superconductivity and for

0097-6156/87/0351-0136$06.00/0

predicting other chemical systems which might also exhibit high
superconducting transition temperatures.

EXPERIMENTAL

Sample Preparation
A sample of $YBa_2Cu_3O_{7-x}$ was prepared by reaction of a
stoichiometric mixture of $BaCO_3$, Y_2O_3 and CuO. The yttrium oxide was
fired at 1000 C before use to remove any carbonate or hydroxide. The
reactants were fired in air in an alumina crucible initially at
846C(48h), then at 900C(24h) and 946C(24h), and finally at 500C(24h).
The sample was reground between each firing and quenched from 500C.
Least squares refinement of the powder X-ray diffraction data from
the final product (see below) gave orthorhombic cell constants a =
3.8228(3), b = 3.8898(5) and c = 11.686(2)Å. The synthesis
conditions used are expected from previous results to lead to the
formation of a slightly oxygen deficient phase. Thermogravimetric
analyses (see below) confirmed this to be the case with a mean oxygen
composition of x = 0.04(3).

Thermogravimetric Analysis
As a step towards the syntheses of samples with other
compositions, the dependence of the oxygen content on temperature and
oxygen partial pressure was investigated by thermogravimetry. Three
different oxygen-containing atmospheres were used, pure oxygen, 5%
oxygen in helium, and helium (which we estimate to contain about
10ppm oxygen). Measurements were made at heating and cooling rates
of 2C/min using a DuPont 1090 thermal analyzer. Typical scans are
shown in Fig.1. In pure oxygen, the parent sample described above (x
= 0.04) begins to oxidize at 350C and reaches a maximum weight at
435C. At this temperature x is very close to zero. Above 435C, the
sample loses weight continuously. At 760C, where x = 0.46, there is
a pronounced discontinuity in the rate of reduction. Above this
temperature, the weight loss continues until at 991C the composition
corresponds to x = 0.63. The reoxidation on cooling is substantially
reversible though some kinetic effects are apparent. Specifically,
the oxidation to x = 0 observed on heating between 350 and 435C is
not observed and the discontinuity in the rate occurs at lower
temperature (690C, x = 0.34). Previous studies of the temperature
dependence of the cell constants in oxygen (18,19,21) show that the
orthorhombic structure becomes tetragonal at about 700C and it seems
likely that this phase transformation is associated with the observed
changes in the rates of oxidation and reduction. The present results
in oxygen are similar to those previously reported (24,25). In all
of these thermogravimetric measurements the stoichiometry is observed
to correspond to x = 0.5 at 800C. The oxidation/reduction behavior
in 5% oxygen is generally similar. The lower partial pressure of
oxygen results, as expected, in a larger value of x at the highest
temperature (x = 0.79 at 975C) and complete reoxidation is not
observed on cooling back to room temperature: x = 0.09. The
occurrence of discontinuities in the slope at 585 and 595C upon
cooling and heating, respectively, presumably correspond to the
changes observed in pure oxygen at 690 and 760C (x = 0.34,0.46).
Above 600C several other small discontinuities in the rate of weight

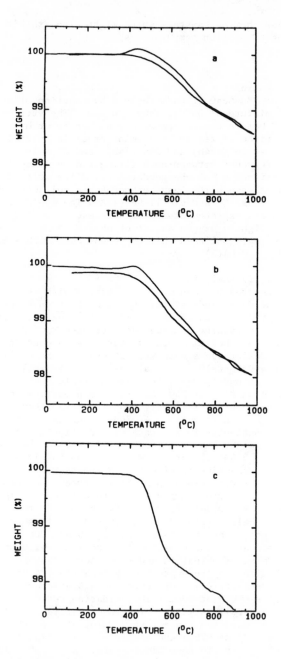

Figure 1.   Weight vs. temperature of $YBa_2Cu_3O_{6.96}$ in (a)pure O_2, (b)5%O_2/He, (c)pure He gas, as the temperature is ramped at 2°C/min.

loss with temperature are also observed. The thermogravimetric data in helium show a weight loss beginning at 400C which is smoothly varying up to 700C. At 850C the composition is close to x =1.0 and X-ray powder diffraction of the sample recovered from the thermogravimetric experiment to this temperature shows that the phase is tetragonal. No attempt was made to reoxidize the sample in helium because of anticipated problems with mass transport at the low oxygen partial pressure. At temperatures above 850C the sample continues to reduce but it has been reported that the structure collapses when x > 1.0 (25).

Differential Scanning Calorimetry

Oxidations of two reduced samples were also investigated by differential scanning calorimetry (DSC) in a DuPont 1090/910 instrument at 10C/min and at an initial oxygen pressure of 34.0 atm. The data are shown in Fig 2. Because the DSC head was sealed during the measurements, the pressure increased with time, reaching 41 atm at 500C. The lower curve corresponds to oxidation of 197 mg of the same batch of material (x=0.04) as used in the thermogravimetric experiments. The data show the oxidation to be centered at 390C and the measured weight gain corresponds to complete oxidation to x=0.00. Oxidation of 748 mg of an initially more reduced sample (x=0.60) shows two distinct peaks centered at 310C and 390C. The major part of the enthalpy change is associated with the first of these while the second corresponds nearly exactly to that observed for the more oxidized sample. The reason for the clear separation of the oxidation into two distinct processes is not yet clear but it does suggest that two distinctly different vacant oxygen sites are being filled. On cooling back to 100C at 10C/min, the exothermic peaks are gone, and only the heat capacity baselines are observed (not shown), consistent with our interpretation of the peaks as arising from oxidation.

Synthesis of other compositions.

Several samples were prepared for X-ray and neutron diffraction studies and magnetic measurements. Two samples were prepared from the initial sample by reduction in helium at 550C(48h) and 840C(48h). The samples were furnace-cooled in helium and their oxygen compositions determined thermogravimetrically by reduction in 5% hydrogen in helium at 10C/min to 1000C to be x = 0.60 and 1.00, respectively. Additional samples were prepared by equilibrating appropriate mixtures of, for example, compositions with x =1.0 and x = 0.04. The mixtures were sealed in silica tubes and heated at 650-670C for 16h and then furnace-cooled. The compositions were confirmed by thermogravimetric analysis in hydrogen as described above. The powder X-ray diffraction patterns of these materials are sharp and, except close to x=0.5 (see below), monophasic. This mode of synthesis exploits the high intrinsic oxygen mobility in these phases (indicated by the substantial reversibility of the TGA measurements and, e.g. ^{18}O exchange experiments, (26,27)) and provides a convenient route to homogeneous intermediate compositions 0 < x < 1.00.

Powder X-ray Diffraction Measurements

Powder X-ray diffraction data for all samples were measured on an automated Siemens D500 diffractometer using Cu Kα radiation. Scans were typically $2 \le 2\theta \le 100°$ in steps of $0.02°$, with $1°$ incident beam slits and count times of 2 sec per point. Data were transferred to a VAX 11/750 and lattice constants derived by least-squares profile fitting of the entire diffraction profiles (28). Atomic coordinates were taken from the neutron diffraction results at similar compositions (see below). Peak shapes were modeled using the pseudo-Voigt function, and cell constants, zeropoint and half-width parameters were optimized using a locally modified version of the DBW3.2 program (29). Results are plotted in Figure 3.

Powder Neutron Diffraction Experiments

Powder neutron diffraction (PND) data were collected from four different samples with respective oxygen contents of x=0.04 (at 296K, 120K and 8K), 0.34 (296K), 0.60 (296K and 8K), and 1.00 (296K) on the powder diffractometer at the Missouri University Research Reactor Facility (30). In each case, approximately 5g of sample was loaded into a 1/4" outside diameter, 0.005" walled vanadium can and centered on the instrument. A wavelength of 1.2893Å was selected from the (220) planes of a Cu monochromator at a take-off-angle of ca. 60.6° and calibrated by cross-comparison against the cell constants for three of the compositions obtained from the X-ray profile refinements (above). Data from four 25° spans of the linear position sensitive detector were each accumulated over some 3-6 hrs and combined to yield the diffraction profile $5 \le 2\theta \le 105°$, rebinned in 0.1° steps. The PND data were analyzed by full matrix least squares profile refinement (28) using DBW3.2 (29) and the pseudo-Voigt peak shape function. The background was treated by linear interpolation between a set of estimated points that was updated periodically during the refinement process. Scattering lengths of 7.750, 5.250, 7.718 and 5.805 fermi (10^{-15}m) for Y, Ba, Cu and O, respectively, were taken from the compilation of Koester and Yelon (31).

In beginning analysis of the first data set (on the x=0.04 sample), structural data for $YBa_2Cu_3O_{7-x}$ materials had not yet been reported. The powder X-ray and neutron diffraction data indicated the structure to be orthorhombic, and analyses commenced in space group Pmmm. This choice was confirmed by the subsequent analyses and by the preprints from other groups that became available during the course of this work (12,13,15,21,32-40). We found no indication of lower symmetry in these materials and satisfactory agreements for all data sets collected from the x=0.04, and 0.34 samples were obtained in Pmmm (although the data collected from samples in the Displex refrigerator suffer from an instrument misalignment which affects the final goodnesses of fit, and, to a small extent, the refined cell constants and half-width parameters). The parameters defining the final convergences are listed in Tables I (overall parameters and residuals), II (final atomic parameters) and III (selected separations and angles). Possible occupancy of other oxygen sites (O5 and O6) were also considered, although in no case was a significant population observed in the refinements. The total oxygen populations are, in all cases, within three esd's of the compositions derived from the TGA analyses. For the x=1.00 material, the O1 occupancy is close to zero, and the site has a large temperature factor. When the

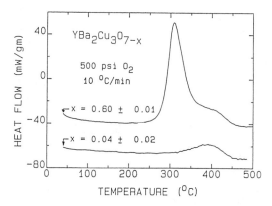

Figure 2. Heat flow vs. temperature obtained with a high pressure differential scanning calorimeter for two samples of YBa$_2$Cu$_3$O$_{7-x}$.

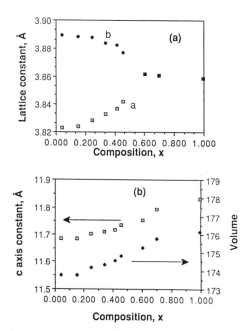

Figure 3. Lattice constants and unit cell volume vs. oxygen stoichiometry x for YBa$_2$Cu$_3$O$_{7-x}$ ((a) - a and b; (b) - c and unit cell volume)

TABLE I. FINAL OVERALL PARAMETERS AND RESIDUALS[a]

Composition,x Temperature	0.04 296K	0.04 120K	0.04 8K	0.34 296K	0.60 296K	0.60 8K	1.00 296K
a Å	3.8240(2)	3.8134(3)	3.8127(2)	3.8336(2)	3.8592(2)	3.8516(2)	3.8577(2)
b Å	3.8879(2)	3.8784(3)	3.8786(3)	3.8844(3)	"	"	"
c Å	11.6901(8)	11.6409(11)	11.6311(10)	11.7126(10)	11.7811(9)	11.7304(12)	11.8274(8)
N reflections	239	237	236	240	143	143	143
N at. var.	15	17	17	15	14	14	14
N tot. var.	27	27	29	25	25	23	23
R_B[b]	1.81	3.73	1.98	1.86	2.03	3.50	1.49
R'_B	1.88	--	--	--	3.10	--	--
R_p	3.35	5.39	4.64	3.81	4.08	6.15	3.98
R_{WP}	4.37	6.74	6.01	4.80	5.13	7.77	5.22
R_E	1.19	1.21	1.39	2.76	1.26	1.51	2.76

a. Standard deviations in parentheses
b. Residuals as defined by Rietveld (28); R'_B refers to fitting of a minor contribution in the pattern from
    the stainless steel sample can end caps.

TABLE II. FINAL ATOMIC PARAMETERS[a] WITH ESTIMATED STANDARD DEVIATION IN PARENTHESES

Composition, x		0.04	0.04	0.04	0.34	0.60	0.60	1.00
Temperature		296K	120K	8K	296K	296K	8K	296K
Y1	B	0.27(5)	0.79(7)	0.60(6)	0.56(6)	0.61(6)	1.09(9)	0.54(6)
Bal	z	0.1854(3)	0.1854(4)	0.1853(4)	0.1888(3)	0.1930(3)	0.1942(3)	0.1954(3)
	B	0.77(6)	0.73(8)	0.66(7)	0.28(7)	0.74(7)	0.57(8)	0.55(7)
Cu1	B	0.70(5)	0.63(8)	0.63(7)	0.74(7)	0.82(6)	0.69(7)	0.97(7)
Cu2	z	0.3557(2)	0.3553(2)	0.3559(2)	0.3572(2)	0.3598(2)	0.3597(3)	0.3608(2)
	B	0.48(4)	0.62(5)	0.56(5)	0.45(5)	0.59(4)	0.49(5)	0.39(4)
O1	B	1.6(1)	1.6(2)	0.7(2)	2.1(3)	1.6(6)	5.6(1.0)	4(3)
	occ.	0.96(1)	0.94(3)	0.82(2)	0.71(2)	0.33(2)	0.43(3)	0.08(3)
O2	z	0.3786(3)	0.3773(4)	0.3776(4)	0.3782(4)	0.3797(2)	0.3785(2)	0.3795(2)
	B	0.41(7)	0.73(11)	0.89(10)	0.45(9)	0.53(4)	0.71(6)	0.51(5)
	occ.	1.94(2)	2.01(3)	2.03(3)	2.02(3)	3.95(3)	4.08(4)	4.04(3)
O3	z	0.3775(3)	0.3778(5)	0.3786(4)	0.3769(5)	---	---	---
	B	0.75(6)	0.45(12)	0.14(10)	0.47(8)	---	---	---
	occ.	2.0	1.93(3)	1.83(3)	2.0	---	---	---
O4	z	0.1582(3)	0.1579(4)	0.1586(3)	0.1572(3)	0.1525(4)	0.1529(4)	0.1528(3)
	B	0.82(6)	0.74(12)	1.34(11)	0.81(8)	2.03(10)	0.85(11)	1.16(9)
	occ.	2.0	1.90(3)	2.11(3)	2.0	2.11(2)	1.89(3)	2.01(2)
Total O occ		6.90(2)	6.78(6)	6.79(6)	6.73(4)	6.39(4)	6.40(6)	6.13(5)

a. The various atomic species occupy sites in space group Pmmm (No. 47) of types Y1 1/2 1/2 1/2, 1h in Wyckoff notation (1d in the tetragonal structures of space group P4/mmm (No. 123)); Bal 1/2 1/2 z, 2t(2h); Cu1 0 0 0, 1a (1a); Cu2 00z, 2q(2g); O1 0 1/2 0, 1e(2f); O2 1/2 0 z, 2s(4i); O3 0 1/2 z, 2r(4i, equivalent to O2); O4 0 0 z, 2q(2g). Occupancies are expressed relative to the structural formula $YBa_2Cu_3O_{7-x}$. The temperature factor, B, is given as Å².

TABLE III. SELECTED SEPARATIONS (Å) AND ANGLES (°) WITH ESTIMATED STANDARD DEVIATIONS IN PARENTHESES

Composition, x Temperature	0.04 296K	0.04 120K	0.04 8K	0.34 296K	0.60 296K	0.60 8K	1.00 296K
Cu1-01	1.944	1.939	1.939	1.942	1.930	1.926	1.929
-04	1.849(3)	1.838(4)	1.845(4)	1.841(4)	1.797(5)	1.794(5)	1.807(4)
mean Cu1-0	1.896(1)	1.888(1)	1.892(1)	1.891(1)	1.864(2)	1.860(2)	1.868(1)
Cu2-02	1.931(1)	1.924(1)	1.923(1)	1.933(1)	1.944(1)	1.939(1)	1.942(1)
-03	1.961(1)	1.957(1)	1.957(1)	1.956(1)	--	--	--
-04	2.309(4)	2.298(5)	2.294(5)	2.342(5)	2.442(5)	2.426(6)	2.461(5)
mean Cu2-(03,04)	1.946(1)	1.940(1)	1.940(1)	1.945(1)	1.944(1)	1.939(1)	1.942(1)
Ba-01	2.890(2)	2.880(3)	2.878(3)	2.926(3)	2.982(2)	2.983(3)	3.010(2)
-02	2.980(4)	2.959(5)	2.960(5)	2.948(5)	2.926(3)	2.896(3)	2.910(3)
-03	2.949(4)	2.941(6)	2.947(5)	2.920(5)	--	--	--
-04	2.745(1)	2.738(1)	2.737(1)	2.754(1)	2.770(1)	2.766(1)	2.774(1)
mean Ba-0	2.862(1)	2.851(2)	2.852(2)	2.860(2)	2.875(1)	2.861(1)	2.876(1)
Y-02	2.407(2)	2.408(3)	2.406(3)	2.410(3)	2.394(1)	2.396(2)	2.398(1)
-03	2.389(2)	2.379(3)	2.372(3)	2.399(3)	--	--	--
mean Y-0	2.398(1)	2.394(2)	2.389(2)	2.404(2)	2.394(1)	2.396(2)	2.398(1)
Cu2-02-Cu2	164.0(2)	164.7(4)	164.9(3)	165.4(3)	166.1(2)	166.9(3)	166.9(2)
-03-Cu2	165.1(3)	164.7(4)	164.5(4)	166.4(4)	--	--	--

occupancy of this site was set to zero, residuals of R_P=4.06 and R_{WP}=5.31, similar to those presented in Table I, were obtained.

Magnetic Measurements
    Magnetic susceptibility (χ) and Meissner effect data were obtained from 4K to 800K using a George Associates Faraday magnetometer in a magnetic field (H) of 50G (Meissner effect) or 6.3kG (χ) at a temperature sweep rate of $\le$ 1K/min. Purified He exchange gas was used in the sample chamber. Ferromagnetic impurity contributions to χ were measured and corrected for via magnetization vs. H isotherms. Additional Meissner effect data were obtained using a P.A.R. vibrating sample magnetometer (VSM) in a field of 10 G.

RESULTS AND DISCUSSION

    Structural variation with composition
    The general form of the unit cell variation with x is similar to that seen in the high-temperature X-ray (18,19) and neutron (21) diffraction experiments, where compositional change is driven by changing temperature; of course, the observed lattice constants in such experiments intrinsically also reflect thermal expansion effects.
    The details of the evolution towards tetragonal symmetry as x approaches 0.5 are somewhat uncertain. The PXD data for the composition closest to x=0.5, x=0.506, were refined as a combination of both orthorhombic and tetragonal forms. The optimized scale factors for the two components were in the approximate ratio 4 (orthorhombic) : 1 (tetragonal). Biphasic behavior for x close to 0.5 has been noted previously (25).
    For reasons of space, the parallel structural results of other groups (12,13,15,21,32-40) are not incorporated in Tables I-III. Cross comparisons between the collated structural data, however, indicate generally excellent consistency, apart from minor differences that can now be ascribed largely to small differences in oxygen composition. Such differences are indicated by the unit cell constants (Table I, Figure 3), by coordinates such as Cul z, or O4 z (Table II), and by certain separations such as Cul-O4 (Table III).
    The compositional variation $0.04 \le x \le 1.00$ is accommodated structurally by changing occupancy of the O1 site. The parent phase of the related $La_3Ba_3Cu_6O_{14-x}$-type materials for which the regime x < 0 is also accessible (41) (and which, too, provide superconducting compositions (42)) also displays increased O1(and equivalently, O5) site occupancy in an again tetragonal phase (43). The composition x =0.0 corresponds to complete filling of the O1 site. The O5 site (1/2 0 0, 1b) is vacant. This unequal distribution gives rise to an increased distance between the Cul atoms separated by O1 (b = 3.888Å) compared to those separated by the vacant O5 site (a = 3.824Å). This ordered arrangement has long range coherence and defines the one-dimensional ... Cul-O1-Cul-O1.... chains along b. Based on these structural data, a genuine orthorhombic to tetragonal transition at some elevated temperature for a fixed composition, 0 < x < 0.5, that would correspond to disordering of these oxygen atoms over the O1 and O5 sites appears possible. Experimentally, there is to date no clear evidence that such a transition has been observed for 0 < x < 0.5.

In previous reports, the transitions observed at higher temperatures have invariably involved concomitant compositional changes (c.f. Fig. 1).

        Reduction of the oxygen content from x = 0.04 to x≈0.4 occurs by steady depopulation of the O1 site as described in the extensive study of Jorgensen et al. ($\underline{21}$). Initially, both the lattice constants and the superconductivity (see below) are little affected (Figures 3 and 5). As x approaches 0.5, the orthorhombic a and b axis lengths begin to converge rapidly, implying that the disparity between the Cu1 - Cu1' (along b) and Cu1 - Cu1' (along a) separations diminishes. At x =0.5 (interestingly the bond percolation threshold for a 2-dimensional net), the long range coherence of the O1 arrangement is lost. Local Cu1-O1-Cu1-O1... and Cu1-O5-Cu1-O5... chains occur with equal probability, but neither achieves long-range coherence or connectivity.

        For x > 0.5, this evolution progresses, with further depopulation of O1/O5 (now, of course, equivalent), leading ultimately to the x=1.0 structure which contains only O4-Cu1-O4 links along c. The interactions giving rise to the difference of 0.06Å between the a and b axis lengths in the orthorhombic phase at x = 0.04 might be expected to persist in a local sense in the tetragonal phase, perhaps giving rise to small static displacement for O1. The temperature factor of O1 is consistently higher than that of the other oxygen atoms, although there is no significant difference in this parameter between the x=0.04 (and x=0.34) and x=0.60 structures (Table III).

        Throughout this marked change in the composition and local structure of the Cu1-O1/O5 layer, the Cu2-O2/O3 'dimpled sheet' is remarkably invariant, exemplified perhaps most notably by the small extent of the changes in bond angles about O2 and O3 (Table III).

    Meissner effect measurements
        Shown in Fig. 5 are Meissner effect data obtained in the Faraday magnetometer (H=50G) for two series of $YBa_2Cu_3O_{7-x}$ samples both prepared from the x=0.04 sample of Figs. 1 and 2. The first set was taken to determine the relevant range of x to be more closely studied; this superconducting range was found to be 0≤x<0.5. The second set of measurements provided more detailed information about the dependence of the superconducting properties on x. A desired value of x was obtained by heating the previously measured sample (in-situ) up to 460-530C, waiting until the desired weight loss was achieved, dropping the temperature to prevent further weight loss and pumping out the sample chamber under high vacuum to remove the evolved oxygen. The sample chamber was then back-filled with clean He exchange gas in preparation for the next Meissner effect/magnetic susceptibility measurements. The two sets of measurements on the x=0.04 sample showed no significant differences in T_c or susceptibility, indicating that any sample degradation during storage in air over several months did not affect these properties. From Fig. 5, the superconducting transition widths remain nearly constant from x=0 to x≈0.4; however, above x≈0.4 the apparent T_c and volume of superconducting material drop dramatically, until no superconductivity is seen at x=0.60.

The variation of the superconducting midpoint temperature and the temperature at which a sample achieves 10% of its maximum diamagnetism from Fig. 4 are plotted vs. oxygen stoichiometry in Fig. 5. Surprisingly, both data sets show an S-shape, and there is no region where the bulk T_c drops smoothly to zero. T_c is nearly independent of x near x=0.0 and x=0.4, and at larger x the superconducting volume, as noted above, goes to zero. The two plateaus are smoothly joined by the S-shape composition dependence of T_c. From Figs. 4 and 5, the transition width is seen to be nearly independent of x between x=0.4 and x=0.0. Combined with the strong variation of T_c which occurs over this range, this indicates that upon increasing x in situ within the magnetometer, oxygen is removed uniformly throughout the sample. To verify this indication and to examine whether the preparation temperature influences T_c, several additional samples were prepared with various x values in flowing He gas or in sealed quartz tubes at 550 to 840C as discussed in the experimental section. Meissner effect measurements using the vibrating sample magnetometer (VSM) showed the superconducting behavior of these samples to be very similar to that shown in Figs. 4 and 5; the T_c data from the VSM are included in Figure 5.

From the evolution of the structure of $YBa_2Cu_3O_{7-x}$ with x as discussed above and as manifested in Fig. 5, the existence of intact Cu-O chains in the central Cu layer is seen not to be a prerequisite for high T_c superconductivity in this compound; if it were, one would have expected to see a $T_c(x)$ with negative curvature only, i.e., with increasingly negative slope with increasing x, contrary to observation. In fact, the data in Fig. 5 suggest instead that the T_c of 50K near x=0.4 is associated with the bounding "dimpled" Cu-O layers in the triple Cu-O layer block, since (i) the isolated Cu-O layers in the $La_{2-x}Sr_xCuO_4$ compounds have a similar maximum T_c of ~50K, (ii) the Cu-O chains in $YBa_2Cu_3O_{7-x}$ are severely disrupted for this region of x (Table 2), (iii) the plateau near x≈0.4 is associated with a mean Cu oxidation state similar to that in optimally superconducting $La_{2-x}(Sr,Ba)_xCuO_4$ compounds, and (iv) the magnetic susceptibilities are also quite similar (see below). We speculate that the role of the Cu-O chains in the central Cu-O layer is to electronically couple the two outer Cu-O layers together, thereby reducing quantum mechanical fluctuation effects associated with low dimensional systems, and allowing the observed T_c to more closely approach the mean-field T_c with decreasing x. The reason that the superconductivity disappears near x=0.5 is probably associated with the fact that at this composition all the Cu is formally Cu^{2+}, forming a Mott-Hubbard insulator/semiconductor, as in the ground state of La_2CuO_4 itself. Why the superconductivity does not reappear for 0.5<x<1 remains an open question.

Magnetic susceptibility studies

After each Meissner effect measurement plotted in Fig. 5, $\chi(T)$ was measured from T_c to T^{max} = 460-530C, where T^{max} was the temperature necessary to remove enough oxygen for the next desired oxygen stoichiometry. T^{max} increased with increasing x in $YBa_2Cu_3O_{7-x}$. Some of the $\chi(T)$ data are shown in Fig. 6, where $\chi(T)$ data for a separately prepared sample of $YBa_2Cu_3O_6$ (see experimental section) are also shown.

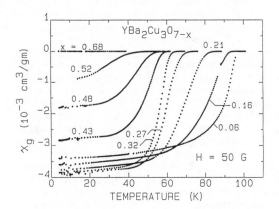

Figure 4.   Meissner effect data in a field of 50G for samples of $YBa_2Cu_3O_{7-x}$ for various oxygen stoichiometries.

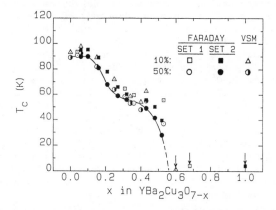

Figure 5.   Superconducting transition temperature T_C vs. oxygen stoichiometry x for $YBa_2Cu_3O_{7-x}$, as determined from Faraday Magnetometer Meissner effect measurements in Figure 4. The (50%) midpoints of the transitions (O,●) and temperatures at which the samples attained 10% of the maximum diamagnetism (□,■), are indicated. Also shown are additional corresponding data (◑,△) from the VSM. The arrows denote temperatures above which no superconductivity was observed.

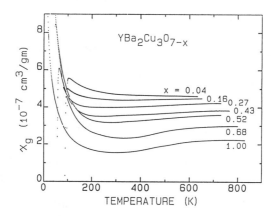

Figure 6.   Magnetic susceptibility χ_g vs. temperature for samples of $YBa_2Cu_3O_{7-x}$ for various values of x.

From Fig. 6, there is a clear and systematic evolution of the shape and magnitude of $\chi(T)$ with oxygen content. Particularly for the samples with $x \geq 0.3$, $\chi(T)$ appears to be the sum of three contributions: (i) a nearly temperature independent contribution, possibly corresponding to the Pauli susceptibility of the conduction carriers; (ii) a low-temperature Curie-Weiss contribution which increases with increasing x and may be associated with ≈1% nearly isolated Cu^{2+} spin 1/2 defects created during oxygen removal; and (iii) a contribution which increases with increasing temperature and exhibits a broad maximum near 800K; this contribution may result from short-range antiferromagnetic ordering of the bulk Cu^{+2} spins, but it may have a more exotic origin. In any case, contribution (iii) is similar to the $\chi(T)$ of certain $La_{2-x}Sr_xCuO_4$ samples and to the high temperature susceptibility of La_2CuO_4 itself. Detailed analysis and interpretation of the data in Fig. 6 will be presented elsewhere.

ACKNOWLEDGMENTS

We thank colleagues at Exxon for many helpful discussions and authors for preprints of unpublished work. We also thank M. Hillpot for her expertise in typing this paper.

REFERENCES
1. Bednorz, J. G.; Müller, K. A. Z. Phys. 1986, B64, 189.
2. Bednorz, J. G.; Müller, K. A.; Takashige, M. Science 1987, 236, 73.
3. Uchida, S.; Takagi, H.; Kitazawa, K.; Tanaka, S. Jpn. J. Appl. Phys. 1987, 26, L1.
4. Takagi, H.; Uchida, S.; Kitazawa, K.; Tanaka, S. Jpn. J. Appl. Phys. 1987, 26, L123.
5. Chu, C. W.; Hor, P. H.; Meng, R. L.; Gao, L.; Huang, Z. J.; Wang, Y. Q. Phys. Rev. Lett 1987, 58, 405.
6. Cava, R. J.; van Dover, R. B.; Batlogg, B.; Rietman, E. A. Phys. Rev. Let. 1987, 58, 408.
7. Tarascon, J. M.; Greene, L. H.; McKinnon, W. R.; Hull, G. W.; Geballe, T. H. Science 1987, 235, 1373.

8. Onoda, M.; Shamoto, S.; Sato, M.; Hosoya, S. Jpn. J. Appl. Phys. 1987, 26, L363.

9. Wu, M. K.; Ashburn, J. R.; Torng, C. J.; Hor, P. H.; Meng, R. L.; Gao, L.; Huang, Z. J.; Wang, Y. Q.; Chu, C. W. Phys. Rev. Lett. 1987, 58, 908.

10. Cava, R. J.; Batlogg, B.; van Dover, R. B.; Murphy, D. W.; Sunshine, S.; Siegrist, T.; Remeika, J. P.; Rietman, E. A. Zahurak, S. M.; Espinosa, G. P. Phys. Rev. Lett. 1987, 58, 1676.

11. Grant, P. M.; Beyers, R. B.; Engler, E. M.; Lim, G.; Parkin, S. S. P.; Ramirez, M. L.; Lee, V. Y.; Nazzal, A.; Vazquez J. E.; Savoy, R. J. Phys. Rev. 1987, B35, 7242.

12. Hazen, R. M.; Finger, L. W.; Angel, R. J.; Prewitt, C. T.; Ross, N. L.; Mao, H. K.; Hadidiacos, C. G.; Hor, P. H.; Meng, R. L.; Chu, C. W. Phys. Rev. 1987, B35, 7238.

13. Beno, M. A.; Soderholm, L.; Capone II, D.W.; Hinks, D. G.; Jorgensen, J. D.; Schuller, I. K.; Segre, C. U.; Zhang K.; Grace, J. D. Appl. Phys. Lett. in press.

14. Rao, C. N. R.; Ganguly, P.; Raychaudhuri, A. K.; Mohan Ram, R. A.; Sreedhar, K. Nature 1987, 326, 856.

15. Beyers, R.; Lim, G.; Engler, E. M.; Savoy, R. J.; Shaw, T. M.; Dinger, T. R.; Gallagher, W. J.; Sandstrom, R. L. Appl. Phys. Lett. 1987, 50, 1918.

16. Kini, A. M.; Geiser, U; Kao, H. I.; Carlson, K. D.; Wang, H. H.; Monaghan, M. R.; Williams, J. M. Inorg. Chem. 1987, 26, 1836-1837.

17. Johnston, D. C.; et. al. to be published.

18. Beyers, R.; Lim, G.; Engler, E. M.; Lee, V. Y.; Ramirez, M. L.; Savoy, R. J.; Jacowitz, R. D.; Shaw, T. M.; LaPlaca, S.; Boehme, R.; Tsuei, C. C.; Park, Sung I.; Shafer, M. W.; Gallagher, W. J.; Chandrashekhar, G. V. Appl. Phys. Lett. in press.

19. Schuller, I. K.; Hinks, D. G.; Beno, M. A.; Capone II, D. W.; Soderholm, L.; Locquet, J. P.; Bruynseraede, Y.; Segre, C. U.; Zhang, K.; Solid State Commun., submitted.

20. Saito, Y.; Noji, T.; Endo, A.; Matsuzaki, N.; Katsumata, M.; Higuchi, N. Jpn. J. Appl. Phys. 1987, 26, L366.

21. Jorgensen, J. D.; Beno, M. A.; Hinks, D. G.; Soderholm, L.; Volin, K. J.; Hitterman, R. L.; Grace, J. D.; Schuller, I. K.; Segre, C. U.; Zhang, K.; Kleefisch, M. S. Phys. Rev. to be published.

22. Ihara, H.; Hirabayashi, M.; Terada, N.; Kimura, N.; Senzaki, K.; Tokumoto, M. Jpn. J. Appl. Phys. 1987, 26, L463.

23. Shamoto, S.; Hosoya, S.; Onoda, M.; Sato, M. Jpn. J. Appl. Phys. 1987, 26, L493.

24. Strobel, P.; Capponi II, J. J.; Chaillout, C.; Marezio, M.; Tholence, J. L. Nature, 1987, 327, 306.

25. Gallagher, P. K.; O'Bryan, H. M.; Sunshine, S. A.; Murphy, D. W. Mater. Res. Bull. 1987, 22, 995.

26. Batlogg, B.; Cava, R. J.; Jayaraman, A.; van Dover, R. B.; Kourouklis, G. A.; Sunshine, S.; Murphy, D. W.; Rupp, L. W.; Chen, H. S.; White, A.; Short, K. T.; Mujsce, A. M.; Rietman, E. A. Phys. Rev. Lett. 1987, 58, 2333.

27. Bourne, L. C.; Crommie, M. F.; Zettl, A.; zur Loye, H.-C.; Keller, S. W.; Leary, K. L.; Stacy, A. M.; Chang, K. J.; Cohen, M. L.; Morris, D. E. Phys. Rev. Lett. 1987, 58, 2337.

28. Rietveld, H. M. J. Appl. Cryst. 1969, 2, 65.
29. Wiles, D. B.; Young, R. A. J. Appl. Cryst., 1981, 14, 149.
30. Thompson, C. W.; Mildner, D. F. R.; Mehregany, M.; Sudol, J.; Berliner, R.; Yelon, W. B. J. Appl. Cryst. 1984, 17, 385.
31. Koester, L.; Yelon, W. B. Summary of Low Energy Neutron Scattering Lengths, Netherlands Energy Research Foundation, 1982.
32. LePage, Y.; McKinnon, W. R.; Tarascon, J. M.; Greene, L. H.; Hull, G. W.; Hwang, D. M. Phys. Rev. 1987, B35, 7245.
33. Siegrist, T.; Sunshine, S.; Murphy, D. W.; Cava, R. J.; Zalurak, S. M. Phys. Rev. 1987, B35, 7137.
34. Capponi, J. J.; Chaillout, G.; Hewat, A. W.; Lejay, P.; Marezio, M.; Nguyen, N.; Raveau, B.; Soubeyroux, J. L.; Tholence, J. L. Tournier, R. Europhys. Lett. 1987, 3, 1301.
35. David, W. I. F.; Harrison, W. T. A.; Gunn, J. M. F.; Moze, O.; Soper, A. K.; Day, P.; Jorgensen, J. D.; Hinks, D. G.; Beno, M. A.; Soderholm, L.; Capone, D. W.; Schuller, I. K.; Segre, C. U.; Zhang, K.; Grace, J. D. Nature, 1987, 327, 310.
36. Beech, F.; Miraglia. S.; Santoro, A.; Roth, R. S. Phys. Rev. 1987, B35, 8778.
37. Greedan, J. E.; O'Reilly, A.; Stager, C. V. Phys. Rev. 1987, B35, 8770.
38. Cox, D. E.; Moodenbaugh, A. R.; Hurst, J. J.; Jones, R. H. J. Phys. Chem. Solids submitted
39. Santoro, A.; Miraglia, S.; Beech, F.; Sunshine, S. A.; Murphy, D. W.; Schneemeyer, L. F.; Waszczak, J. V. Mater. Res. Bull. 1987, 22, 1007.
40. Bordet, P.; Chaillout, C.; Capponi, J. J.; Chenavas, J.; Marezio, M. Nature 1987, 327, 687.
41. Er-Rakho, L.; Michel, C.; Provost, J.; Raveau, B. J. Solid State Chem. 1981, 37, 151.
42. Mitzi, D. B.; Marshall, A. F.; Sun, J. Z.; Webb, D. J.; Beasley, M. R.; Geballe, T. H.; Kapitulnik, A. preprint.
43. Newsam, J. M.; Mitzi, D. B. et al. to be published.

RECEIVED July 8, 1987

Chapter 15

Structure–Property Relationships
for $RBa_2Cu_3O_x$ Phases

C. C. Torardi[1], E. M. McCarron[1], M. A. Subramanian[1], H. S. Horowitz[1], J. B. Michel[1], Arthur W. Sleight[1], and D. E. Cox[2]

[1]Central Research and Development Department, E. I. du Pont de Nemours and Company, Experimental Station, Wilmington, DE 19898
[2]Physics Department, Brookhaven National Laboratory, Upton, NY 11973

The structures of several 1-2-3 compounds, e.g. $YBa_2Cu_3O_7$, have been refined from powder neutron diffraction data collected at 298 K. Orthorhombic symmetry was found for the superconducting phases $NdBa_2Cu_3O_{6.85}$, $YBa_2Cu_3O_{6.91}$, and $ErBa_2Cu_3O_{6.99}$. The short Cu-O distance of about 1.85 Å decreases with increasing size of the rare earth cation, but the average Cu-O distance in the chains remains nearly constant. Refinement of a reduced $YBa_2Cu_3O_{7-x}$ phase shows that a tetragonal $YBa_2Cu_3O_6$ composition can be achieved before structural collapse. This phase is semiconducting and not superconducting. We have also prepared a sample of $YBa_2Cu_3O_7$ with the tetragonal 1-2-3 structure where oxygens have not ordered in a way to produce the linear $+(Cu-O)_n$ chains. This form of $YBa_2Cu_3O_7$ is not superconducting above 4.2 K. The structure of $La_{1.5}Ba_{1.5}Cu_3O_{7.33}$ with the tetragonal 1-2-3 structure was refined and again there are no linear $+(Cu-O)_n$ chains and no superconductivity. On the basis of these six structural refinements, we tentatively conclude that the $+(Cu-O)_n$ chains are required for superconductivity.

Ever since the identification of the 1-2-3 compound (i.e. $RBa_2Cu_3O_x$ where R = Y or nearly any lanthanide) as a high

0097–6156/87/0351–0152$06.00/0

temperature superconductor, considerable effort has been expended to understand the relationship between the superconductivity and the structural features of these compounds. Although these compounds are structurally related to the ideal AMO_3 perovskite structure, there are extensive oxygen vacancies and there are two distinct A cation sites. Furthermore, compounds with the basic 1-2-3 structure may have either tetragonal or orthorhombic symmetry depending on the oxygen vacancy ordering.

We have used the high resolution Guinier x-ray powder diffraction technique to determine whether a given compound is tetragonal or orthorhombic. Powder neutron diffraction data have been obtained for refinement by the Rietveld profile technique. Since crystals of orthorhombic 1-2-3 compounds seem to be invariably twinned, there is little advantage to single crystal techniques. Furthermore, the greater relative scattering power of oxygen with neutrons compared to x-rays allows for determination of the oxygen content even when impurity phases are present.

We have used neutron diffraction to elucidate the structure of various 1-2-3 compounds as a function of R cation type and ratio of R to Ba. We have also refined various tetragonal and orthorhombic structures as a function of oxygen content. One objective was to determine the range of oxygen content over which the 1-2-3 structure can exist. Of particular interest was the question of whether or not superconductivity exists if the linear $\pm(Cu-O)_n$ chains, a prominent structural feature of the orthorhombic $YBa_2Cu_3O_7$ structure, are disrupted.

Synthesis

Most of the 1-2-3 compounds were prepared from appropriate ratios of reagent grade CuO, $BaCO_3$, La_2O_3, Nd_2O_3, Er_2O_3 and/or Y_2O_3. The rare earth oxides were fired at 1000°C before weighing. After grinding together in a mortar, the intimate mixture was heated in air to 950°C and held for at least 24 hours. This was followed by cooling in air to room temperature at about 150°/hour.

One sample of $YBa_2Cu_3O_x$ was prepared by an oxalate precursor technique. Appropriate quantities of reagent grade nitrates of Y^{3+}, Ba^{2+} and Cu^{2+} were dissolved in water and added dropwise to 1M oxalic acid using about 50% excess over that required to convert all metals to oxalates. The resulting blue slurry was spray dried in a Buchi No. 190 mini spray dryer operated with N_2 as the atomizing gas. The inlet temperature was 230°, and the outlet temperature was 105°C. The chamber atmosphere was air. The resulting fine blue powder was heated in air at 800°C for about 20 hours. The sample was then cooled to room temperature at a cooling rate of about 150°/hour.

Neutron Powder Diffraction

Neutron diffraction data were collected at the Brookhaven National Laboratory High-Flux Beam Reactor. For these experiments, pressed and sintered samples of powder (~10-15g) in the shape of a cylinder approximately 0.9-1.3 cm in diameter and 3.0-3.5 cm in height were used. Full sets of data were collected at ambient temperature at 0.1° step-scan intervals over a 2θ range of 5 to 140°. The experimental configuration consisted of a pyrolytic graphite monochromator and analyzer in the (002) and (004) settings, respectively. Collimation was set at 20', 40', 40' and 20', respectively, for the in-pile, monochromator-sample, sample-analyzer, and analyzer-detector locations. The neutron wavelength was 2.3695 Å, and higher order components were suppressed with a graphite filter.

Non-superconducting Tetragonal Compounds

Consistent with Guinier x-ray powder photographs, the neutron diffraction patterns for $YBa_2Cu_3O_6$, $YBa_2Cu_3O_7$ prepared at 800°C, and $La_{1.5}Ba_{1.5}Cu_3O_{7.3}$ were found to have tetragonal symmetry. For the latter sample, a larger tetragonal unit cell ($a\sqrt{2}$) has been reported ([1]). We found no evidence in the x-ray or neutron data to support this larger cell. The data for the above compounds were fitted by refinement of the structures in the space group P4/mmm (No.123) using the Rietveld profile method ([2]). All perovskite-like oxygen-atom positions were initially included. Starting atomic positions were as follows: Ba (site 2h) $\frac{1}{2},\frac{1}{2},z$; Y (site 1d) $\frac{1}{2},\frac{1}{2},\frac{1}{2}$; Cu(1) (site 1a) 0,0,0; Cu(2) (site 2g) 0,0,z; O(1) (site 2g) 0,0,z; O(2) (site 4i) 0,1/2,z; O(3) (site 2f) 0,$\frac{1}{2}$,0; O(4) (site 1b) 0,0,$\frac{1}{2}$. Because the oxygen content was not accurately known, the site occupancies for the O atoms were also refined along with the positional and isotropic thermal parameters. Only oxygen atoms O(1) and O(2) refined to full occupancy for the three compounds. A check on the metal atom multipliers showed these sites to be fully occupied. The stoichiometry obtained from the profile analyses are given in Tables I and II.

The tetragonal sample of $YBa_2Cu_3O_7$ contained a small amount of $BaCO_3$ impurity that complicated the profile background correction and contributed to the intensity of several peaks thereby resulting in a relatively high weighted profile R-factor. From the present structural refinement, this compound has been tentatively formulated as $YBa_2Cu_3O_{7.3(1)}$. A clean sample of this compound has now been prepared and will soon be reexamined.

In the unit cell of $La_{1.5}Ba_{1.5}Cu_3O_7$, there are available three sites for the lanthanum and barium ions. Because the La/Ba ratio is 1:1, two models for the

Table I. Atomic Positional and Thermal Parameters for
the Tetragonal (P4/mmm) Compounds $YBa_2Cu_3O_6$,
$YBa_2Cu_3O_{7.3(1)}$, and $La_{1.5}Ba_{1.5}Cu_3O_{7.33(4)}$

		$YBa_2Cu_3O_6$	$YBa_2Cu_3O_{7.3(1)}$	$La_{1.5}Ba_{1.5}Cu_3O_{7.33(4)}$
Ba	x	0.50	0.50	0.50
or	y	0.50	0.50	0.50
Ba/La(1)	z	0.1952(4)	0.1886(9)	0.1814(4)
	$B(\mathring{A}^2)$	0.9(1)	1.0(3)	0.4(1)
Y	x	0.50	0.50	0.50
or	y	0.50	0.50	0.50
La(2)	z	0.50	0.50	0.50
	$B(\mathring{A}^2)$	0.6(1)	0.4(4)	0.8(2)
Cu(1)	x	0.00	0.00	0.00
	y	0.00	0.00	0.00
	z	0.00	0.00	0.00
	$B(\mathring{A}^2)$	1.0(1)	1.9(4)	0.5(2)
Cu(2)	x	0.00	0.00	0.00
	y	0.00	0.00	0.00
	z	0.3605(3)	0.3559(7)	0.3465(3)
	$B(\mathring{A}^2)$	0.3(1)	0.5(2)	0.3(1)
O(1)	x	0.00	0.00	0.00
	y	0.00	0.00	0.00
	z	0.1518(4)	0.1567(9)	0.1562(6)
	$B(\mathring{A}^2)$	1.5(1)	1.3(5)	2.3(3)
O(2)	x	0.00	0.00	0.00
	y	0.50	0.50	0.50
	z	0.3794(2)	0.3759(6)	0.3657(3)
	$B(\mathring{A}^2)$	0.7(1)	1.3(3)	1.0(1)
O(3)	x	0.00	0.00	0.00
	y	0.50	0.50	0.50
	z	0.00	0.00	0.00
	$B(\mathring{A}^2)$	1.00	12(2)	7.1(7)
	atom/site	0.02(1)	0.63(5)	0.64(1)
O(4)	x	--	0.00	0.00
	y	--	0.00	0.00
	z	--	0.50	0.50
	$B(\mathring{A}^2)$	--	1.50	1.50
	atom/site	--	0.08(3)	0.08(2)
$a(\mathring{A})$		3.8519(1)	3.8657(3)	3.9024(2)
$c(\mathring{A})$		11.8037(4)	11.6015(17)	11.6908(9)
R_{NUC} (%)		3.9	7.6	2.5
R_{WP} (%)		11.5	19.1	11.0
R_E (%)		5.8	7.9	6.5

Table II. Bond Distances in the Tetragonal (P4/mmm) Compounds $YBa_2Cu_3O_6$, $YBa_2Cu_3O_{7.3(1)}$, and $La_{1.5}Ba_{1.5}Cu_3O_{7.33(4)}$

	$YBa_2Cu_3O_6$	$YBa_2Cu_3O_{7.3(1)}$	$La_{1.5}Ba_{1.5}Cu_3O_{7.33(4)}$
Cu(1)-O(1)	1.792(5)	1.818(10)	1.826(7)
Cu(1)-O(3)[a]		1.933	1.951
Cu(2)-O(1)	2.463(4)	2.311(9)	2.225(6)
Cu(2)-O(2)	1.939(1)	1.947(2)	1.964(1)
Cu(2)-O(4)[a]		1.672(3)	1.795(2)
Ba-O(1)	1.772(2)	2.758(3)	2.775(3)
Ba-O(2)	2.905(4)	2.908(9)	2.907(6)
Ba-O(3)[a]		2.919(7)	2.882(3)
La/Y-O(2)	2.395(1)	2.410(3)	2.504(2)
La/Y-O(4)[a]		2.733	2.759

[a]These oxygen atom sites are partially occupied, see Table I.

large-cation ordering were examined. One model placed
the ions at half occupancy in all three sites. The
second model placed a full La at $\frac{1}{2},\frac{1}{2},\frac{1}{2}$ and the remaining
lanthanum along with all of the barium statistically
distributed over the $\frac{1}{2},\frac{1}{2},z$ sites. It was evident from
the refinement that the second model was best. In fact,
the occupancies of the La and Ba ions at $\frac{1}{2},\frac{1}{2},z$ were
refined with the constraint that the sum of the
occupation factors must give a fully occupied site. To
within one-half of a standard deviation unit, these sites
were found to contain 75% Ba and 25% La. During these
refinements, the z coordinates for Ba and La were
constrained to be equal as were the isotropic B values.
Atoms O(1) and O(3) exhibited relatively high isotropic B
values, 2.3(3) and 7.1(7) $Å^2$, respectively. They were
then refined with anisotropic terms. Both showed very
large B_{11} and B_{22} values on the order of 4 $Å^2$ for O(1)
and 9 $Å^2$ for O(3). These are believed to be an artifact
due to the statistical disorder of barium and lanthanum
on the 0,0,z sites. Both oxygen atoms bond to these
metal ions, and for any given M-O (M = Ba, La) bond, the
La-O bond lengths are expected to be shorter than the
Ba-O distances (e.g., the respective sums of the ionic
radii (3) for La-O and Ba-O are 2.67 and 2.92 Å). This
means that in some or all of the unit cells these O atoms
actually are located slightly away from their 'ideal'
positions (4).

Further details on the profile analyses will be
published in the future. Table I compares the results of
these refinements. Interatomic distances in the three
tetragonal materials are given in Table II.

Superconducting Orthorhombic Compounds

Rietveld profile refinements were done on the compounds
$NdBa_2Cu_3O_{6.9}$, $YBa_2Cu_3O_{6.9}$, and $ErBa_2Cu_3O_7$ which all have
orthorhombic symmetry. The data for these compounds were
fitted by refinement of the structures in the space group
Pmmm (No. 47). The lower crystallographic symmetry,
relative to the tetragonal symmetry described above,
splits some of the oxygen atom sites into two independent
groups. Following an analogous procedure used in the
refinements of the tetragonal materials, all
perovskite-like oxygen atom positions were initially
included and the site occupancies for the oxygen atoms
were also refined. The variable z coordinates were taken
from the recently described structures of $YBa_2Cu_3O_7$.
Oxygen atoms O(1), O(2), and O(3) (Table III) refined to
full occupancy for the three compounds. The metal atom
sites were also found to be fully occupied. Comparisons
of the refined structural parameters and of the
interatomic distances are given in Tables III and IV,
respectively. The stoichiometry obtained from the
refinements of these compounds is also given in the
Tables.

Table III. Atomic Positional and Thermal Parameters for the
            Superconducting Orthorhombic (Pmmm) Compounds
            $NdBa_2Cu_3O_{6.85(6)}$, $YBa_2Cu_3O_{6.91(4)}$, and
            $ErBa_2Cu_3O_{6.99(6)}$.

		$NdBa_2Cu_3O_{6.85(6)}$	$YBa_2Cu_3O_{6.91(4)}$	$ErBa_2Cu_3O_{6.99(6)}$
Ba	x	0.50	0.50	0.50
	y	0.50	0.50	0.50
	z	0.1846(5)	0.1847(4)	0.1838(6)
	$B(\text{Å}^2)$	0.9(2)	0.5(1)	0.3(2)
Nd,Y	x	0.50	0.50	0.50
or Er	y	0.50	0.50	0.50
	z	0.50	0.50	0.50
	$B(\text{Å}^2)$	0.1(1)	0.2(1)	0.0(2)
Cu(1)	x	0.00	0.00	0.00
	y	0.00	0.00	0.00
	z	0.00	0.00	0.00
	$B(\text{Å}^2)$	1.1(2)	0.4(1)	0.2(2)
Cu(2)	x	0.00	0.00	0.00
	y	0.00	0.00	0.00
	z	0.3505(4)	0.3556(3)	0.3563(4)
	$B(\text{Å}^2)$	0.3(1)	0.2(1)	0.0(1)
O(1)	x	0.00	0.00	0.00
	y	0.00	0.00	0.00
	z	0.1559(6)	0.1586(4)	0.1596(5)
	$B(\text{Å}^2)$	1.1(2)	0.6(2)	0.4(2)
O(2)	x	0.00	0.00	0.00
	y	0.50	0.50	0.50
	z	0.3721(7)	0.3784(5)	0.3790(6)
	$B(\text{Å}^2)$	0.8(2)	0.3(1)	-0.1(2)
O(3)	x	0.50	0.50	0.50
	y	0.00	0.00	0.00
	z	0.3715(6)	0.3782(4)	0.3791(6)
	$B(\text{Å}^2)$	0.5(1)	0.3(1)	0.4(2)
O(4)	x	0.00	0.00	0.00
	y	0.50	0.50	0.50
	z	0.00	0.00	0.00
	$B(\text{Å}^2)$	1.9(7)	1.6(5)	1.8(6)
	atom/site	0.71(4)	0.87(3)	0.89(4)
O(5)	x	0.50	0.50	0.50
	y	0.00	0.00	0.00
	z	0.00	0.00	0.00
	$B(\text{Å}^2)$	1.50	1.50	1.50
	atom/site	0.14(2)	0.05(1)	0.10(2)
$a(\text{Å})$		3.8687(2)	3.8179(1)	3.8123(2)
$b(\text{Å})$		9.9150(2)	3.8801(2)	3.8756(2)
$c(\text{Å})$		11.7477(8)	11.6655(6)	11.6576(7)
R_{NUC} (%)		3.7	2.2	3.5
R_{WP} (%)		12.8	11.1	13.6
R_E (%)		6.9	6.1	9.9

Table IV. Bond Distances in the Orthorhombic (Pmmm)
Superconducting Compounds, $NdBa_2Cu_3O_{6.85(6)}$,
$YBa_2Cu_3O_{6.91(4)}$, and $ErBa_2Cu_3O_{6.99(6)}$

	$NdBa_2Cu_3O_{6.85(6)}$	$YBa_2Cu_3O_{6.91(4)}$	$ErBa_2Cu_3O_{6.99(6)}$
Cu(1)-O(1)	1.831(7)	1.850(5)	1.859(7)
Cu(1)-O(4)[a]	1.957	1.940	1.938
Cu(1)-O(5)[a]	1.934	1.909	1.906
Cu(2)-O(1)	2.286(6)	2.298(5)	2.295(6)
Cu(2)-O(2)	1.974(2)	1.958(1)	1.955(2)
Cu(2)-O(3)	1.950(2)	1.927(1)	1.924(2)
Ba-O(1)	2.773(2)	2.739(1)	2.733(2)
Ba-O(2)	2.931(5)	2.958(4)	2.967(5)
Ba-O(3)	2.942(5)	2.976(4)	2.988(5)
Ba-O(4)[a]	2.906(4)	2.879(3)	2.868(4)
Ba-O(5)[a]	2.921(4)	2.899(3)	2.889(4)
Nd/Y/Er-O(2)	2.449(4)	2.378(3)	2.373(4)
Nd/Y/Er-O(3)	2.472(4)	2.405(3)	2.398(4)

[a]These oxygen atom sites are partially occupied, see Table III.

Structural Description

The structures of tetragonal $YBa_2Cu_3O_6$, $YBa_2Cu_3O_{7.3}$, and $La_{1.5}Ba_{1.5}Cu_3O_{7.3}$ and of orthorhombic $NdBa_2Cu_3O_{6.9}$, $YBa_2Cu_3O_{6.9}$, and $ErBa_2Cu_3O_7$ are very closely related. All six structures contain layers consisting of a copper-oxygen framework. Each layer is constructed from two copper-oxygen sheets with composition CuO_2 obtained by corner sharing of nearly-planar CuO_4 units. The two sheets are connected by linear O-Cu-O groups so that the copper atoms within the sheets, Cu(2) in Figure 1, are five-coordinated in a square pyramidal configuration. $YBa_2Cu_3O_6$, which contains no more oxygen atoms, has this structure. In the orthorhombic $YBa_2Cu_3O_7$-type structures, additional oxygen atoms connect the linear O-Cu-O groups along the b axis making this copper atom, Cu(1), planar and four-coordinated, and creating infinite O-Cu-O strings along the b direction. In the tetragonal $YBa_2Cu_3O_{7.3}$ and $La_{1.5}Ba_{1.5}Cu_3O_{7.3}$ compounds, the additional oxygen atoms connect the linear O-Cu-O groups along the a_1 and a_2 tetragonal axes in a random fashion. These two phases have vacant approximately forty percent of the oxygen-atom sites along the chains. They are also the only ones to show a small amount of oxygen in the Y or La layer thereby forming a Cu-O-Cu connection between the copper-oxygen layers. The La-Ba material has all of the barium and one-third of the lanthanum ions situated within the Cu-O layer network and the remaining lanthanum ions located between the layers. The other compounds have only Ba within the copper-oxygen framework and Nd, Y, or Er between the layers.

In the rare-earth layer, La, Nd, Y, and Er are bonded to essentially eight oxygen atoms arranged to form a square prism around the cation. Barium is bonded to eight oxygen atoms in $YBa_2Cu_3O_6$, ~10.5 oxygen atoms in the tetragonal Y and La-Ba compounds, and almost 10 oxygen atoms in the superconducting orthorhombic phases.

Discussion

There has been considerable speculation about the significance for superconductivity of the linear $\{Cu-O\}_n$ chains in the $RBa_2Cu_3O_7$ compounds. It should be noted that a prominent structural feature of the old T_c record holders, e.g. Nb_3Ge, was strongly bonded, infinite linear chains of Nb. From our neutron diffraction studies, it appears that the $\{Cu-O\}_n$ chains are a necessary requirement for superconductivity in the $RBa_2Cu_3O_x$ family of compounds where x is about seven. The evidence is particularly compelling in the case of orthorhombic vs. tetragonal $YBa_2Cu_3O_x$. The composition and formal oxidation states for copper are essentially the same, yet the orthorhombic form with chains is superconducting with a T_c of about 90 K whereas the tetragonal form without

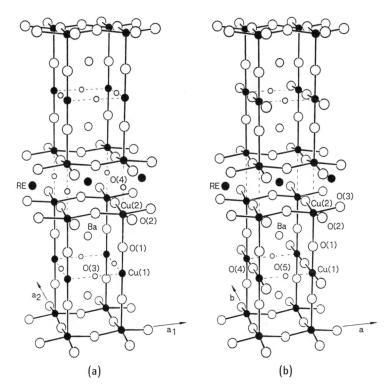

(a)   (b)

Figure 1. (a) Structure of tetragonal $YBa_2Cu_3O_6$, $YBa_2Cu_3O_{7.3}$, and $La_{1.5}Ba_{1.5}Cu_3O_{7.3}$. O(3) and O(4) sites are either empty or partially occupied depending on the compound. (b) Structure of orthorhombic $NdBa_2Cu_3O_{6.9}$, $YBa_2Cu_3O_{6.9}$, and $ErBa_2Cu_3O_7$. O(4) and O(5) sites are partially occupied.

chains is not superconducting above 4.2 K. We also see that superconductivity disappears in the $RBa_2Cu_3O_x$ series as x decreases towards 6.5 and the chains disappear. Again in the $La_{1.5}Ba_{1.5}Cu_3O_{7.3}$ compound, there are no chains and no superconductivity above 4.2 K.

The Cu(1)-O distance along the linear chains is simply one half the b axis. Thus, this Cu-O distance increases monotonically with the increasing size of the rare earth cation. In view of the apparent importance of the linear chains, it seems unexpected that T_c does not change significantly in a regular manner with this changing Cu-O distance. However, there is a compensation effect. As the Cu-O distance along the chains increases, the shorter Cu-O distances perpendicular to this chain are decreasing. Thus, the average Cu(1)-O distance in the chains does not change much in $RBa_2Cu_3O_7$: 1.89, 1.89, and 1.90Å for the Nd, Y and Er compounds, respectively. Therefore, the copper-oxygen framework can be visualized as being expanded in the ab plane and compressed along the c axis as the rare-earth ion gets larger.

The situation for $La_{1+y}Ba_{2-y}Cu_3O_x$ compositions with the 1-2-3 structure is very complex since there are two compositional variables which play a critical role in whether or not superconductivity is observed. The value of y apparently never reaches zero but should be as low as possible for optimum superconducting properties. The value of x must be controlled through thermal treatment and oxygen pressure. We deal only with the case of y = 0.5 in this paper. There has been partial La substitution for Ba at the Ba site, and the value of x is about 7.3. There is more than enough oxygen to form chains along one axis only. The remaining oxygen might have been disordered along one axis only, but instead it is disordered along both axes. There are now no infinite $\{Cu-O\}_n$ chains. It is also important to note that the average copper oxidation in $La_{1.5}Ba_{1.5}Cu_3O_{7.3}$ is +2.37 which is nearly the same as in $YBa_2Cu_3O_{7.0}$, i.e. +2.33.

A structure for $La_{1.5}Ba_{1.5}Cu_3O_{7.3}$ had been previously proposed on the basis of powder x-ray diffraction data ([1]). The proposed structure had the same basic cation positions as in the 1-2-3 structure, but the oxygen vacancy ordering was much different. Furthermore, it was proposed that the cell edge was a√2 despite the fact that there was apparently no evidence for this larger cell. A 100 reflection was listed in a table, but no observed d value was given. Our attempts to refine our data with this proposed structure met with no success, and we thus conclude that the proposed structure ([1]) is incorrect.

We have now established that the range of oxygen content in $RBa_2Cu_3O_x$ phases with the 1-2-3 structure is x = 6 to 7. However, in the $La_{1+y}Ba_{2-y}Cu_3O_x$ system, the value of x can clearly exceed 7. It appears that the value of x must be reasonably close to seven in order to have oxygen vacancy ordering to produce the orthorhombic structure.   Too much oxygen or too little oxygen causes a type of oxygen disorder which results in a tetragonal structure.

Furthermore, although $RBa_2Cu_3O_x$ compounds are normally orthorhombic and superconducting for values of x close to seven, we observe semiconducting behavior as x decreases below 6.5.   Thus for $YBa_2Cu_3O_6$, it seems logical to assume localized electron behavior and to assign an oxidation state of +1 to the linearly coordinated copper and an oxidation state of +2 to the copper in the sheets (5).

Literature Cited

1.   Er-Rakho, L.; Michel, C.; Provost, J.; Raveau, B. *J. Solid State Chem.* 1981, *37*, 151.
2.   Rietveld, H. M. *J. Appl. Crystallogr.* 1965, *2*, 65.
3.   Shannon, R. D. *Acta Crystallogr.* 1976, *A32*, 751.
4.   Torardi, C. C.; McCarron, E. M.; Subramanian, M. A.; Sleight, A. W.; Cox, D. E., in preparation
5.   Torardi, C. C.; McCarron, E. M.; Bierstedt, P. E.; Sleight, A. W.; Cox, D. E. *Solid State Commun.*, submitted.

RECEIVED July 6, 1987

Chapter 16

Oxygen-Defect Perovskites and the 93-K Superconductor

N. L. Ross, R. J. Angel, L. W. Finger, R. M. Hazen, and C. T. Prewitt

Geophysical Laboratory, Carnegie Institution of Washington, 2801 Upton Street, NW, Washington, DC 20008

The structure of the tetragonal phase of $YBa_2Cu_3O_6$ is presented and discussed in relation to both the orthorhombic structure of the 93K superconductor and the known structures of other oxygen defect perovskites. The reduction of oxygen content from the ideal ABO_3 stoichiometry of perovskites reduces the primary coordination of both the A and B cation sites. In particular, with decreasing oxygen content the octahedral B site first becomes a square-based pyramid (5 coordinating oxygens), then square-planar or pseudo-tetrahedral (4) and finally linearly coordinated by two oxygens. Such behaviour is especially prevalent in those phases containing variable oxidation state transition-metal cations such as Cu, Mn, and Fe, which can be accommodated in these variable-coordination sites. The superconducting phase $Ba_2YCu_3O_{7-x}$ is a defect perovskite which exhibits many of these structural variations. Samples with high oxygen content ($x=0$) have two types of oxygen vacancies. One is the removal of all the oxygen atoms from layers level with the Y atoms (in A sites). This reduces the coordination of the two adjacent layers of Cu sites to 5-fold square-based pyramids. The second set of oxygen vacancies is adjacent to the other copper atoms, reducing their coordination to square-planar. Further reduction of the oxygen content to O_6 ($x=1$) results in the removal of oxygen from this square-planar array to produce linear O-Cu-O groups; these remaining oxygens also form the apices of the square-based pyramids.

The discovery by Wu et al. (1) of superconductivity at 93K in a mixed-phase sample in the Y-Ba-Cu-O system has stimulated an unprecedented amount of research effort directed at solving the structures of the phase(s) responsible for superconductivity. In this paper we will present the results of the first single-crystal structure analyses of the phases present in these samples to be carried out (2), and then review the additional structural information now available in relation to the previously determined structures of other oxygen defect perovskites.

Y_2BaCuO_5 Structure

The polycrystalline sample in which superconductivity at temperatures in excess of 93K was first observed consisted of two major phases. Electron microprobe analyses indicated that a green transparent phase, which comprised the majority of the early samples, contained Ba, Cu and Y in the ratios 1:1:2. A single crystal diffraction study and structure solution by direct methods indicated that this green phase has an ideal composition Y_2BaCuO_5, and a structure similar to that reported on the basis of powder diffraction (3). Crystallographic data for this structure are reported in Table I.

0097–6156/87/0351–0164$06.00/0
© 1987 American Chemical Society

Table I. Crystallographic data for Y_2BaCuO_5

Space group: Pbnm (determined by diffraction absences and structure determination)
Unit Cell data: a=7.123 Å, b=12.163 Å, c=5.649 Å, V_{cell} = 489.4 Å^3, Z = 4, ρ_{calc} = 6.22

Positional Parameters

Atom	Mult.	x/a	y/b	z/c
Ba	4	0.93	0.90	0.25
Y1	4	0.12	0.29	0.25
Y2	4	0.40	0.07	0.25
Cu	4	0.71	0.66	0.25
O1	8	0.16	0.43	0.00
O2	8	0.36	0.23	0.50
O3	4	0.03	0.10	0.25

This structure, shown in Figure 1, is unusual in that it contains Y_3+ in trigonal pyramid coordination, as well as copper in capped square-planar coordination (alternatively described as a square-based pyramid), the latter feature being shared with the structure of the superconducting phase.

$YBa_2Cu_3O_{7-x}$

Structure Analysis. The superconducting phase comprised about one-third of the earliest samples received from the University of Houston. Electron microprobe analyses indicated a cation ratio of Ba:Cu:Y of 2.1:2.9:1.0. Unit cell parameters were determined on a number of different crystals, and indicated that the unit cell was tetragonal with a=3.859 Å, c=3.904 Å. Some crystals in later batches deviated from exact tetragonal symmetry, indicating that ordering within the structure was reducing the symmetry to orthorhombic. X-ray diffraction studies showed that single crystals had 4/mmm Laue symmetry, and diffraction symbol 4/mmmP..., allowing as possible space groups P422, P4mm, P$\bar{4}$m2, P$\bar{4}$2m, and P4/mmm. With all of the cations within the unit cell occupying the special positions of the perovskite structure, these four possible space groups are only distinguished by the position of the oxygen atoms, which are comparatively weak scatterers of X-rays. Thus an N(z) test of X-ray diffraction data is not a reliable indicator of the presence or absence of a center of symmetry in the structure. A combination of direct methods and least squares refinements was used to solve the structure, and crystallographic data are reported in Table II for the two most likely space groups.

The structure of this tetragonal phase is shown in Figure 2a and, like a number of oxygen-defect perovskites discussed below, exhibits variable co-ordination of cations by the oxygen anions. The key to understanding the structure lies in the ordering of the barium and yttrium cations over the 'A' sites of the perovskite structure. In cubic and metrically cubic perovskites these sites are coordinated to 12 anion positions at a distance of $\sqrt{2}a_c$. However, Y^{3+} is a much smaller cation than Ba^{2+}, (Shannon-Prewitt radii of 1.02 Å and 1.42-1.60 Å respectively (4)), and the oxygens surrounding the yttrium sites can be thought as being 'drawn in' towards the central Y site. As a consequence the ideally cubic sub-cell around the Y site at 1/2,1/2,1/2 is shortened along the c axis of the structure, while the cells around the Ba sites are expanded. This shortening of the central sub-cell of the structure results in a distance of 3.30 Å between pairs of Cu2 sites at 0,0,0.36 and 0,0,0.64. Oxygen anions are therefore excluded from the site at 0,0,1/2 as their presence would require Cu-O distances of 1.65 Å. Thus we see that this phase is required to have at least one vacant oxygen site by crystal chemical constraints. The resulting coordination of the Cu2 site can be described as slightly distorted versions of a square-based pyramid, or a capped square-planar array.

In contrast to the vacancies at z=1/2, the oxygen sites at z=0 are not required to be vacant by crystal chemistry. It is these sites that show the variable occupancy that gives rise to the variable stoichiometry and complicated structural phase relations to be discussed below. In our tetragonal phase however these sites are also almost completely devoid of scattering density, indicating that the oxygen content of this material is very close to O_6. The coordination of the Ba sites is accordingly reduced from the value of twelve for an ideal structure, to eight. The off center distribution of the remaining oxygens (Figure 2a) accounts for the displacement of the barium atoms from an ideal z-coordinate of $1/6$ to the observed $z=0.19$. The Cu1 sites are also reduced in coordination, from six in an ideal perovskite, to two. The presence of this linear O-Cu-O group could account for some of the similarities to cuprite, Cu_2O, noted in the Raman spectra of these samples (5).

Table II. Crystallographic data for $YBa_2Cu_3O_6$

Unit cell parameters: a=b=3.859 Å, c=11.71 Å, $V_{cell} = 174.4$ Å^3, $\rho_{calc} = 6.19$

Positional Parameters

| Atom | P$\bar{4}$m2 | | | | P4/mmm | | | |
	Mult.	x/a	y/b	z/c	Mult.	x/a	y/b	z/c
Ba	2	0.50	0.50	0.19	2	0.50	0.50	0.19
Y	1	0.50	0.50	0.50	1	0.50	0.50	0.50
Cu1	1	0.00	0.00	0.00	1	0.00	0.00	0.00
Cu2	2	0.00	0.00	0.36	2	0.00	0.00	0.36
O1	2	0.50	0.00	0.38	4	0.50	0.00	0.38
O2	2	0.50	0.00	0.61				
O3	2	0.00	0.00	0.15	2	0.00	0.00	0.15

Phase transitions and Average Structures. It is clear from the large quantity of structural information now available for the $YBa_2Cu_3O_{7-x}$ compound that there are a number of distinct structures that differ in structural detail. Both the local structure and the apparent long-range structure of these materials are determined by the thermal history and the oxygen content of the structure, which appears to vary between O_6 and O_7, although there is one report (Beyers, R. et al., Appl. Phys. Letts., submitted) of an $O_{7.4}$ composition. At the lowest oxygen contents, O_6, to which the structure reported above approximates, the structure is truly tetragonal both on local and long-range scales. This is because both of the oxygen sites at $z=0$ $(1/2,0,0$ and $0,1/2,0)$ are vacant, and there is nothing to distinguish the a and b axes of the unit cell. When the oxygen content of the structure is increased these sites may either be equally and partially occupied, or one may be occupied in preference to the other. To take an end-member case, with an oxygen content of O_7 the crystal may be tetragonal with both of these oxygen sites being half occupied, or one may be fully vacant and the other fully occupied. In the latter case the a and b axes are now distinguishable, and the structure is orthorhombic (6) with the Cu1 sites in square planar coordination (Figure 2b).

It is also important to distinguish the type of disorder within the structure, which may be dynamic or static. Static disorder is the apparent random occupancy of one site or the other by oxygen atoms, while dynamic disorder involves the rapid exchange of oxygen atoms between the two sites. In both cases the average structure 'seen' by X-ray or neutron single crystal diffraction techniques remains tetragonal. On the other hand, if the distribution is static, a powder diffraction experiment will result in the true symmetry independent of the domain distribution, provided that the domains are larger than the correlation length. It is for this reason that powder diffraction techniques have been successful in determining the orthorhombic structure (7; Beech, F. et al., Phys. Rev. Letts. , submitted; Beno M.A. et al., Appl. Phys. Letts. , submitted). However, in the static case local ordering of the oxygens and vacancies may give rise to orthorhombic domains within the crystals, which then appear to be (110) twins (8; Beyers, R. et al., Appl. Phys. Letts., submitted; Sueno, S. et al., Jap. J. Appl. Phys., in press). The size of these domains is clearly dependent upon both the oxygen content and upon the

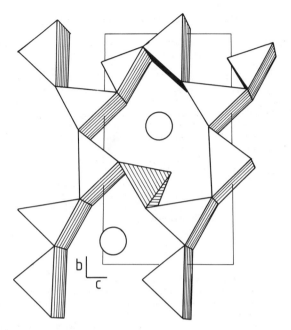

Figure 1. Polyhedral representation of the Y_2BaCuO_5 structure. The large circles represent Ba sites.

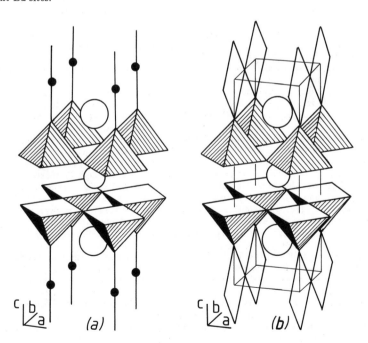

Figure 2. Polyhedral representation of (a) $YBa_2Cu_3O_6$ and (b) $YBa_2Cu_3O_7$. The large circles represent Ba sites, intermediate circles Y sites.

thermal history of the samples, and in the limit single orthorhombic crystals with space group Pmmm can be grown.

Dynamic disordering is responsible for the phase transition between the low temperature orthorhombic structure and a high temperature tetragonal phase in which all of the anion sites at z=0 are statistically and equally occupied by oxygen atoms. The wide range of reported temperatures for this transition (Schuller, I.K. et al., Solid State Comm., submitted; Sueno, S. et al., Jap. J. Appl. Phys. , in press) clearly represent a true variation of the transition temperature with oxygen content and the scale of ordering within the structure (9). The space group of this dynamically disordered phase must be a tetragonal supergroup of the Pmmm symmetry of the orthorhombic phase - and is therefore P4/mmm. However, this does not necessarily imply that a structure with an oxygen content of O_6 must possess this space group. Indeed, the displacements of the fully occupied anion sites are quite likely to be different in the two cases, and could lead to different space groups for the disordered phase and the O_6 tetragonal phase. It should also be noted that neither the orthorhombic - tetragonal phase transition, nor the varying scales of ordering within the orthorhombic phase, give rise to a structure similar to that proposed for $La_3Ba_3Cu_6O_{14}$ (10), as suggested recently by some groups (11,12). In fact, a review of the available data (Ross, N.L. et al., Nature, submitted) suggests that $La_3Ba_3Cu_6O_{14}$ actually possesses the tetragonal structure reported here for the superconductor.

Oxygen-Defect Perovskites

We have now discussed in detail the structure of $YBa_2Cu_3O_{7-x}$ with respect to the thermal history and oxygen content of the phase. These two variables determine the local structure and apparent long-range structure present in the phase, that is, the distribution of oxygen vacancies in the structure. Since $YBa_2Cu_3O_{7-x}$ is a true derivative of the ideal perovskite structure, a comparison of this compound with other oxygen-defect perovskites provides a basis from which to gain a greater understanding of this class of materials. In the following sections, the structure of the high T_c superconductor will be discussed in relation to other oxygen-defect perovskites derived from removal of oxygen atoms from the ideal perovskite prototype.

Polyhedral Elements . The perovskite prototype, ABO_3, consists of a framework of corner-linked BO_6 octahedra extending infinitely in three dimensions, with a large A cation in 12-fold coordination occupying the cavity created by the linkage of eight octahedra at the corners of a cube outlining the unit cell of the structure. The reduction of oxygen content from the ideal ABO_3 stoichiometry by substitution of low-valence metal cations for high-valence metal cations reduces the primary coordination of the A and B cation sites. With decreasing oxygen content, the octahedral B site first becomes a capped square plane or square-based pyramid with five coordinating oxygen atoms. $CaMnO_{2.5}$(13,14) and $SrMnO_{2.5}$(15), for example, have five-coordinate Mn^{3+} cations with approximately square-pyramidal coordination. The Cu2 atoms in both the orthorhombic and tetragonal structures of the high T_c superconductor are also in square-pyramidal coordination (Figure 2). Further removal of oxygen atoms results in cations that are in 4-fold coordination. We have already discussed the example of the $Cu1^{2+}$ atoms in square-planar coordination in the orthorhombic structure of the high T_c superconductor (Figure 2b). There are, however, examples of metal cations that prefer tetrahedral coordination. $Ca_2Fe_2O_5$(16) and $Sr_2Fe_2O_5$(17) crystallize in structures closely related to that of brownmillerite, Ca_2FeAlO_5 (18). In all of these structures one half of the Fe^{3+} cations are in tetrahedral coordination and one half are in octahedral coordination (Figure 3). Finally, the oxygen content may be reduced even further, resulting in B sites that are linearly coordinated by two oxygens. To the authors' knowledge, the coordination of the Cu1 sites in the tetragonal high T_c superconductor (Figure 2a) is the first example of an oxygen-defect perovskite with this type of coordination. It is also the most oxygen-deficient perovskite ($ABO_{2.0}$) known.

From the examples given above, it is apparent that the coordination geometry adopted by the metal cations in the B-sites depends on the particular transition metal cations present in the structure. Mn^{3+} prefers square pyramidal coordination while Cu^{2+} is found in both square pyramidal and square planar coordination. Fe^{3+}, on the other hand, is commonly found in tetrahedral and octahedral coordination, but not in square pyramidal coordination. Thus although $CaMnO_{2.5}$ and $SrFeO_{2.5}$ have the same oxygen stoichiometry, they crystallize with totally different structures.

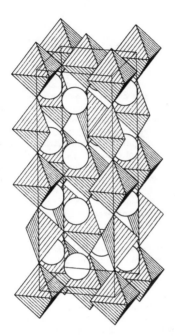

Figure 3. Polyhedral representation of the brownmillerite structure.

<u>Oxygen Vacancy Arrangements</u>. The structure of individual oxygen-defect perovskites therefore depends not only on the overall oxygen content of the structure, but also on the polyhedral coordination preference of the metal cations present in the structure. This section will focus on the oxygen vacancy arrangements present in specific oxygen-defect perovskites resulting from organization of the polyhedral units.

The structures of $CaMnO_{2.5}$ and $SrMnO_{2.5}$ are characterized by orthorhombic cells with $a \cong \sqrt{2}a_c$, $b \cong 2\sqrt{2}a_c$ and $c \cong a_c$. The host lattice is derived from the perovskite prototype by ordering of oxygen vacancies on the (001) planes of the cubic perovskite that form pseudo-hexagonal tunnels running along $[001]_c$. Every second $[110]_c$ row of oxygen sites has oxygen vacancies alternating with oxygen atoms. Successive defect rows are displaced by $a/2$, so as to generate square-based pyramidal coordination for all of the Mn^{3+}. The Ca^{2+} and Sr^{2+} cations, which are surrounded by 10 rather than 12 nearest neighbors, are believed to play a secondary role in maintaining the parallel channels of oxygen vacancies.

The structures of $CaFeO_{2.5}$ and $SrFeO_{2.5}$, on the other hand, are characterized by orthorhombic cells with $a = c \cong \sqrt{2}a_c$ and $b \cong 4a_c$. Layers of perovskite-like FeO_6 octahedra alternate with (001) layers in which two oxygen atoms have been removed from every octahedron. As with the manganate structures, oxygen vacancies propagate along $[110]_c$. Unlike the manganate structures, <u>every</u> oxygen atom along particular $[110]_c$ strings is missing, leaving Fe^{3+} with only four nearest neighbors. The ordering of vacancy strings is such that half the iron atoms remain octahedrally coordinated and half are tetrahedrally coordinated. The Ca^{2+} and Sr^{2+} cations occupy the large interstices between the layers, surrounded by eight nearest neighbors.

Tofield et al.(19) proposed a model for oxygen vacancy ordering in $SrFeO_{2.75}$ which is of interest because of its close relationship to the structures described above. This model structure has either orthorhombic or body-centered tetragonal symmetry with cell dimensions of $a = c \cong 2\sqrt{2}a_c$ and $b \cong 2a_c$. Similar to the ferrite and manganate oxygen-defect perovskites, the oxygen vacancies in $SrFeO_{2.75}$ also propagate along $[110]_c$. Every other oxygen, however, is missing along these $[110]_c$ vacancy strings as in $CaMnO_{2.5}$ and $SrMnO_{2.5}$, but not in $CaFeO_{2.5}$ and $SrFeO_{2.5}$. The manganate structure can, in fact, be simply derived from the structure proposed for $SrFeO_{2.75}$. Let A represent one of the oxygen-vacancy strings along $[110]_c$ which has the oxygen(O)/vacancy(v) arrangement, ...v-O-v-O-v..., and let B represent a $[110]_c$ string with no vacancies, ...O-O-O-O..., and let C represent a string with the pattern,...O-v-O-v-O..., in which the vacancies are displaced relative to A. Adjacent $[110]_c$ strings in $CaMnO_{2.5}$ structure have the pattern, ...ABCBABCBABCBA..., and the corresponding pattern in the model structure proposed for $SrFeO_{2.75}$ is ...ABBBABBBABBBA... . Thus the structure of $CaMnO_{2.5}$ can be derived directly from the proposed structure upon substitution of strings with oxygen vacancies (C) for every second fully-occupied $[110]_c$ oxygen chain (B). The $SrFeO_{2.5}$ structure, however, cannot be directly derived from the proposed structure without involving rearrangement of the oxygen atoms and vacancies concommittant with removal of oxygen atoms.

In Tofield et al.'s model for $SrFeO_{2.75}$, half of the iron atoms remain in octahedral coordination and half are in five-fold coordination. In view of the observed preference of Fe^{3+} for tetrahedral and octahedral coordination, and of Mn^{3+} for pyramidal coordination, we would expect that this proposed structure type would be more appropriate for manganate $O_{2.75}$ perovskites than for those containing iron. Indeed, Grenier et al.(20) have reported a ferrite with the same oxygen stoichiometry, $Ca_4Fe_2Ti_2O_{11}$, which has a structure related to that of brownmillerite, but with a 1:3 ratio of tetrahedral to octahedral layers.

The oxygen vacancies in the oxygen-defect perovskites described above all propagate along the $[110]_c$ of the cubic perovskite cell. The arrangement of the oxygen vacancies at $z=0$ within the orthorhombic structure of the high T_c superconductor follows a similar pattern. However, as a consequence of the low oxygen content of the unit cell ($O_{2.33}$ on an O_3 basis) <u>every</u> one of these $[110]$ rows has alternate occupied and vacant sites. As in $CaMnO_{2.5}$ and $SrMnO_{2.5}$ each successive defect containing row is displaced relative to its neighbours. In $YBa_2Cu_3O_7$ this arrangement results in the square planar coordination of the Cu1 sites. The total absence of oxygen from the $z=1/2$ planes, however, is a new perovskite feature and, as discussed above, is a crystal-chemical consequence of the ordering of Ba^{2+} and Y^{3+} on the A sites.

<u>Static and Dynamic Ordering</u>. As with the 93 K superconductor, the local and long-range ordering present in oxygen-defect perovskites is controlled, in large part, by the oxygen content of the samples. This is illustrated very clearly by the work of Grenier et al.(20,21). Grenier et al.(21) proposed a general model for oxygen ordering in intermediate compositions between

$Ca_2Fe_2O_5$ and ABO_3 ($A=Ca,Sr,Ba,Y,La,Gd$; $B=Fe,Ti$). Based on the relationship between the brownmillerite structure and the perovskite structure, they proposed that the structures of intermediate compositions would consist of a succession of perovskite-like sheets of corner-linked $[BO_6]$ octahedra separated by sheets of $[BO_4]$ tetrahedra. Grenier et al.(20) found, however, in the $Ca_2Fe_{2x}Ti_{2-2x}O_{6-x}$ solid solutions ($0 \leq x \leq 1$) that ordered structures with a succession of octahedral and tetrahedral sheets exist only for compositions with $0.50 \leq x \leq 1$. The most oxygen-deficient endmember, $Ca_2Fe_2O_5$, crystallizes with the brownmillerite structure which has a 1:1 ratio of tetrahedral to octahedral layers. With increasing oxygen content up to $x=0.5$, they observed three more structures with ordered oxygen vacancies, $Ca_5[Fe_2Ti]_O[Fe_2]_TO_{13}$, $Ca_3[FeTi]_O[Fe]_TO_8$ and $Ca_4[FeTi_2]_O[Fe]_TO_{11}$. These compounds have ratios of octahedral to tetrahedral sheets of 3:2, 2:1 and 3:1, respectively, thus corroborating their model. Near compositions of $x=0.5$, ordered microdomains appeared which later gave rise to a long-range ordered structure. For x less than 0.5, the average structure determined by x-ray diffraction had cubic symmetry, corresponding to a disordered perovskite structure. The Mössbauer spectra, however, showed that some of the Fe^{3+} cations were still in tetrahedral coordination suggesting that, even in the absence of superlattice ordering, the oxygen vacancies still tend to combine in pairs with Fe^{3+} inducing a change from octahedral to tetrahedral coordination.

In addition to these examples of static ordering of oxygen and vacancies, even in structures without long-range order, it is possible for these structures to undergo dynamic disordering of oxygens and vacancies to produce a structure with higher symmetry. Such a transition has been observed in both $YBa_2Cu_3O_{7-x}$ and in $Sr_2Fe_2O_5$ (22). The low temperature form of the latter has the orthorhombic symmetry of the brownmillerite structure with statically ordered oxygen vacancies. Above 700°C, the phase has the cubic symmetry of an oxygen-deficient perovskite with all anion sites equally and statistically occupied by oxygen atoms. Given that the transition to a disordered cubic structure involves a complete randomization of the oxygen sublattice, substantial disorder would be expected to exist just below the transition temperature. Similar order-disorder transitions would be expected to occur at high temperatures in intermediate structures in the $CaFeO_{2.5}$ - $CaTiO_3$ system since they have similar oxygen vacancy ordering schemes.

Conclusions

Our review of the structural variations found in a number of oxygen defect perovskites has demonstrated that the structure of the high T_c superconductor $YBa_2Cu_3O_{7-x}$ has both some normal and unusual crystal chemical features. The ordering of oxygen atoms around the Cu1 sites is similar to patterns found in a wide variety of other perovskite derivatives, and gives rise to similar order-disorder behaviour. The novel feature of the oxygen-vacancy ordering arises as a consequence of the ordering of Ba^{2+} and Y^{3+} over the A sites. This ordering forces a complete plane of oxygen atoms to be missing from the structure, but with the interesting result that the B site coordination of Cu2 is pyramidal, a coordination found in other oxygen-defect perovskites.

Acknowledgments

This work has been supported in part by National Science Foundation grants EAR8419982, EAR8608941 and EAR8319504, by a NATO fellowship to RJA, and by the Carnegie Institution of Washington.

Literature Cited

1. Wu, M.K.; Ashburn, J.R.; Torng, C.J.; Hor, P.H.; Meng, R.L.; Gao, L.; Huang, Z.J.; Wang, Y.Q.; Chu, C.W. Phys. Rev. Letts. 1987, 58, 908-910.
2. Hazen, R.M.; Finger, L.W.; Angel, R.J.; Prewitt, C.T.; Ross, N.L.; Mao, H.K.; Hadidiacos, C.G.; Hor, P.H.; Meng, R.L.; Chu, P.W. Phys. Rev. B 1987, 35, 7238-7241.
3. Michel, C.; Raveau, B. J. Solid State Chem. 1982, 43, 73-80.
4. Shannon, R.D.; Prewitt, C.T. Acta Cryst. 1969, B25, 925-946.
5. Hemley, R; Mao, H.K. Phys. Rev. Letts. 1987, 58, 2340-2342.
6. Siegrest, T.; Sunshine, S; Murphy, D.W.; Cava, R.J.; Zahurak, S.M. Phys. Rev. B 1987, 35, 7137-7139.

7. Capponi, J.J.; Chaillout, C.; Hewat, A.W.; Lejay, P.; Marezio, M.; Nguyen, N.; Raveau, B.; Soubeyroux, J.L.; Tholence, J.L.; Tournier, R. Europhys. Letts. 1987, 3, 1301-1308.
8. LePage, Y.; McKinnon, W.R.; Tarascon, J.M.; Greene, L.H.; Hull, G.W.; Hwang, D.W. Phys. Rev. B. 1987, 35, 7245-7248.
9. Strobel, P.; Capponi, J.J.; Chaillout, C.; Marezio, M.; Tholence, J.L. Nature 1987, 327, 306-308.
10. Er-Rakho, L.; Michel, C.; Provost, J.; Raveau, B.J. J. Solid State Chem. 1981, 37, 151-157.
11. Hirabayashi, M.; Ihara, H.; Terada, N.; Senzaki, K.; Hayashi, K.; Waki, S.; Murata, K.; Tokumoto, M.; Kimura, Y. Jap. J. Appl. Phys. 1987, L454-455.
12. Qadri, S.B.; Toth, L.E.; Osofsky, M.; Lawrence, S.; Gubser, D.U.; Wolf, S.A. Phys. Rev. B 1987, 35, 7235-7237.
13. Poeppelmeier, K.R.; Leonowicz, M.E.; Longo, J.M. J. Solid State Chem. 1982, 44, 89-98.
14. Poeppelmeier, K.R.; Leonowicz, M.E.; Scanlon, J.C.; Longo, J.M. J. Solid State Chem. 1982, 45, 71-79.
15. Caignaert, V.; Nguyen, N.; Hervieu, M.; Raveau, B. Mat. Res. Bull. 1985, 20, 479-484.
16. Hughes, H.; Ross, P.; Goldring, D.C. Min. Mag. 1967, 36, 280-291.
17. Greaves, C.; Jacobson, A.J.; Tofield, B.C.; Fender, B.E.F. Acta. Cryst. 1975, B31, 641-646.
18. Colville, A.A.; Geller, S. Acta. Cryst. 1971, B27, 2311-2315.
19. Tofield, B.C.; Greaves, C.; Fender, B.E.F. Mat. Res. Bull. 1975, 10, 737-746.
20. Grenier, J.C.; Menil, F.; Pouchard, M.; Hagenuller, P. Mat. Res. Bull 1978, 13, 329-337.
21. Grenier, J.C.; Darriet, J.; Pouchard, M.; Hagenmuller, P. Mat. Res. Bull. 1976, 11, 1219-1226.
22. Shin, S.; Yonemura, M.; Ikawa, H. Mat. Res. Bull. 1978, 13, 1017-1021.

RECEIVED July 6, 1987

Chapter 17

Oxide Ion Vacancies, Valence Electrons, and Superconductivity in Mixed-Metal Oxides

J. Thiel[1,2], **S. N. Song**[2,3], **J. B. Ketterson**[2,3], **and K. R. Poeppelmeier**[1,2]

[1]Department of Chemistry, Northwestern University, Evanston, IL 60201
[2]Materials Research Center, Northwestern University, Evanston, IL 60201
[3]Department of Physics and Astronomy, Northwestern University, Evanston, IL 60201

High temperature superconductivity in $YBa_2Cu_3O_{7-\delta}$ is dependent on the mixed oxidation states of copper (Cu^{3+}/Cu^{2+}). Replacement of the d^8 Cu^{3+}-ion with the isoelectronic d^8 Ni^{2+}-ion and Ba^{2+} with La^{3+} for charge compensation has been studied. In the composition series $YBa_{2-x}La_xCu_{3-x}Ni_xO_{7\pm\delta}$; $x = 0.0$, 0.1, 0.2, 0.3, 0.4, 0.5, 0.7, and 1.0 a systematic change from orthorhombic to tetragonal symmetry was observed along with a decrease in the resistive transition temperature from 92K → 72K → 29K. Magnetic susceptibility measurements indicate that several phases are present. They are closely related in structure to $YBa_2Cu_3O_7$ and likely contain some fraction of Ni^{3+}/Ni^{4+} and more than seven oxygen atoms per unit cell.

The concept that a significant fraction of oxygen atoms may be removed by reduction from metallic oxide lattices with little or no structural change was first studied (1,2) and debated (3) over twenty-five years ago. In particular the perovskite lattice has been shown to persist over the limits ABO_{3-x}; $0 \geq x \geq 0.5$ for many transition and some main group metal cations. The first compounds with the composition $ABO_{2.5}$ and ordered oxide ion vacancies that were recognized (4,5) to be related to perovskite were $Ca_2Fe_2O_5$ (6) and the mineral brownmillerite, Ca_2AlFeO_5 (7). The vacancy ordering in the former can best be described if the perovskite structure is viewed as a series of AO and BO_2 planes along one [001] direction. Sheets of Fe^{3+} with tetrahedral coordination result from removal of alternate [110] rows of oxide ions from alternate BO_2 planes. The new unit cell is related to the simple cubic dimension(a_c) by $a_o \simeq \sqrt{2}\ a_c$, $c_o \simeq \sqrt{2}\ a_c$ and $b_o \simeq 4a_c$. The series of compounds $A_nB_nO_{3n-1}$ ($n \geq 2$), for example $Ca_2Fe_2O_5$ ($n = 2$) and $CaFeO_3$ ($n = \infty$), have been examined in detail (8). In general the superstructure formed depends upon the electronic configuration (d^n) and preferred coordination of the smaller B-site cation and the ionic radius of the larger, electropositive A-cation (9). Complex pseudocubic and oxygen-deficient structures based on

0097–6156/87/0351–0173$06.00/0

perovskite are known for manganese ($\underline{10}$), iron ($\underline{11}$) and copper ($\underline{12}$). Oxygen deficient compounds based on the related K_2NiF_4 structure, e.g. $Ca_2MnO_{3.5}$, are also known ($\underline{13}$) although they have not been as thoroughly investigated.

In addition to pseudocubic structures, ABO_3 compounds form an extraordinary number of structures based on mixed cubic (c) and hexagonal (h) close-packed AO_3 layers ($\underline{14}$). Some of this polytypism has been shown to be associated with the oxygen composition ($\underline{15,16}$). The combination of mixed sequences (c,h), cation composition, and oxygen stoichiometry can give rise to a large number of complex structures and a range in properties ($\underline{17}$).

Following the recent report of Bednorz and Müller ($\underline{18}$) on possible high T_c (>30K) superconductivity in the Ba-La-Cu-O mixed phase system, the compound $La_{2-x}M_xCuO_{4-x/2+\delta}$ (M = Ba^{2+}, Sr^{2+}, or Ca^{2+}) with the tetragonal K_2NiF_4 structure was identified ($\underline{19,20}$) to be the superconducting phase. Soon thereafter Chu et al. ($\underline{21,22}$) reported superconductivity above 90K in the Ba-Y-Cu-O system. The compound has been identified by numerous groups ($\underline{23-26}$) and has the composition $YBa_2Cu_3O_{7-\delta}$ and a structure (Beno, M. A.; Soderholm, L.; Capone, D. W.; Jorgensen, J. D.; Schuller, I. K.; Segre, C. U.; Zhang, K.; Grace, J. D. Appl. Phys. Lett., in press.) related to the tetragonal structure reported for $La_3Ba_3Cu_6O_{14.1}$ ($\underline{27}$).

In this paper we report on our efforts to replace the d^8 Cu^{3+}-ion in $YBa_2Cu_3O_7$ with the isoelectronic d^8 Ni^{2+}-ion. Lanthanum was substituted in equimolar amounts with nickel to provide the necessary charge compensation.

Experimental

Samples of composition $YBa_{2-x}La_xCu_{3-x}Ni_xO_{7\pm\delta}$ with x = 0.0, 0.1, 0.2, 0.3, 0.4, 0.5, 0.7 and 1.0 were prepared by solid state reaction of Aldrich yttrium oxide (99.999%), cupric oxide (99.999%), barium carbonate (99.999%) lanthanum oxide (99.999%) and nickel oxide (99.999%). Powders were ground with a mortar and calcined in air at 900°C for 8 hours. The samples were reground daily and calcined for 16 hours at 950°C in air for 5 days. The compounds were ground and heated in an oxygen atmosphere first for 6 hours at 900°C and then for 12 hours at 700°C. Disc-shaped specimens 1.25 cm in diameter and 1.5 mm thick were isostatically pressed at 7.2 kbar at room temperature. The oxygen treatment at 900°/700°C was repeated to anneal the pellet. After cooling, the discs were cut into rectangular specimens with cross section of 1.5 x 3 mm^2 and four leads were attached with silver paint for 4-point resistivity measurements.

X-ray diffraction (XRD) and magnetic susceptibility measurements were carried out on polycrystalline materials that had been calcined in oxygen. XRD powder patterns were recorded with Cu Kα radiation using a Ni filter on a Rigaku diffractometer. For reference all patterns were recorded with an internal NBS Si standard. After correcting the peak position based on the observed silicon line positions, the lattice constants were then refined by a least squares method weighted proportional to the square root of the height and inversely with the square of the width of the peak.

A   VTS-50   susceptometer   was   used   for   the   magnetization measurements.   When equipped with a specially designed transport probe   it   could   also   be   used   for   resistivity   measurements. Temperature was controlled in the range 1.7-400K and magnetic fields in the range 0-5T.   The susceptometer was calibrated with NBS   aluminum   and   platinum   standards.   For   the   transport measurements temperatures could be read with either the thermometer in the VTS or with a carbon glass thermometer mounted on the probe; both were calibrated to an accuracy of 0.010K.   Field-cooling (H = 100 Oe) was used to trace the temperature dependence of the susceptibility.   The samples were cooled slowly, passing through the transition point to lower temperature at a fixed field.   The contribution from the quartz basket was carefully subtracted.   The DC   resistivity   was   measured   with   care   to   insure   thermal equilibrium.   Currents of 1 mA were employed and voltages were determined with a Keithley 181 nanovoltmeter.

Results

Sample Characterization

Figure 1 shows the X-ray powder diffraction of the compositions x = 0.0, 0.1, 0.2, 0.3, 0.4, 0.5, 0.7 and 1.0 from $32.00 \leq 2\theta$ (deg) $\leq$ 33.50.   In this region the 013 (d calcd./(Å) 2.750) and 103/110 (d calcd./(Å)   2.727,   2.726)   diffraction   peaks   occur   for   the orthorhombic phase $YBa_2Cu_3O_{7-\delta}$.   The equimolar substitution of La^{3+}/Ni^{2+} for Ba^{2+}/Cu^{3+} results   in   a   systematic   change   from orthorhombic to tetragonal symmetry.   Small amounts of a phase, green in color, with a diffraction pattern similar to Y_2BaCuO_5 were observed for x > 0.5.   The data are summarized in Table I.

Table I.   Comparison of lattice constants

	a/Å	b/Å	c/Å
x=0.0	3.823(1)	3.888(1)	11.670(2)
0.1	3.821(1)	3.886(1)	11.666(2)
0.2	3.819(1)	3.891(1)	11.678(2)
0.3	3.824(1)	3.885(1)	11.653(2)
0.4	3.840(2)	2.879(1)	11.638(3)
0.5	3.837(2)	3.879(2)	11.642(4)
0.7	3.863(1)	-	11.592(4)
1.0	3.858(1)	-	11.574(2)

This result implies that the oxygen occupancy in the xy plane adjacent to the Cu^{3+} ion changes from essentially no oxide ions along the a-axis and full occupancy along the b-axis (orthorhombic case) to where the occupancy is more equal for x > 0.5 (tetragonal case).   The extreme case for x = 1.0 would have no vacancies in the xy plane around a Ni^{4+} (d^6—ion) if oxidation of the nickel has occurred.   Gravimetric studies are presently underway to clarify the oxygen composition, along with TEM to study the structural variations of these complex compositions.

Our susceptibility results are summarized in Figure 2 where normalized signal intensity at 5.2K and H = 100 Oe are plotted for the compositions x = 0.1, 0.2, 0.3, 0.4, 0.5, 0.7 and 1.0. As we have reported earlier (28,29), the magnetic susceptibility is particularly sensitive to the presence or absence of a superconducting phase. The intensity of the diamagnetic signal should scale with the volume fraction and therefore could be used to locate phase boundaries. We find that substitution of La^{3+}/Ni^{2+} greatly reduces the diamagnetic signal, and for x = 0.7, 0.8 and even 1.0 a small remnant diamagnetic signal was detected.

In Figure 3 R(T)/R(300K) plots show two distinct regions. For x = 0.0, 0.1, 0.2, and 0.3 the midpoint of the resistive transition decreases from 92K to 72K in a nearly linearly fashion. This can be seen quite clearly in Figure 4. For x = 0.4 and 0.5 a drop in resistance is detected between 70-75K and a resistive transition occurs near 28-29K. In Figure 5 the much larger resistance observed for the x = 0.7 composition and semiconducting behavior is plotted. Similar behavior was observed for x = 0.8 and x = 1.0.

Discussion

In the compound $YBa_2Cu_3O_{7-\delta}$ when $\delta = 0$ all oxygen sites are fully occupied. When $0.0 \leq \delta \leq 1.0$ there are oxygen vacancies, but these are thought to be located specifically on oxygen atoms located along the b-axis. Along the a-axis additional oxygen could potentially be found although probably not in $YBa_2Cu_3O_7$ because this would require further oxidation of Cu^{3+} to Cu^{4+}. We have attempted to form the isoelectronic compound $YBaLaNiCu_2O_7$ and solid solutions with $YBa_2Cu_3O_7$. However we have prepared all our compositions in oxygen and additional oxygen in the structure up to $YBaLaNiCu_2O_8$ is entirely possible. Given that $LaNiO_3$ (Ni^{3+}) and $BaNiO_3$ (Ni^{4+}) are known compounds this possibility seems all the more likely. In our thinking we assume that the planes that contain yttrium will not allow additional oxygen into the lattice, and therefore nine oxygen atoms per unit cell are not possible and the limiting oxygen stoichiometry is eight. We also assume that any additional oxygen incorporated into the lattice along the a-axis will result in the oxidation of Ni^{2+} (substituting for Cu^{3+}).

From the X-ray diffraction data, resistivity measurements, and susceptibility measurements we observe three regions of distinct behavior:

Region   I: x = 0.0 to x = 0.3
Region II: x = 0.3 to x = 0.7
Region III: x = 0.7 to x = 1.0

In region I (x = 0.0 to x = 0.3) the midpoint of the resistive transition changes from greater than 90K to approximately 72K. The X-ray diffraction patterns of these compositions show an orthorhombic structure from which we infer that the oxygen occupancy along the a- and b-axes are still very unequal in at

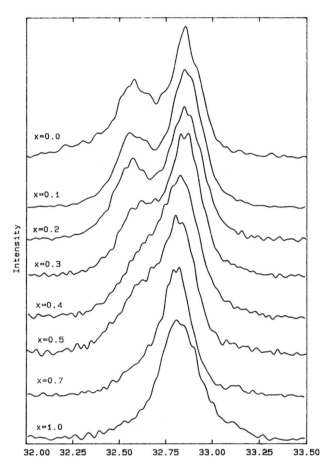

Figure 1. The X-ray powder diffraction pattern for the region $32° \leq 2\theta \leq 33.5°$ using Cu Kα radiation.

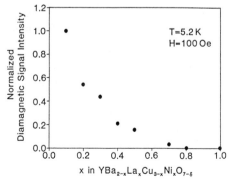

Figure 2. The normalized shielding signal intensity vs. composition x in $YBa_{2-x}(LaNi)_xCu_{3-x}O_{7\pm\delta}$. The signal intensity was normalized with respect to that from $YBa_{1.9}(LaNi)_{0.1}Cu_{2.9}O_{7\pm\delta}$.

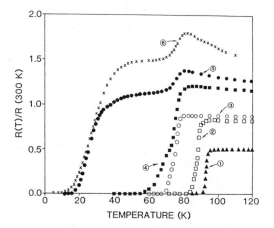

Figure 3. The normalized resistance vs. temperature for $YBa_{2-x}(LaNi)_x Cu_{3-x}O_{7+\delta}$. (1) x = 0.0; (2) x = 0.1; (3) x = 0.2; (4) x = 0.3; (5) x = 0.4; and (6) x = 0.5.

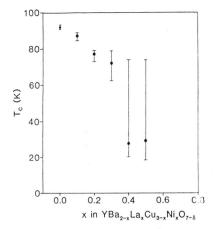

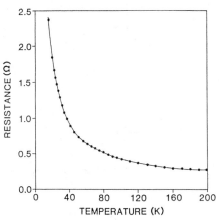

Figure 4. The relation between resistive transition temperature and composition x in $YBa_{2-x}(LaNi)_x Cu_{3-x}O_{7+\delta}$. The error bars show the corresponding 0.9 - 0.1 R_n transition widths. For x > 0.7, no resistive transition was observed.

Figure 5. The temperature dependence of the resistance of $YBa_{1.3}(LaNi)_{0.7}Cu_{2.3}O_{7+\delta}$. The solid curve is a guide to the eyes.

least one of the phases present. At most five phases can coexist at equilibrium in the $BaO-Y_2O_3-La_2O_3-CuO-NiO$ system when the temperature and oxygen partial pressure (1 atm) are fixed, although fewer are likely due to the chemical similarity of the components. The fraction of oxide superconductor drops significantly with substitution of La^{3+}/Ni^{2+} for Ba^{2+}/Cu^{3+} and indicates that at least one additional non-superconducting phase is present.

In region II the resistive behavior of x = 0.4 and 0.5 show a small fraction of the 72K material and a new transition near 29K. Similarly the X-ray diffraction patterns show the major phases present have a much smaller orthorhombic lattice distortion. Small amounts of $La_{2-x}Ba_xCuO_{4-x/2+\delta}$ could be responsible for the transition near 30K, but this phase was not observed in the diffraction patterns. The susceptibility data indicate that these samples may be largely bi- or triphasic depending on the amount of non-superconducting oxide present.

Region III (x = 0.7, 0.8 and 1.0) does not show any evidence for the presence of a superconducting phase based on resistance measurements. A weak diamagnetic signal is found for all three samples up to approximately 60K which indicates that a small amount of the oxide superconductor from Region I still remains. This implies that the Region II superconductor has reacted which is consistent with the fairly sharp X-ray diffraction patterns observed of a tetragonal phase. In this case the tetragonal phase is probably due to an excess of oxygen and Ni^{4+} e.g. $YBaLaNi^{4+}(Cu^{2+})_2O_8$. The intermediate region II material may contain the d^7-ion Ni^{3+}, e.g. $YBa_{1.5}La_{0.5}Ni_{0.5}Cu_{2.5}O_{7.25}$. The excess oxygen in the lattice would oxidize the filled $(d_z2)^2$ states and create bridges between the one-dimensional chains in the structure. Experiments and calculations are underway to verify and understand the role of excess oxygen on the superconducting transition temperature.

Acknowledgments

We acknowledge support for this research from NSF (DMR-8610659) and from the Materials Research Center of Northwestern University, NSF (MRL-8520280). We are grateful to S. Massidda, J. J. Yu and A. J. Freeman for helpful discussions.

Literature Cited

1. Jonker, G. H. <u>Physica</u>. 1954, <u>20</u>, 1118.
2. Kestigian, M.; Dickenson, J. G.; Ward, R. <u>J. Am. Chem. Soc</u>. 1957, <u>79</u>, 5598.
3. Anderson, S.; Wadsley, A. D. <u>Nature</u> 1960, <u>187</u>, 499.
4. Watanabe, H.; Sugimoto, M.; Fukase, M.; Hirone, T. <u>J. Appl. Phys</u>. 1965, <u>36</u>, 988.
5. Geller, S.; Grant, R. W.; Gonser, U.; Wiedersich, H.; Espinosa, G. P. <u>Phys. Lett</u>. 1966, <u>20</u>, 115.
6. Bertaut, E. F.; Blum, P.; Sagnieres, A. <u>Acta Crystallogr</u>. 1959, <u>12</u>, 149.
7. Hansen, W. C.; Brownmiller, L. T. <u>Amer. J. Sci</u>. 1928, <u>15</u>, 224.

8. Grenier, J. C.; Parriet, J.; Pouchard, M. Mat. Res. Bull.
    1976, 11, 1219.
9. Poeppelmeier, K. R.; Leonowicz, M. E.; Longo, J. M. J. Solid
    State Chem. 1982, 44, 89.
10. Poeppelmeier, K. R.; Leonowicz, M. E.; Scanlon, J. C.; Longo,
    J. M.; Yelon, W. B. J. Solid State Chem. 1982, 45, 71.
11. Grenier, J. C.; Pouchard, M.; Hagenmuller, P. Structure and
    Bonding 1981, 47, 1.
12. Michel, C.; Raveau, B. Revue de Chimie Minerale 1984, 21, 407.
13. Leonowicz, M. E.; Poeppelmeier, K. R.; Longo, J. M. J. Solid
    State Chem. 1985, 59, 71.
14. Katz, L.; Ward, R. Inorg. Chem. 1964, 3, 205.
15. Negas, T.; Roth, R. S. J. Solid State Chem. 1970, 1, 409.
16. Negas, T.; Roth, R. S. J. Solid State Chem. 1971, 3, 323.
17. Goodenough, J. B.; Longo, J. M. Landolt-Bornstein, Group
    III/Vol.4a, Springer-Verlag, New York/Berlin 1970.
18. Bednorz, J. G.; Müller, K. A. Z. Phys. 1986, B64, 189.
19. Takagi, H.; Uchida, S.; Kitazawa, K.; Tanaka, S. Jpn. J.
    Appl. Phys. Lett. 1987, 26, L123.
20. Cava, R. J.; van Dover, R. B.; Batlogg, B.; Rietman, E. A.
    Phys. Rev. Lett. 1987, 58, 408.
21. Wu, M. K.; Ashburn, J. R.; Torng, C. J.; Hur, P. H.; Meng, R.
    L.; Gao, L.; Huang, Z. J.; Wang, Y. Q.; Chu, C. W. Phys. Rev.
    Lett. 1987, 58, 908.
22. Hur, P. H.; Gao, L.; Meng, R. L.; Huang, Z. J.; Forster, K.;
    Vassilious, J.; Chu, C. W. Phys. Rev. Lett. 1987, 58, 911.
23. Hwu, S.-J.; Song, S. N.; Thiel, J.; Poeppelmeier, K. R.;
    Ketterson, J. B.; Freeman, A. J. Phys. Rev. B. 1987, 35,
    7119.
24. Cava, R. J.; Batlogg, B.; van Dover, R. B.; Murphy, D. W.;
    Sunshine, S.; Siegrist, T.; Remcika, J. P.; Reitman, E. A.;
    Zahurak, S.; Espinosa, G. P. Phys. Rev. Lett. 1987, 58, 1676.
25. Stacy, A. M.; Badding, J. V.; Geselbracht, M. J.; Ham, W. K.;
    Holland, G. F.; Hoskins, R. L.; Keller, S. W.; Millikan, C.
    F.; zur Loye, H.-C. J. Am. Chem. Soc. 1987, 109, 2528.
26. Steinfink, H.; Swinnea, J. S.; Sui, Z. T.; Hsu, H. M.;
    Goodenough, J. B. J. Am. Chem. Soc. 1987, 109, 3348.
27. Er-Rakho, L.; Michel, C.; Provost, J.; Raveau, B. J. Solid
    State Chem. 1981, 37, 151.
28. Hwu, S.-J.; Song, S. N.; Ketterson, J. B.; Mason, T. O.;
    Poeppelmeier, K. R. Commun. Am. Ceram. Soc. 1987, 70, C-165.
29. Wang, G.; Hwu, S.-J.; Song, S. N.; Ketterson, J. B.; Marks, L.
    D.; Poeppelmeier, K. R.; Mason, T. O. Adv. Ceram. Mater.
    1987, 2, 313.

RECEIVED July 6, 1987

Chapter 18

Effects of Oxygen Stoichiometry on Structure and Properties in $Ba_2YCu_3O_x$

D. W. Murphy, S. A. Sunshine, P. K. Gallagher, H. M. O'Bryan, R. J. Cava, B. Batlogg, R. B. van Dover, L. F. Schneemeyer, and S. M. Zahurak

AT&T Bell Laboratories, Murray Hill, NJ 07974

Results of a series of studies on the oxygen stoichiometry in $Ba_2YCu_3O_x$, $6.0 \leq x \leq 7.0$, are reported. The structure at x=7 contains layers of corner-shared, square pyramidal copper and chains of square planar copper. Oxygen is removed from the chains leaving two-fold coordinate copper at x=6. Superconductivity is optimal for x≈7 and for x=6 the material is a semiconductor. A valence description ascribes Cu^{+2} to the Cu in the layers for all x with the Cu in the chains going from Cu^{+3} at x=7 to Cu^{+1} at x=6. A model for twinning based on oxygen defects similar to those in $Ba_2YCu_3O_6$ is discussed.

The recent discovery of superconductivity above 90K in a number of cuprate perovskites (1-6) has made these materials the focus of intense scientific effort. The general class of compounds exhibiting these high T_c's is $Ba_2MCu_3O_x$ (M=Y, La, Nd, Sm, Eu Gd, Dy, Ho, Er, Tm, Yb; x≃ 7) (7-10). Soon after their discovery it became evident that oxygen stoichiometry plays a vital role in the structure and properties of these materials (11-15). Two of these studies on the prototype phase $Ba_2YCu_3O_x$ have revealed a stability range of $6.0 \leq x \leq 7.0$ (11,15). Reversible changes in crystallographic symmetry and physical properties have been found to occur over this range of composition. Detailed studies of the structures, oxygen mobility, and properties of $Ba_2YCu_3O_x$ for $6.0 \leq x \leq 7.0$ have provided insight into the chemical nature of these materials and into possible

0097-6156/87/0351-0181$06.00/0
© 1987 American Chemical Society

mechanisms for superconductivity and for crystal twinning. The results of these studies are summarized below.

STRUCTURES OF $Ba_2YCu_3O_x$ PHASES

The crystallographic unit cells of the end members $Ba_2YCu_3O_7$ and $Ba_2YCu_3O_6$ are presented in Figure 1. Neutron powder diffraction profile analysis indicates that the unit cell of $Ba_2YCu_3O_7$ is orthorhombic (Pmmm) with a=3.8198(1)Å , b=3.8849(1)Å , and c=11.6762(3)Å (17-19) while that of $Ba_2YCu_3O_6$ is tetragonal (P4/mmm) with a=3.8570(1)Å   and c=11.8194(3)Å (20,21). These structures are conveniently described as oxygen deficient perovskites with tripled unit cells due to Ba-Y ordering along the c axis. The ordered oxygen vacancies in these structures result in a reduction in the coordination numbers of Cu from the ideal six-fold octahedral coordination of a stoichiometric perovskite. For $Ba_2YCu_3O_7$, the two oxygen vacancies result in four-fold coordinate (Cu(1)) and five-fold coordinate (Cu(2)) Cu atoms. This compound can be thought of as being made of layers and chains. The Cu(1) atoms form linear chains of corner shared square planes oriented along the b axis and the Cu(2) atoms form two-dimensional layers of corner shared square pyramids as shown in Figure 2. The O(1) atom from the chain also serves as the apical oxygen atom for the square pyramidal Cu(2). The relative importance to superconductivity of the layers and chains is one of the central questions that must be answered to understand superconductivity in these materials and to aid in the design of new phases. We discuss later the dependence of T_c on oxygen stoichiometry which may provide some insight into this question.

The structure of $Ba_2YCu_3O_6$ differs from that of $Ba_2YCu_3O_7$ by the removal of O(4) from along the b axis. This results in a change in the coordination about Cu(1) from square planar to a linear two-fold coordination. The five-fold coordination about Cu(2) is maintained. The Cu(1)-O(1) distance decreases and the Cu(2)-O(1) distance increases making Cu(2) more square planar as seen in Figure 3. The structure of $Ba_2YCu_3O_6$ can be described as being made of the same type of layers as in the x=7 phase, but with isolated, two-coordinate Cu(1) atoms replacing the chains (Figure 4).

A neutron structure study of $Ba_2YCu_3O_{6.8}$ (20) showed that the additional vacancies were randomly distributed on the O(4) atoms along the b axis. Another study on a composition with x=6.5 showed the same random oxygen vacancy on O(4) (22), but as we indicate later, this may depend on the details of sample preparation.

The structures of the x=6 and x=7 phases lend themselves to a valence description of the Cu on the different crystallographic sites. The key to this description is that the linear two-fold coordination observed for Cu(1) at x=6 is typical of Cu^{+1} compounds (23,24), whereas Cu^{+2} and Cu^{+3} may be four, five, or six coordinate. The average oxidation state of Cu in

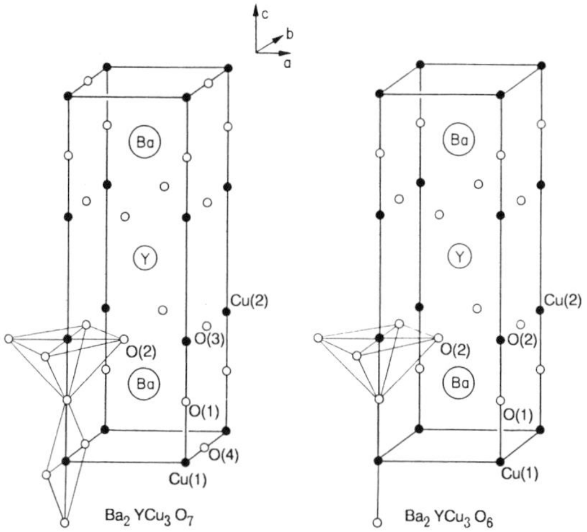

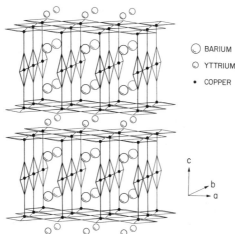

Figure 1. Schematic drawings of the unit cells of $Ba_2YCu_3O_7$ and $Ba_2YCu_3O_6$.

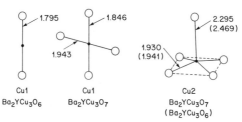

Figure 2. A perspective view of $Ba_2YCu_3O_7$ emphasizing the chains of Cu(1) and layers of Cu(2).

Figure 3. The coordination environments and bond lengths (Å) for Cu(1) and Cu(2) in $Ba_2YCu_3O_7$ and $Ba_2YCu_3O_6$.

the end members is 2.33 for x=7 and 1.67 for x=6. Since Cu(2) is relatively unchanged between end members and there are two Cu(2) per formula unit we assign it as Cu^{+2} in both end members. This reasoning leads to the limiting formulas $Ba_2YCu^{+3}Cu_2^{+2}O_7$ and $Ba_2YCu^{+1}Cu_2^{+2}O_6$ for the end members and $Ba_2YCu_{(1-x)}^{+3}Cu_x^{+1}Cu_2^{+2}O_{7-x}$ for intermediate oxygen stoichiometries. Thus, Cu^{+3} atoms are associated with the chains oriented along the b axis and Cu^{+2} atoms with the layers. This further indicates that the reduction that occurs in going from O_7 to O_6 is primarily a reduction of Cu^{+3} to Cu^{+1} at Cu(1).

The change in weight of a sample of $Ba_2YCu_3O_7$ when heated in pure oxygen is shown in Figure 5. At high temperatures this compound begins to lose oxygen and at 686°C the orthorhombic to tetragonal transition occurs. This corresponds to an oxygen stoichiometry of 6.6 (11,14,16). As noted in Figure 5, both the oxygen loss and symmetry change are reversible even under rapid cooling (100°C/min). Quenching a sample from high temperatures may preserve oxygen deficiency, but a more controlled way to assure a particular stoichiometry is to heat to the desired weight in an inert gas. The evolution of lattice parameters and cell volume in air as a function of temperature is displayed in Figure 6. It can be seen that the orthorhombic to tetragonal transition occurs at a lower temperature (610°C) in air than in oxygen but the transition still occurs at the same oxygen stoichiometry. The stoichiometry at which the transition occurs does, however, depend on how the sample is prepared. For example, a sample of $Ba_2YCu_3O_{6.5}$ prepared by combining equal mole portions of $Ba_2YCu_3O_6$ and $Ba_2YCu_3O_7$ and heating in a sealed tube at 450°C and a series of samples with 6.6 < x < 6.3 prepared by a low temperature Zr gettering technique were orthorhombic at room temperature (25). These compositions are in the tetragonal region at high temperature.

Dramatic changes in the electrical properties occur with changing oxygen stoichiometry. The $Ba_2YCu_3O_7$ phase is a bulk superconductor ($\simeq 78\%$ Meissner effect) (4) whereas $Ba_2YCu_3O_6$ is semiconducting (20) (ρ_{300K}=27 Ωcm with an effective energy gap of 0.21eV over the range 125-300K). The semiconducting to superconducting behavior is reversible and varies directly with oxygen stoichiometry. The superconducting T_c is lowered and broadened by removal of oxygen. A second, relatively sharp T_c near 60K is observed for compositions near x=6.6 (13,25). This may suggest a further oxygen vacancy ordering at this composition.

As T_c depends on the oxygen stoichiometry, changes that occur with reduction may provide the key to superconductivity in these materials. The most dramatic structural effect of oxygen removal is the loss of connectivity in the 1-D chains that run along the b axis. This may indicate that conductivity occurs along these chains. An alternate explanation (26)

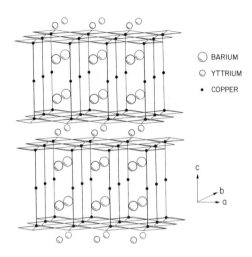

Figure 4. Schematic drawing of $Ba_2YCu_3O_6$ showing linear two-coordination about Cu(1) and layers of Cu(2).

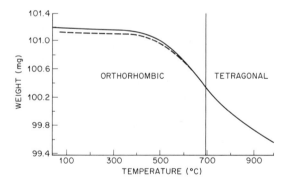

Figure 5. Weight as a function of temperature (TGA data) for $Ba_2YCu_3O_7$ heated in O_2. (Data from ref. 11). The solid line is on heating at 1 °C/min; dashed line is on cooling at 100 °C/min cooling. The transition from orthorhombic to tetragonal symmetry occurs at 686 °C.

involves valence fluctuations of the copper atoms (26). Such fluctuations depend on the displacement of O(1) between Cu(1) and Cu(2). This model has many common features with the bipolaron ideas that originally motivated the search for superconductivity in ternary copper oxides by Bednorz and Müller (27). The movement of O(1) toward Cu(1) upon reduction will make these fluctuations less likely. Regardless of the exact mechanism, it is clear that the 1-D chains play a vital role in superconductivity, either by modulating the 2-D layers or by carrying the superconducting current.

Individual crystallites of $Ba_2YCu_3O_x$ are heavily twinned (28-31). This twinning results in an apparent disorder of O(4) atoms on both the a and b axes. The above description of oxygen stoichiometry may provide insight into crystal twinning. The twin boundaries are readily apparent by either electron microscopy or optical microscopy. They occur around the (110) reflection plane with an average spacing of 50-3000Å (30) depending on the material preparation. Within each twin domain the chains along the b axis are parallel. Across a twin boundary, the b axes of adjacent domains are mutually perpendicular. One possible model of the twin boundary, presented in Figure 7, arises from ordering of oxygen vacancies adjacent to Cu(1) along (110). Along the twin boundary each Cu(1) would have linear, two-coordination as in $Ba_2YCu_3O_6$. The chain direction could propagate in either orientation from such a boundary. The simplest model assumes that between twin boundaries the O(4) sites would be fully occupied. This model suggests that lower oxygen stoichiometries would result in more frequent twin boundaries. Assuming oxygen defects only at such twin boundaries, the oxygen stoichiometry is given by $x=6+\dfrac{n-2}{n}$ where n is the average number of unit cells between twin boundaries. As n approaches infinity (i.e. no twin boundaries) x=7 while n=2 gives x=6, the lower limit of possible oxygen stoichiometries. A typical x for O_2 annealed samples is 6.98 implying that the twins should be about 100 unit cells (390Å) apart. This distance is within the observed range. Additional disordered O vacancies may still occur within a domain.

An important consequence of the oxygen mobility and variable stoichiometry in $Ba_2YCu_3O_7$ is facile exchange between [16]O and [18]O (32,33). Interestingly, mass spectrometry, Raman, and Rutherford backscattering experiments indicated that exchange occurs at all sites not just O(4) (32). Most importantly, the BCS theory of superconductivity predicts that such an exchange should lead to a lowering of T_c by 3.5-4K due to lower phonon frequencies. Figure 8 shows Meissner effect data for an [18]O exchanged sample and an [16]O control sample which indicate that no isotope effect is observed (32). This strongly suggests that a mechanism other than the traditional electron-phonon coupling is operative.

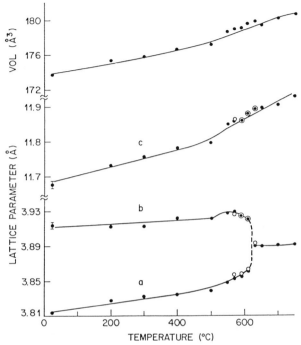

Figure 6. Lattice parameters and cell volume versus temperature in air for $Ba_2YCu_3O_7$.

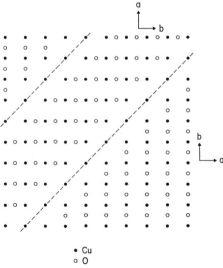

Figure 7. Proposed nature of twin boundaries in $Ba_2YCu_3O_7$. The unit cell orientation on either side of the twin boundary is labeled.

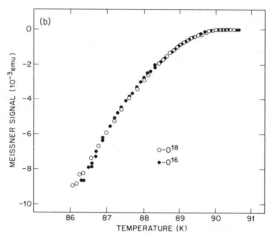

Figure 8. Meissner signal for ^{18}O substituted and ^{16}O control samples of $Ba_2YCu_3O_7$ indicating no isotope effect.

COMPARISON WITH OTHER CUPRATES

There is an interesting contrast in the compositional ranges found for the two superconducting cuprate systems $(La,Sr)_2CuO_4$ and $Ba_2YCu_3O_7$. The former has the K_2NiF_4 type structure and exists over a wide range of La/Sr ratios, but has a limited range of oxygen stoichiometry. The latter system is close to a line phase with respect to the cations, but has a wide oxygen stoichiometry range. In both cases, superconductivity is optimal at a particular stoichiometry of the variable component and one end member is a semiconductor. In many respects it is an advantage to have a fixed cation line phase, since oxygen can be adjusted in an annealing step. It is also fortunate that in the casse of $Ba_2YCu_3O_7$ the optimal oxygen stoichiometry can be achieved by annealing in the convenient range of one atmosphere oxygen. Chemical variability is at the core of attempts to tailor the properties in any system. These two systems, each with one chemical variable, have presented enormous challenges to a complete understanding. It does not require great imagination to think that there might be other superconducting systems where both the cations and oxygen can independently vary.

An example of a system in which both cations and oxygen can vary is $Ba_3La_3Cu_6O_{14+y}$ (34). This phase was reported to exist for 0.05 < y < 0.43. The phase progressively becomes more conductive going from semiconducting to semimetallic with increasing y. A recent powder neutron refinement (35) indicates that the structure of this compound is like that of $Ba_2YCu_3O_7$ with several small differences. First, the Y site is occupied only by La and the extra La atoms are disordered with Ba, and second, the cell is tetragonal rather than orthorhombic. The likely cause for the increased symmetry is that the oxygen stoichiometry in this material is 14.5 (i.e. $Ba_{1.5}La_{1.5}Cu_3O_{7.25}$) which necessitates occupation of the oxygen site along the a axis. The solid solution $Ba_{2-x}La_{1-x}Cu_3O_{7+\delta}$ $0.0 \leq x \leq 0.5$ exhibits superconductivity for $0.0 < x < 0.3$ (7,36). As the amount of La increases, δ also increases to maintain a nearly constant average Cu oxidation state.

A second oxidized cuprate perovskite is $La_4BaCu_5O_{13.2}$ which exhibits metallic behavior, but, does not lose oxygen in an oxygen atmosphere (37). Preliminary experiments in this lab indicate, however, that oxygen can be removed under reducing conditions (N_2) without changing the structure. As oxygen is removed the compound becomes less conductive so that $La_4BaCu_5O_{12.9}$ shows semimetallic behavior. This demonstrates that the presence of Cu^{+3} alone is not sufficient to insure facile oxygen loss.

SUMMARY

In this paper we have addressed the question of the nature and importance of oxygen stoichiometry in cuprates of the $Ba_2YCu_3O_x$ family. Within this structure type, oxygen is the key chemical variable which can be changed in order to give insight into structure-bonding-property relationships. The description of this phase in chemical valence terms is useful and leads to the conclusion that the principal redox that occurs as the oxygen content varies is in the chains along the b axis. The electronic properties are very sensitive to x and suggest that the chains are crucial to superconductivity, either directly by carrying the current or indirectly by providing the source for the electron coupling mechanism. Finally, the easy exchange of oxygen has allowed for isotopic exchange studies that give strong evidence that the phonon coupling mechanism is not applicable to this material. We have high hopes that understanding the mechanism of superconductivity in these materials will lead to new superconductors.

LITERATURE CITED

[1] Wu, M. K, Ashburn, J. R., Torng, C. J., Hor, P. H., Meng, R. L., Gao, L., Huang, Z. J., Wang, Y. Q., and Chu, C. W., *Phys. Rev. Lett. 58,* 908 (1987).
[2] Hikami, S., Hirai, T., and Kagoshima, S., *Jpn. J. Appl. Phys. 26,* L314 (1987).
[3] Takagi, H., Uchida, S., Kishio, K., Kitazawa, K., Fueki, K., and Tanaka, S., *Jpn. J. Appl. Phys. 26,* L320 (1987).
[4] Cava, R. J., Batlogg, B., van Dover, R. B., Murphy, D. W., Sunshine, S., Siegrist, T., Remeika, J. P., Rietman, E. A., Zahurak, S., and Espinosa, G. P., *Phys. Rev. Lett. 58,* 1676 (1987).
[5] Tarascon, J. M., Green, L. H., McKinnon, W. R., and Hull, G. W., *Phys. Rev. B 35,* 7115 (1987).

[6] Grant, P. M., Beyers, R. B., Engler, E. M., Lim, G., Parkin, S. S., Ramirez, M. L., Lee, V. Y., Nazzal, A., Vasquez, J. E., and Savoy, R. J., *Phys. Rev. B 35,* 7242 (1987).

[7] Murphy, D. W., Sunshine, S. A., van Dover, R. B., Cava, R. J., Batlogg, B., Zahurak, S. M., and Schneemeyer, L. F., *Phys. Rev. Lett. 58,* 1888 (1987).

[8] Hor, P. H., Meng, R. L., Wang, Y. Q., Gao, L., Huang, Z. J., Bechtold, J., Forster, K., and Chu, C. W., *Phys. Rev. Lett. 58,* 1891 (1987).

[9] Tarascon, J. M., McKinnon, W. R., Greene, L. H., Hull, G. W., and Vogel, E. M., submitted for publication.

[10] Schneemeyer, L. F., Waszczak, J. V., Zahurak, S. M., van Dover, R. B., and Siegrist, T., submitted for publication.

[11] Gallagher, P. K., O'Bryan, H. M., Sunshine, S. A., and Murphy, D. W., *Mater. Res. Bull.,* in press.

[12] Strobel, P., Capponi, J. J., Chaillout, C., Marezio, M., and Tholence, J. L., *Nature, 327,* 306 (1987).

[13] Tarascon, J. M., McKinnon, W. R., Greene, L. H., Hull, G. W., Bagley, B. G., Vogel, E. M., and LePage, Y., *MRS meeting,* Anaheim, April 1987.

[14] Manthiram, A., Savinnea, J. S., Sue, Z. T., Steinfink, H., and Goodenough, J. B., submitted for publication.

[15] Kubo, Y., Yoshitake, T., Tabuchi, J., Nakabayashi, Y., Ochi, A., Utsumi, K., Igaraski, H. Yonezawa, M., *Jpn. J. Appl. Phys.,* in press.

[16] O'Bryan, H. M., and Gallagher, P. K., *Adv. Ceram. Matls.,* in press.

[17] Beech, F., Miraglia, S., Santoro, A., and Roth, R. S., Phys. Rev. B., submitted.

[18] Copponi, J. J., Chaillout, C., Hewat, A. W., Lejay, P., Marezio, M., Nguyen, N., Raveau, B., Sorbezroux, J. L., Tholence, J. L., and Tournier, R., *Europhys. Lett.,* submitted.

[19] David, W. I. F., Harrison, W. T. A., Gunn, J. M. F., Moze, O., Soper, A. K., Day, P., Jorgensen, J. D., Hinks, D. G., Beno, M. A., Soderholm, L., Capone II, D. W., Schuller, I. K., Segre, C. U., Zhang, K., and Grace, J. D., *Nature, 327,* 310 (1987).

[20] Santoro, A., Miraglia, S., Beech, F., Sunshine, S. A., Murphy, D. W., Schneemeyer, L. F., and Waszczak, J. V., *Mater. Res. Bull,* in press.

[21] Swinnea, J. S., and Steinfink, H., *J. Mat. Sci.,* submitted.

[22] Santoro, A., Miraglia, S., Beech, F., Sunshine, S. A., and Murphy, D. W., *Mater. Res. Bull,* in press.

[23] Ishiguro, T., Ishizawa, N., Mizutani, N., and Kato, M., *J. Sol. State Chem., 49,* 232 (1983).

[24] Haas, H. and Kordes, E., *Z. Krist, 129,* 259 (1969).

[25] Cava, R. J., Batlogg, B., Chen, C. H., Rietman, E. A., Zahurak, S. M., and Werder, D., submitted.

[26] Whangbo, M. H., Evain, M., Beno, M. A., and Williams, J. M., *Inorg. Chem., 26,* 1832 (1987).

[27] Bednorz, J. G. and Müller, K. A., *Z. Phys. B, 64,* 189 (1986).

[28] Chen, C. H., Werder, D. J., Liou, S. H., Kwo, J. R., and Hong, M., submitted.

[29] Syono, Y., Kikuchi, M., Oh-ishi, K., Hiraga, K., Arai, H., Matsui, Y., Koboyashi, N., Sasaoka, T., and Muto, Y., *Jpn. J. Appl. Phys., 26,* L498 (1987).

[30] Nakahara, S., Boone, T., Yan, M. F., Fisanick, G. J., and Johnson, Jr., D. W., *J. Appl. Phys.* submitted.

[31] Beyers, R., *MRS Meeting,* Anaheim, April 1987.

[32] Batlogg, B., Cava, R. J., Jayaraman, A., van Dover, R. B., Kourouklis, G. A., Sunshine, S. A., Murphy, D. W., Rupp, L. W., Chen, H. S., White, A., Short, K. T., Mujsce, A. M., and Rietman, E. A., *Phys. Rev. Lett., 58,* 2333 (1987).

[33] Bourne, L. C. Crommie, M. F., Zettl, A., Zurloye, H-C., Keller, S. W., Leary, K. L., Stacy, A. M., Chang, K. J., Cohen. M. L., and Morris, D. E., *Phys. Rev. Lett., 58,* 2337 (1987).

[34] Provost, J., Studer, F., Michel, C., and Raveau, B., *Synth. Met., 4,* 157 (1981).

[35] Sunshine, S. A., Schneemeyer, L. F., Murphy, D. W., Waszczak, J. V., Santoro, A., Miraglia, S., and Beech, F., to be published.

[36] Mitzi, D. B., Marshall, A. F., Sun, J. Z., Webb, D. J., Beasley, M. R., Geballe, T. H., and Kapitulnik, A., submitted.

[37] Michel, C., Er-Rakho, L., Hervieu, M., Pannetier, J., and Raveau, B., *J. Solid State Chem., 68,* (1987).

RECEIVED July 6, 1987

Chapter 19

Comparison of the 1-2-3 Phase and the 3-3-6 Phase in the La-Ba-Cu-O Superconductor Series

John P. Golben, Sung-Ik Lee, Yi Song, Xiao-Dong Chen, R. D. McMichael,
J. R. Gaines, and D. L. Cox

Department of Physics, Ohio State University, Columbus, OH 43210

The $La_1Ba_2Cu_3O_{9-\delta}$ (1-2-3) and $La_{3-x}Ba_{3+x}Cu_6O_{14+y}$ (3-3-6) phases are correlated to the variation in T_c in the La-Ba-Cu-O series. These phases are both oxygen deficient perovskites of the form $ABO_{3-\delta}$. From x-ray diffraction the two phases are nearly indistinguishable due to the similar scattering powers of the isoelectronic La^{+3} and Ba^{+2} ions and the low scattering power of the O atoms. The major difference between them is the La-Ba ratio and the oxygen occupancy conditions. The implications of the similarities and differences of these phases is also discussed.

Since the discovery of high temperature superconductivity in La-Ba-Cu-O (<u>1</u>), an interest in this series as well as other super-conductors has accelerated. The first reported high T_c (zero resistance) phase in this series was isolated to be $La_{1.85}Ba_{.15}CuO_4$ (<u>2,3</u>) (hereafter referred to as the 2-1-4 phase), with a T_c of $12^\circ K$. However, recently several groups (<u>4-7</u>) have reported higher zero resistances in the phase $La_1Ba_2Cu_3O_{9-\delta}$ (hereafter referred to as the 1-2-3 phase); with one report as high as 75 K. (<u>4</u>) This has renewed interest (<u>8-11</u>) on the compound $La_3Ba_3Cu_6O_{14+y}$ (hereafter referred to as the 3-3-6 phase) where $0 < y \leq 1$. This latter compound has demonstrated metallic to insulating electrical properties as a function of y. (<u>9,10</u>) The La-Ba series is thus emerging as a potent combination where the relationship between the structure and the superconducting mechanism can be completely studied.

The 1-2-3 compound of $Y_1Ba_2Cu_3O_{9-\delta}$ has been found to have a sequential Ba-Y-Ba ordering along the c-axis of the unit cell. (<u>12-16</u>) The similar lattice constants and x-ray diffraction data of the compounds $M_1Ba_2Cu_3O_{9-\delta}$ (M = Nd, Sm, Eu, Gd, Ho, Er, Lu, and Eu)

0097-6156/87/0351-0192$06.00/0

($\underline{4,5,17,18}$) strongly indicate that these compounds have the same
sequential ordered structure. An additional ordering of oxygen
vacancies in the outer copper planes of the cell, as suggested by
neutron diffraction data for $Y_1Ba_2Cu_3O_{9-\delta}$ ($\underline{19}$) is responsible for
the slight orthorhombic distortion that has been observed for
compounds of this type. Though the observed x-ray pattern for
$La_1Ba_2Cu_3O_{9-\delta}$ indicates that this structure is tetragonal, orthohombic shouldering on the peaks has been observed in the samples with
the highest T_c's, i.e. those samples with T_c's closer to the "90 K"
plateau set by the other 1-2-3 superconductors. ($\underline{17,18}$) This suggests
that the 1-2-3 structure of La-Ba-Cu-O belongs in the same group as
these other 1-2-3 compounds.

The $La_3Ba_3Cu_6O_{14+y}$ phase has a similar sequential ordering of
the form (La,Ba)-La-(La,Ba) along the c axis, i.e. the La atom still
exclusively occupies the central position of the unit cell. ($\underline{8}$) The
major difference between this 3-3-6 phase and the 1-2-3 phase lies
with the random placement of the extra La atoms within the Ba-rich
layer of the cell, and the amount and location of the oxygen vacancies in the cell. These two differences may be inherently related
since the replacement of the Ba^{+2} ion with the La^{+3} ion should alter
the oxygen structure and the Cu valences. In the 1-2-3 structure
there is a square planar Cu-O arrangement between adjacent unit cells
along the c-axis. In the 3-3-6 structure this arrangement is replaced by octahedral Cu-O coordinations between the unit cells
along the c-axis, while the oxygen deficiency is relocated in the Ba
planes. This new placement of the oxygen deficiency requires that
the unit cell be redefined as tetragonal with a new lattice constant
$a^* \approx \sqrt{2}a$. The 3-3-6 unit cell is then about twice the size of the
1-2-3 cell, and now has the square planar Cu-O arrangement between
adjacent cells in the a-b plane.

Figure 1 shows the transformation of the orthorhombic 1-2-3 phase
to the tetragonal 3-3-6 phase. In all the representations oxygen is
indicated at the corners of the square planar, square pyramid, and
octahedral units. The substitution of the extra La atoms in the
3-3-6 structure on the Ba sites is not shown. In "a" the 1-2-3 phase
is shown with a stacked ordering of the Ba-La-Ba atoms. In this
cell, two oxygens are missing from the $(ABO_3)_3$ ideal perovskite form.
Specifically, one is missing from site "1" and another is missing
from site "2" where this latter site represents 4 equivalent sites of
1/4 oxygen each in the unit cell. In "b" the 3-3-6 cell with twice
the volume of the 1-2-3 cell is heavily outlined. The dashed line
retains the 1-2-3 form. In "c" the 3-3-6 unit cell is featured. The
dashed lines highlight the Ba-La-Ba layers. In "d" the oxygen
ordering is changed to form the final tetragonal form. Four oxygens
are missing from the $(ABO_3)_6$ ideal perovskite form. One is from site
"1" and the other three are from sites 3, 4, and 5; each representing
4 equivalent sites of 1/4 oxygen each in the unit cell. Note that
site 2 is now filled, increasing the symmetry of the cell and negat-
ing the orthorhombic distortion.

The visual effect of the oxygen vacancies is further
illustrated in Figure 2, where the 1-2-3 (a) and the 3-3-6 (b)
structures are compared with added depth. In the 1-2-3 structure,
ordered oxygen deficiencies on the outer Cu planes set up "lines" of
Cu-O bonds. Cu-O "sheets" exist between the La and Ba layers. In

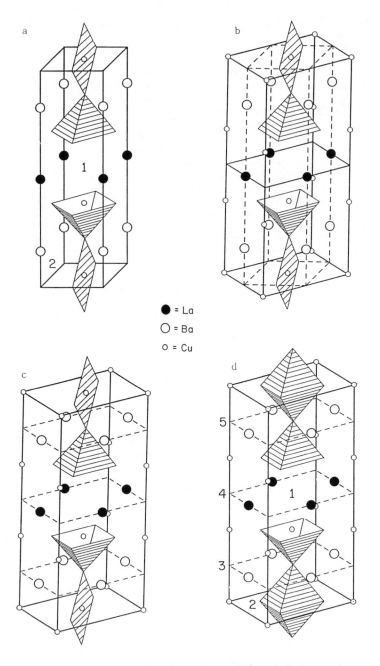

Figure 1. The transformation of the orthorhombic 1–2–3 phase to the tetragonal 3–3–6 phase. In all of the representations, oxygen is indicated at the corners of the square planar, square pyramid, and octahedral units. The substitution of the extra La atoms in the 3–3–6 structure on the Ba sites is not shown. A detailed description of the figure appears in the test.

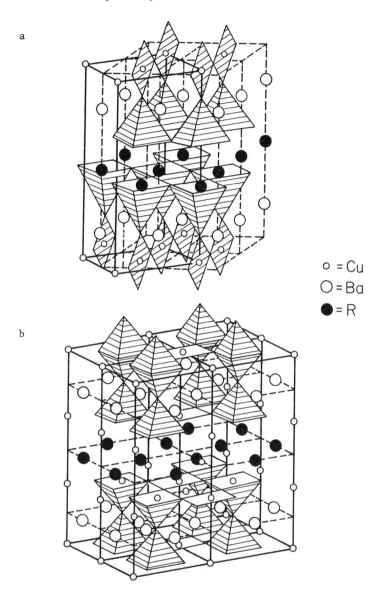

Figure 2. The 1-2-3 (a) and the 3-3-6 (b) structures are compared with depth added. The Cu-O "lines" and "sheets" are disrupted in the 3-3-6 structure by the redefined oxygen positions.

the 3-3-6 structure, the outer Cu-O lines are gone and the repeated square pyramidal coordination along the a-b plane is disrupted. Since superconductivity in the 1-2-3 compounds is generally associated with either the sheets or the lines (or both), an intensive study of the oxygen arrangement and content in these phases is important to the understanding of the superconducting mechanism.

Detailed investigation of the La-Ba structure is difficult by using x-ray techniques alone. Since the La^{+3} and Ba^{+2} ions are isoelectronic, their x-ray scattering power is nearly equal. The O atom has a low atomic number and therefore contributes little to the x-ray spectrum. And because the metallic ion locations are nearly identical, the 1-2-3 and 3-3-6 phases are basically indistinguishable by x-ray diffraction data (We believe that the lone 010/100 x-ray superlattice peak for the 3-3-6 structure is too weak to be of value in discriminating between the two structural possibilities.).

Recently, it was reported that in single phase samples of $La_1Ba_2Cu_3O_{9-\delta}$ there was a variance in the value of the zero resistance T_c from about 27 K to 57 K. ($\underline{6,7}$) Two possible explanations were given for this observation. The first explanation was that the amount and location of the oxygen deficiencies varied between these samples. This caused a movement of the Ba and Cu atoms in the cell, a corresponding change in the intensity of the (014)/(005) line in the x-ray spectrum, and the change in T_c. The other explanation was that there was a sequentially ordered (Ba-La-Ba) phase that competed with a "random cubic" phase in these samples. The former phase was postulated to be a 1-2-3 "90 K" superconductor while the latter phase was possibly non-superconducting. The superconducting properties and the value of T_c of the sample as a whole would then depend upon the relative presence of these competing phases in the sample.

The inclusion of the 3-3-6 phase is consistent with both of these interpretations. As far as the first explanation goes, it has already been mentioned that oxygen doping can change the electrical properties of this phase from insulating to metallic. Further doping of oxygen, or an alteration of the oxygen structure by doping for La or Ba, may produce a superconductor. Certainly the 1-2-3 phase can be considered an extreme form of "Ba for La" doping that results in a superconductor, i.e. both the 3-3-6 and 1-2-3 phases settle into the $ABO_{3-\delta}$ perovskite form where A = (La,Ba) and B = Cu.

It appears that the second explanation is a bit more feasible if the notion of a non-superconducting "random cubic" phase is replaced by the non-superconducting 3-3-6 phase. The superconducting "sequentially ordered" 1-2-3 phase would then compete with this phase in the sample, and could alter the value of T_c. X-ray diffraction alone cannot distinguish the exact presence of the 1-2-3, 3-3-6, or "random cubic" phases, if indeed the latter phase exists.

Neutron scattering and thermogravimetric analysis will help to determine the relationship between these phases. The oxygen content and structural comparison of these phases to the 2-1-4 superconducting phase is also intriguing. The financial support of the National Science Foundation through a grant DMR 83-16989 to the Ohio State University Materials Research Laboratory and grant DMR 84-05403 is gratefully acknowledged.

Literature Cited

1. J.G.Bednorz and K.A.Muller, Z.Phys. B 1986, 64, 189.
2. J.G. Bednorz, M. Takashige, and K.A. Muller, Europhys. Lett 1987, 3, 379.
3. H. Takagi, S. Uchida, K. Kitazawa, and S. Tanaka, Jpn. J. Appl. Phys. (to be published).
4. P.H. Hor, R.L. Meng, Y.Q. Wang, L. Gao, Z.J. Huang, J. Bechtold, K. Forster, and C.W. Chu, Phys. Rev. Lett. 1987, 58, 1891.
5. D.W. Murphy, S. Sunshine, R.B. van Dover, R.J. Cava, B. Batlogg, S.M. Zahurak, and L.F. Schneemeyer, Phys. Rev. Lett. 1987, 58, 1888.
6. Sung-Ik Lee, John P. Golben, Sang Young Lee, Xiao-Dong Chen, Yi Song, Tae W. Noh, R.D. McMichael, Yue Cao, Joe Testa, Fulin Zuo, J.R. Gaines, A.J. Epstein, D.L. Cox, J.C. Garland, T.R. Lemberger, R. Sooryakumar, Bruce R. Patton, and Rodney T. Tettenhorst, Extended Abstract Materials Research Society, 1987.
7. Sung-Ik Lee, John P. Golben, Sang Young Lee, Xiao-Dong Chen, Yi Song, Tae W. Noh, R.D. McMichael, J.R. Gaines, D.L. Cox, and Bruce R. Patton, Phys. Rev. B (to be published).
8. L. Er-Rakho, C. Michel, J. Provost, and B. Raveau, J. Sol. State Chem. 1981, 37, 151.
9. J. Provost, F. Studer, C. Michel, and B. Raveau, Synthetic Metals 1981, 4, 147.
10. C. Michel and B. Raveau, Rev. Chimie Minerale 1984, 21, 407.
11. Y. Le Page, W.R. Mckinnon, J.M. Tarascon, L.H. Greene (submitted to Phys. Rev. Lett.).
12. R. Beyers, G. Lim, E.M. Engler and R.J. Savoy, T.M. Shaw, T.R. Dinger, W.J. Gallagher and R.L. Sandstrom (unpublished).
13. P.M. Grant, R.B. Beyers, E.M. Engler, G. Lim, S.S.P. Parkin, M.L. Ramirez, V.Y. Lee, A. Nazzal, J.E. Vazquez and R.J. Savoy (unpublished).
14. T. Siegrist, S. Sunshine, D. W. Murphy, R.J. Cava and S.M. Zahurak (submitted to Phys. Rev. Lett.).
15. M. Hirabayashi, H. Ihara, N. Terada, K. Senzaki, K. Hayashi, S. Waki, K. Murata, M. Tokumoto, and Y. Kimura, Jap. J. Appl. Phys. 1987, 26, 1454.
16. Sung-Ik Lee, John P. Golben, Yi Song, Sang Young Lee, Tae W. Noh, Xiao-Dong Chen, Joe Testa, J.R. Gaines and Rodney T. Tettenhorst, Appl. Phys. Lett. (to be published).
17. John P. Golben, Sung-Ik Lee, Sang Young Lee, Yi Song, Tae W. Noh, Xiao-Dong Chen, J.R. Gaines and Rodney T. Tettenhorst, Phys. Rev. B (to be published).
18. J.E. Greedan, A. O'Reilly and C.V. Stager, (submitted to Phys. Rev. Lett.).

RECEIVED July 6, 1987

Chapter 20

High-Temperature Superconducting Oxide Synthesis and the Chemical Doping of the Cu-O Planes

J. M. Tarascon, P. Barboux, B. G. Bagley, L. H. Greene, W. R. McKinnon, and G. W. Hull

Bell Communications Research, 331 Newman Springs Road, Red Bank, NJ 07701-7020

Different synthesis techniques for the preparation of dense superconducting ceramics are discussed, and a sol-gel process is shown to be very promising. The effect of oxygen content, and the effect of substitution of Ni and Zn for copper, on the structural, transport and superconducting properties of the La-Sr-Cu-O and Y-Ba-Cu-O systems are presented. We find that substitution on the copper sites destroys T_c in the La-Sr-Cu-O system and decreases it in the Y-Ba-Cu-O system, and this effect is insensitive as to whether the 3d metal is magnetic (Ni) or diamagnetic (Zn). A detailed study of the $YBa_2Cu_3O_{7-y}$ system as a function of oxygen content (y) shows that superconductivity can be destroyed in these materials by the removal of oxygen and restored by reinjecting oxygen; either thermally at 500°C or at temperatures (80°C) compatible with device processing by means of a novel plasma oxidation process. Of scientific interest, the plasma process induces bulk superconductivity in the undoped La_2CuO_4.

For the past decade the maximum superconducting critical temperature (T_c) was 23K as obtained with a Nb_3Ge alloy ([1]). Despite the large amount of materials synthesis, which resulted in a new family of superconducting compounds containing either chalcogenides (Chevrel phases) ([2]) or oxides ($LiTi_2O_4$ ([3]), $BaPb_{1-x}Bi_xO_3$ ([4])), the Nb_3Ge T_c of 23K was never exceeded. In late 1986, the discovery of superconductivity at 36K in the La-Sr-Cu-O system by Bednorz and Muller ([5]) has stimulated the field of superconductivity, with physicists and chemists now in a race towards new high T_c materials. The T_c upper limit was rapidly increased to 40K by several groups ([6-8]) using the same oxide system (with Sr instead of Ba) and then pushed above liquid nitrogen temperatures by Wu et al. ([9]) with a multiphase Y-Ba-Cu-O oxide. Superconductivity at 93K was independently confirmed by other researchers ([10-11]) and the compound of formula $YBa_2Cu_3O_{7-y}$ was identified as the superconducting phase ([12]) and its crystal structure, consisting of a stack of Y,CuO_2,BaO,CuO,BaO,CuO_2 and Y planes perpendicular to the c-axis with Cu-O chains running along the b-axis, was established ([13]). During this exciting time the physics of these materials received more attention than their chemistry because no major problems were encountered in their preparation. The materials science of these oxides is, therefore, poorly understood and in this paper

0097–6156/87/0351–0198$06.00/0

we present different novel synthetic routes to obtain homogeneous and compact ceramics.

Band structure calculations performed on these oxides by L. F. Mattheis et al. (14) led to the conclusion that the Cu3d-O2p electrons govern the superconducting properties of these materials. To further our understanding of the mechanism of superconductivity in these oxides it was important to perform several chemical substitutions. We recently reported that the substitution of rare earths for La in the La-Sr-Cu-O system or for Y in the Y-Ba-Cu-O system has only a minor effect on T_c (15-16). In this paper we report on the chemical doping of the Cu-O planes with the replacement of magnetic (Ni) or diamagnetic (Zn) ions for Cu in both the La-Sr-Cu-O and Y-Ba-Cu-O systems, which in the following will be denoted as 40K and 90K materials, respectively. Independent of the nature of the 3d metal substitution superconductivity is rapidly depressed.

These new High T_c oxides belong to a large family of perovskite materials which have been studied over the years because of their ability to absorb or lose oxygen reversibly (i.e. for their non-stoichiometry in oxygen). It is well known in the field of superconductivity that T_c can be extremely sensitive to problems of stoichiometry. Thus it was of special interest to study how processing conditions (temperature, ambient) affect the oxygen content in these materials and thereby the superconducting properties. We have undertaken a detailed study of the $YBa_2Cu_3O_{7-y}$ system and report herein that the changes in oxygen content (y) and T_c in these materials are completely reversible. T_c is a maximum for the maximum oxygen content (y = 0) whereas the compound is semiconducting for y = 1. We show that both high oxygen content and a high T_c for these oxides can be reached at temperatures (80°C) attractive for high-speed microelectronic device processing using a novel plasma oxidation technique. This technique is used to produce bulk superconductivity at 40K in La_2CuO_4.

Synthesis

a) Solid state reaction

The 40K doped materials $La_{1.85}Sr_{0.15}Cu_{1-x}M_xO_{4-7}$ (M = Ni,Zn) were prepared by reacting appropriate amounts of $SrCO_3$ and the corresponding metal oxide powders, each 99.999%. The mixed powders, pressed into pellets and placed into platinum crucibles, were reacted at 1140°C under an oxygen ambient for 48 hours and then cooled to 500°C before being removed from the furnace. After such a treatment the pellets were well sintered and the material was single phase. We observe that lower reaction temperatures result in a sample contaminated by traces of unreacted La_2O_3 (moisture sensitive), so that after several days the pellet disintegrates.

The undoped and nickel doped $YBa_2Cu_3O_{7-y}$ phases were prepared in a similar way to the 40K materials with the difference that, in this case, loose powders were placed in an alumina crucible and reacted at temperatures of 950°C for 48 hours. Then the reacted powders were ground, pressed into a pellet and reheated at the same temperature for 48 hours in an oxygen ambient and then cooled slowly (taking 6 hours) to room temperature. The resulting materials are single phase (as indicated by powder x-ray diffraction using a Bragg-Brentano geometry with CuK_α radiation), but poorly sintered. A similar synthesis was made using nitrates instead of carbonates, and in the following the samples made from the carbonate and the nitrate will be denoted A and B, respectively. The powders resulting from the above process contain crystals of reasonable size for some single crystal studies but not suitable for other physical measurements.

Larger crystals are obtained either by changing the annealing temperature and cooling rate or by using an excess carbonate as follows: A small amount (up to 400mg) of Rb or K carbonate is added to 1.5g of the starting mixture prior to reaction. After reaction at 890°C, followed by several days treatment at 960°C, the powders are then slowly cooled. In the liquid state (T>880°C) the K or Rb carbonates wet the crucible and tend to creep up the crucible wall leaving behind a sparkling black powder with a few crystals of about 1mm size. X-ray diffraction of the resulting powder indicates a single phase material (no extra Bragg peaks) whose lattice parameters are similar to those of pure $YBa_2Cu_3O_{7-y}$ and which exhibits (measured by means of a SQUID magnometer) a 65% Meissner effect. With the same preparation heat treatment, the Meissner effect was always greater for materials made with the flux than without it. We thus believe that the potassium or rubidium carbonate fluxes act as a getter for impurities and thereby produce a cleaner material. Single crystals, isolated from this preparation show a resistive superconducting transition at 90K.

One disadvantage encountered in preparing the $YBa_2Cu_3O_{7-y}$ ceramic by a solid state reaction at 950°C is that the resulting material is poorly sintered. In an attempt to overcome this problem we developed a sol-gel (i.e. controlled precipitation and growth) process as we describe next.

b) Sol-Gel process

The gels are obtained by adding, to a known amount of barium hydroxide, appropriate amounts of yttrium nitrate and copper acetate. During this process the pH of the solution is constantly adjusted so as to provide a homogeneous precipitation. A few minutes after mixing the solutions, colloidal particles nucleate and continue to grow until they form a viscous blue gel (after a few hours). The gels are then freeze-dried and the resulting material ,consisting of a very fine amorphous powder, is then fired in a manner similar to that for the ceramics prepared by solid state reactions. A more detailed description of the sol-gel process will be reported elsewhere (17).

The ceramic material obtained by this method (sample C) is compared to those obtained by solid state reactions using either a carbonate (A) or nitrate (B) as the source of barium. Micrographs for the three samples are shown in Figure 1. Note that sample C is homogeneous with a particle size of about 1 micron whereas about 100 and 20 micron particle sizes are observed for samples A and B respectively. Resistivity measurements on the same set of samples, were performed using a standard four-probe method with silver paint contacts in an exchange gas cryostat with a Si-diode thermometer. A T_c of 91K with a transition width of 0.5K is observed for sample C whereas samples B and A exhibit widths of 2 and 1K respectively. The Meissner effect was observed to be 35, 65 and 50% for samples A, B and C respectively.

From this study we conclude that the sol-gel technique produces the most homogeneous and dense ceramics with the sharpest superconducting transitions. Also, the use of barium nitrate instead of of barium carbonate or the flux method (with $RbCO_3$ or KCO_3) allows a more complete reaction giving rise to a better Meissner effect and sharper transitions.

Results and Discussion

a) Substitution for copper

Nickel and zinc doped $La_{1.85}Sr_{0.15}Cu_{1-x}M_xO_{4-y}$ compounds were investigated (18) with the materials being single phase over the range of stoichiometry (x) studied. The x-ray powder diffraction patterns are completely indexed on a basis of a tetragonal unit cell. The crystal data are summarized in table I as a function of x for the nickel and zinc doped 40k materials. Note that, independent of the substituted element, the "a" lattice parameter increases while the "c" parameter decreases. For the La-Sr-Cu-O phase the CuO_6 octahedron is distorted, being elongated along the c axis. The c/a ratio can be used to characterize this distortion and its variation with x. Table I shows that, for both Ni and Zn substitutions, c/a

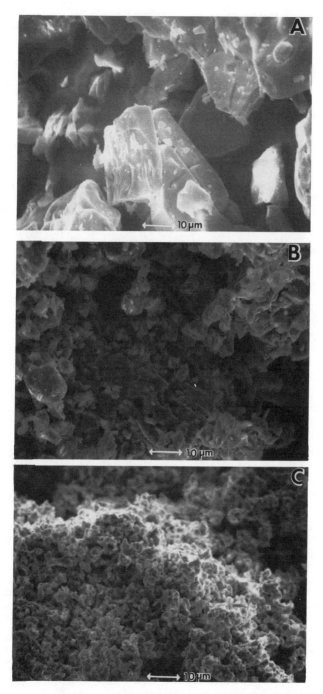

Figure 1. Scanning electron microscopy pictures of samples made from carbonate (A), nitrate (B) and by sol-gel process (C).

decreases with increasing x (i.e. the octahedron becomes less distorted) in a similar way. This indicates that the Jahn-Teller distortion is removed by substituting Zn and Ni for copper. This is expected since there is no Jahn-Teller effect for d^{10} (Zn^{2+}) or d^8 (Ni^{2+}) ions whereas it is a maximum for d^9 ions (Cu^{2+}).

The variation of the unit cell volume is also shown in Table I. The volume increases as a function of x when M = Zn and decreases when M = Ni as expected from size considerations, since the ionic radius of Ni^{2+} is smaller than that of Cu^{2+} which is smaller than that of Zn^{2+} either in an octahedral (0.83 < 0.87 < 0.88Å) or tetrahedal (0.69 < 0.71 < 0.74Å) environment (19).

TABLE I: Crystallographic lattice parameters and superconducting critical temperatures (T_{cm}) for the $La_{1.85}Sr_{0.15}Cu_{1-x}M_xO_4$(M=Ni,Zn) and $YBa_2Cu_{3-x}Ni_xO_{7-y}$ series

Compounds	a	b	c	c/a	V	T_{cm}
$La_{1.85}Sr_{0.15}Cu_{1-x}M_xO_4$	(Å)	(Å)	(Å)	(Å)	(Å)^3	(K)
x = 0 M = Ni	3.777(3)		13.226(1)	3.501	188.60(1)	39.3
X = 0.025	3.7749(2)		13.222(1)	3.502	188.42(2)	22.6
x = 0.05	3.7761(2)		13.2078(9)	3.497	188.33(2)	< 4.2
x = 0.075	3.7789(4)		13.187(1)	3.489	188.31(4)	no
x = 0.1	3.7815(3)		13.1671(3)	3.481	188.29(6)	no
x = 0.125	3.7833(3)		13.1582(3)	3.478	188.36(3)	no
x = 0.2	3.7873(4)		13.1138(2)	3.462	188.09(6)	no
x = 0.3	3.7971(3)		13.0460(1)	3.436	188.10(3)	no
x = 0.025 M = Zn	3.7772(2)		13.2213(1)	3.50	188.663(4)	15
x = 0.05	3.7809(1)		13.135(3)	3.494	188.86(3)	no
x = 0.075	3.7827(3)		13.2014(2)	3.489	188.892(2)	no
x = 0.10	3.7850(4)		13.1940(1)	3.485	189.017(5)	no
x = 0.2	3.7936(3)		13.1638	3.470	189.442(6)	no
$YBa_2Cu_{3-x}Ni_xO_7$						
x = 0	3.8237(8)	3.8874(3)	11.657(3)	---	173.28(6)	91.6
x = 0 (VA)	3.8589(8)		11.800(4)	3.051	175.72(8)	no
x = 0.25	3.8191(8)	3.8857(8)	11.6571(3)	---	172.98(7)	63.9
x = 0.5	3.8197(7)	3.8832(4)	11.6470(5)	---	172.75(8)	52.3

In contrast to these results for the 40K material, the substitution of Ni for Cu in the 90K phase did not produce significant structure changes for concentrations up to 0.5. At higher Ni contents, single phase materials cannot be made.

The variation of the superconducting critical temperatures (as determined by resistivity measurements) as a function of x for the doped 40K materials is shown in Table I. Note first that T_c decreases from 39K (x = 0) to 22K and then to about 4.2K for x = 0.05. It is tempting to ascribe the reduction in T_c to a magnetic effect. An unexpected result, however, is that the substitution of diamagnetic Zn decreases T_c even faster than does the magnetic Ni. Thus we conclude that the suppression of T_c in these oxides is not a magnetic effect, but more likely a disorder effect. The zinc substitution is expected to create more structural disorder than does the nickel since the end member phase exists with nickel (La_2NiO_4) whereas the equivalent phase "La_2ZnO_4" exists, but with a different structure. Thus T_c should be reduced more efficiently with zinc than with nickel, as we observe.

For the 90K material the nickel substitution affects T_c less than it does for the 40K material since T_c (Table I) is still 50K at x = 0.5. There are two types of copper sites in the 90K structure, and our x-ray data does not allow us to determine whether the nickel substitution occurs within the Cu-O chains or within the Cu-O planes. Thus the interpretation of our results on the effect of Ni being on the copper site are less conclusive for the 90K compound that for the 40K compound. Besides substitutions at the copper site, the Cu-O planes can be modified by removing or adding oxygen and next we show that oxygen doping dramatically affects the physical properties of the $YBa_2Cu_3O_{7-y}$ superconducting phase.

b) Oxygen doping the 90K material

The $YBa_2Cu_3O_{7-y}$ series with $0<y<1$ in steps of 0.1 have been studied ([20]). The materials were prepared as follows: A starting material was prepared in the manner described in the synthesis section and its oxygen content was evaluated to be $6.96+0.1$ by a thermogravimetric analysis method (TGA) described elsewhere ([15]). This value will be taken as 7.0 (i.e. $y=0$) in the following. Different values of y were then obtained by a contolled heat treatment of rectangular bars of the compound in an argon ambient.

X-ray powder diffraction patterns of the resulting materials are shown in Figure 2 for the range of two theta 35-50°. Note that, independent of the oxygen content (y), the Bragg peaks remain sharp suggesting that our materials are homogeneous. A striking feature of Figure 2 is that, as oxygen is removed from the as-grown material (orthorhombic unit cell), the 020 and 200 Bragg peaks (hatched peaks) become closer and finally merge together beyond $y=0.59$ becoming a single peak (tetragonal unit cell). The variation of the lattice parameters a and b as a function of y are shown in Figure 3a. Note that the orthorhombic-tetragonal transition occurs at about $y=0.6$. Figure 3b shows the variation of the unit cell length, c, and the unit cell volume, V, with oxygen content and we note that both c and V increase with decreasing oxygen content. Single crystal x-ray diffraction and neutron diffraction studies have shown that the oxygen is removed from the chains; and from bond length considerations it has been suggested that, upon removal of oxygen, monovalent copper ions are formed ([21-22]). Indeed, recent photoemission studies provide evidence for the formation of Cu^{+1} upon removal of oxygen thereby confirming the structural studies ([23]). From the smaller average valence of copper in the reduced material one would expect the unit cell volume to increase in going from $y=0$ to 1, as we observe.

The superconducting transition temperatures for the $YBa_2Cu_3O_{7-y}$ series were determined (using a SQUID magnetometer) by cooling the sample in zero field and warming it in a field of 10G (shielding measurements). Figure 4a shows the shielding measurement traces. Note first that the orthorhombic - tetragonal transition is correlated with the superconducting transition, since for y greater than 0.6 the compound is both tetragonal and not superconducting. It is difficult to believe that this correlation is coincidental. On the other hand there are reports that the compound $LaBa_2Cu_3O_{7-y}$ may be tetragonal and superconducting ([24]). Another interesting feature of Figure 4a is that T_c does not change continuously with oxygen content; we observe that the transition is broad at small y, sharpens for values of y ranging from 0.25 to 0.45 and broadens again at larger values of y. This suggests the existence of a bulk superconducting transition at 55K in the Y-Ba-Cu-O system. A two T_c phase system is even more evident in the data shown in Figure 4b where ac-susceptibility traces are shown for a vacuum annealed sample in which oxygen was restored at 300°C. Note that samples prepared deliberately inhomogenous (See Figure Caption 4b) can contain both the 55K and 90K phases. Reintercalation of oxygen into these materials at even lower temperatures (80° C) produces the 55K phase which transforms discontinuously to the 90K phase ([25]). These two phases likely correspond to a different ordering of the oxygen atoms (or vacancies) within the central Cu-O plane (i.e. Magneli type phases). Previous reports ([26]) have shown that T_c changes continuously with oxygen content in samples prepared at temperatures of about 700°C and quenched. It is possible that these intermediate T_c's

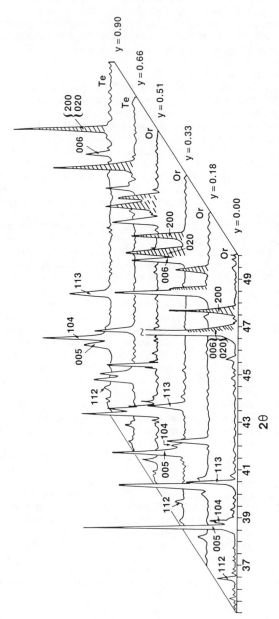

Figure 2. X-ray powder diffraction patterns as a function of y for the $YBa_2Cu_3O_{7-y}$ series. Indices are shown above each peak. Several peaks are hatched as a guide to the reader to better emphasize the structural changes occurring upon removal of oxygen. Or and Te refer to orthorhombic and tetragonal unit cell respectively.

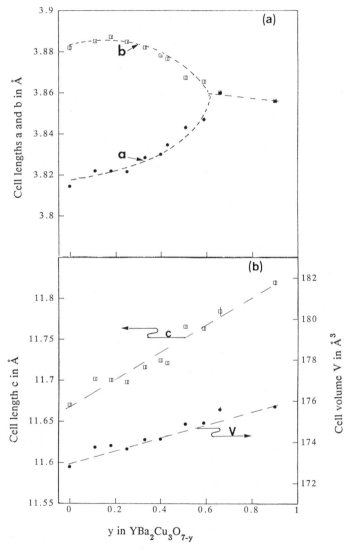

Figure 3. Variation of the orthorhombic lattice paramenters a, b, c and V as a function of y for the YBa$_2$Cu$_3$O$_{7-y}$ series.
a) The unit cell lengths a and b are shown.
b) The unit cell length c and unit cell volume V are shown.

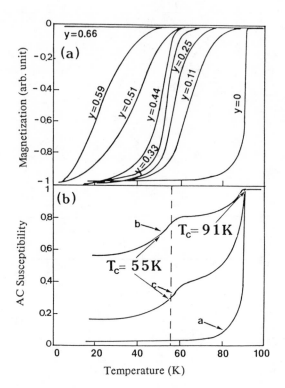

Figure 4. Effect of oxygen content on the superconducting properties of $YBa_2Cu_3O_{7-y}$.

a) Temperature dependent magnetization for the $YBa_2Cu_3O_{7-y}$ series ($0<y<1$) collected while warming the samples in a field of 10G after they have been cooled in zero field. The curves shown have been normalized by defining the baseline before and after the superconducting transition as 0 and -1, respectively.

b) The ac-susceptibility vs. temperature is shown for the as-grown $YBa_2Cu_3O_{7-y}$ (a) and for two vacuum annealed samples (b,c) heated to 300°C under 10^{-2} Torr of oxygen for 5 and 15 hours respectively. Note the two superconducting transitions at 55 and 91K.

(not 55 nor 90K) are achieved because the concentration of vacancies is variable and the ordering kinetics necessary to produce distinct 55 or 90K material are too slow to track the quenching rate.

The above results show that the superconducting properties of these materials are sensitive to oxygen content. The maximum oxygen content (i.e. maximum T_c) can be obtained by annealing the samples at 500°C in an oxygen ambient. Such high annealing temperatures may clearly limit the use of these new high T_c oxides in the microelectronics industry. This problem is solved by putting the oxygen back into the material by means of a low temperature process, namely a plasma oxidation(25), that we present next.

Plasma oxidation

We demonstrate the efficacy of this technique on a non-superconducting $YBa_2Cu_3O_6$ sample (i.e. vacuum annealed sample). The sample is plasma oxidized in a cylindrical reactor at a power density of less than 0.005 watts/ cm³ so as to avoid sample heating (< 80° C). The oxygen pressure is 0.7Torr and the frequency 13.56MHz. The sample, in the form of a rectangular bar (0.5x4x8mm), is placed in such a way that all surfaces are exposed to the plasma. The effect of the process is shown in Figure 5a. Prior to the plasma oxidation the sample exhibits semiconducting-like behavior (the resistance increasing with decreasing temperature). After 117 hours in the plasma the sample is still semiconducting, but with a notable decrease in the resistivity. An additional 103 hours in the plasma results in a further decrease in the resistivity of the sample and, in addition, it becomes superconducting at 92K with a resistive foot at 55K due to the oxygen deficient ordering discussed previously. After a total of 280 hours in the oxygen plasma the sample recovers its metallic-like temperature dependence and shows a superconducting transition about 2K greater than for the initial material. The Meissner effect for the plasma oxidized sample was observed to be 24%, confirming the ability of the plasma process to produce bulk property changes. Even though this process takes place at temperatures of about 80°C we observe that as the material goes from semiconducting to superconducting, it undergoes a tetragonal-orthorhombic transition similar to that discussed previously. In addition to its potential importance with respect to high-speed electronic device processing, this low-temperature process could lead to the discovery of new metastable phases (e.g. rich in oxygen) that cannot be made via high temperature reactions. The La_2CuO_4 system is used to demonstrate this last point.

Early reports on the La_2CuO_4 phase indicated that this compound was semiconducting and its behavior was ascribed to the presence of a half filled band which favored a Peierls instability (i.e. opening gap). However, recent work has shown that $La_2 CuO_4$ exhibits signs of superconductivity when prepared at low temperatures (950°C) or becomes fully superconducting when annealed under oxygen pressure (27). We induced this electronic transition by plasma oxidation and Figure 5b shows our results. The starting $La_2 CuO_4$ specimen, made at a temperature of 1140°C, was semiconducting-like (resistivity decreasing with increasing temperature). After a plasma exposure of this sample for only 18 hours we observe a marked decrease in the resistivity (note the scale change) and, in addition, residual semiconducting type behavior above the 38K temperature at which the sample will become a superconductor. Note, however, that the resistance does not go to zero. Further treatment for a period of 6 days produces an additional decrease in the resistivity. The resistivity shows a metallic-like temperature dependence above 50K, drops abruptly at 38K, and becomes equal to zero at 25K. To prove bulk superconductivity in the plasma oxidized $La_2 CuO_4$ sample, Meissner and shielding effects were measured. The Meissner effect is observed to be 18% for the plasma oxidized sample demonstrating the existence of bulk superconductivity for $La_2 CuO_4$.

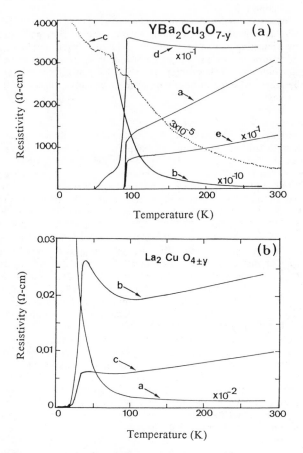

Figure 5. The ability of the plasma oxidation process to restore oxygen and promote superconductivity in several oxides is shown.

a) The resistivity vs. temperature is shown for $YBa_2Cu_3O_{7-y}$ as-grown (a), vacuum annealed (b), and after treatments of 15 (c), 117 (d) and 210 hours (e).

b) The resistivity vs. temperature is shown for "La_2CuO_4" as grown (a) and after plasma oxidation for 10 hours (b) and 6 days (c).

At first sight these findings are intriguing and appear to be contrary to theoretical models which stress the importance of having Cu^{3+} in these materials for the occurance of superconductivity, since, from simple valence considerations and assuming the material is stoichiometric, the valence of copper in the superconducting La_2CuO_4 phase is 2. A simple explanation to resolve this discrepancy is to assume that " La_2 CuO_4 " is not stoichiometric and has vacancies on the La and/or Cu sites. Both types of vacancies will increase the average valence of copper. The presence of copper vacancies will produce defects in the Cu-O plane and we argued previously that such defects (i.e. disordering) should decrease T_c, although it is possible that a small fraction of vacancies can be present without affecting T_c perceptibly. Since the compound does become superconducting we believe that the most likely assumption is that the material is deficient in lanthanum. This is also consistent with the experimental observation that attempts to make stoichiometric La_2CuO_4 at 950°C result in a material which is extremely moisture sensitive due to the presence of unreacted La_2O_3 which transforms to $La(OH)_3$ destroying the pellets. A final possibility is that, during the synthesis of the phase at 1140°C, vacancies are created in the La, Cu and oxygen sites so that the sample formula should be written as $La_{2-x} Cu_{1-x} O_{4-3x}$. In that case, during the plasma oxidation process, oxygen will be introduced into the oxygen vacancies leading to a compound of formula $La_2 CuO_{4+y}$ in which the valence of copper is then greater than 2. Distinguishing between these two possibilities may be resolved through careful analytical work. Finally we point out that this plasma oxidation process has also been used to restore oxygen, and superconductivity at 40K, in a semiconducting vacuum annealed La-Sr-Cu-O phase (25).

Conclusions

We report and compare different synthesis techniques for the preparation of the high T_c compound $YBa_2Cu_3O_{7-y}$ and conclude that sol-gel is very promising because dense and homogeneous ceramics with transitions as sharp as 0.5K are obtained. This technique is not specific to the yttrium compound, but can be used to make homogeneous chemical substitutions.

We show that the substitution of Ni and Zn for copper in these new high T_c oxides destroys T_c faster with diamagnetic Zn than with magnetic Ni and we conclude that the observed decrease in T_c is not of a magnetic origin but more likely is related to a disordering effect. For BCS type superconductors, T_c should be extremely sensitive to magnetic ions because of their interaction with the superconducting electrons thereby breaking the Cooper pairs and suppressing T_c. The indifference of T_c to as whether the dopant is magnetic or diamagnetic suggests that the new mechanism of superconductivity in these materials could be, in contrast to BCS theory, insensitive to magnetic impurities.

We observe that removal of oxygen from the Cu-O plane destroys T_c and that these changes in T_c and oxygen content are completely reversible. We show that superconductivity can be restored in these materials by annealing them at temperatures of 500°C in an oxygen ambient or at low temperatures by a plasma oxidation. This novel low temperature technique could be very useful for the technical application of these materials, as well as for the synthesis of new metastable phases (i.e. rich in oxygen content) that cannot be made via high temperature solid state reactions.

Acknowledgments

We wish to thank J. M. Rowell, N. G. Stoffel, E. M. Vogel, J. H. Wernick and E. Yablonovitch for valuable discussions, and W. L. Feldman and W. E. Quinn for technical assistance.

Literature Cited

1) L. R. Testardi, R. L. Meek, J. M. Poate, W. A. Royer, A. R. Storm and J. H. Wernick, Phys. Rev. B11, 4304 (1975).
2) M. Chevrel, M. Sergent, J. Prigent, J. Solid State Chem. 3, 515 (1971).
3) D. C. Johnston, H. Prakash, W. H. Zachariasen and R. Viswanathan, Mater. Res. Bull. 8, 777 (1973).
4) A. W. Sleight, J. L. Gillson, F. E. Biersled, Solid State Commun. 17, 27 (1975).
5) J. G. Bednorz and K. A. Muller, Z. Phys. B 64 189 (1986).
6) K. Kishio, K. Kitazawa, S. Kanbe, I. Yasuda, N. Sugil, H. Takagi, S. Uchida, K. Fueki and S. Tanaka, Chem. Lett. (Japan) 429 (1987).
7) R. J. Cava, R. B. van Dover, B. Batlogg and E. A. Rietman, Phys. Rev. Lett. 58 408 (1987).
8) J. M. Tarascon, L. H. Greene, W. R. McKinnon, G. W. Hull and T. H. Geballe, Science 235 1373 (1987).
9) M. K. Wu, J. R. Ashburn, C. J. Torng, P. H. Hor, R. L. Meng, L. Gao, Z. J. Huang, Y. Q. Wang, and C. W. Chu, Phys. Rev. Lett. 58 908 (1987).
10) People's Daily, China 25 Feb 1987.
11) J. M. Tarascon, L. H. Greene, W. R. McKinnon and G. W. Hull, Phys. Rev. B 235, 7115 (1987).
12) R. J. Cava, B. Batlogg, R. B. van Dover, D. W. Murphy, S. Sunshine, T. Siegrist, J. P. Remeika, E. A. Rietman, S. Zahurak and G. Espinosa, Phys. Rev. Lett. 58 1676 (1987).
13) Y. Le Page, W. R. McKinnon, J. M. Tarascon, L. H. Greene, G. W. Hull and D. M. Hwang, Phys. Rev. B35 7245 (1987).
14) L. F. Mattheis and D. R. Hamann, Solid State Commun. (in press).
15) J. M. Tarascon, W. R. McKinnon, L. H. Greene, G. W. Hull and E. M. Vogel, Phys. Rev. B36 July 1, 1987, (in press)
16) J. M. Tarascon, W. R. McKinnon, L. H. Greene, G. W. Hull, B. G. Bagley, E. M. Vogel and Y. LePage, Proceedings of the Materials Research Society 1987 Spring Meeting, Symposium S: High Temperature Superconductors with T_c over 30K. (D. V. Gubser and M. Schluter, eds.) in press.
17) P. Barboux, J. M. Tarascon et al (to be published)
18) J. M. Tarascon, L. H. Greene, W. R. McKinnon, G. W. Hull and P. Barboux, Solid State Commun.
19) R. D. Shannon, Acta Crystallogr., A32, 7511 (1965).
20) W. R. McKinnon, J. M. Tarascon, L. H. Greene and G. W. Hull, Phys. Rev. B (to be submitted).
21) J. E. Greedan, A. O'Reilly and C. V. Stager, Phys. Rev. B 36 June 1, 1987 (in press).
22) A. S. Santoro, S. Miraglia, F. Beech, S. A. Sunshine, D. W. Murphy, L. F. Schneemeyer and J. W. Waszczak, (preprint).
23) N. Stoffel, J. M. Tarascon, Y. Chang, M. Onellion, D. W. Niles and G. Margarito, Phys. Rev. B
24) D. B. Mitzi, A. F. Marshall, J. Z. Sun, D. J. Webb, M. R. Beasley, T. H. Geballe and A. Kapitulnik, Phys. Rev. B (submitted).
25) B. G. Bagley, L. H. Greene, J. M. Tarascon and G. W. Hull, App. Phys. Lett. (submitted).
26) I. K. Schuller, D. G. Hinks, M. A. Beno, D. W. Capone, L. Soderholm, J. P. Locquet, Y. Bruynseraede, C. V. Segre, K. Zhang (Preprint).
27) J. Beille, R. Cabanel, C. Chaillout, B. Chevalier, G. Demazeau, F. Deslandes, J. Elourneau, P. Lejay, C. Michel, J. Provost, B. Raveau, A. Sulpice, J. Tholence and R. Tournier, C. R. Acad. Sc. Paris 18, 304 (1987).

RECEIVED July 8, 1987

SURFACES AND INTERFACES

Chapter 21

High-Temperature Superconductors
Occupied and Unoccupied Electronic States, Surface Stability, and Interface Formation

Y. Gao[1], T. J. Wagener[1], D. M. Hill[1], H. M. Meyer III[1], J. H. Weaver[1], A. J. Arko[2], B. K. Flandermeyer[2], and D. W. Capone II[2]

[1]Department of Chemical Engineering and Materials Science, University of Minnesota, Minneapolis, MN 55455
[2]Division of Materials Research, Argonne National Laboratory, Argonne, IL 60439

The energy distribution of the occupied and unoccupied electronic states of the high temperature superconductors $La_{1.85}Sr_{0.15}CuO_4$ and $YBa_2Cu_3O_{6.9}$ have been identified using x-ray photoemission and inverse photoemission. These results show low state density within several eV of the Fermi energy, E_F, well-defined Cu-O occupied levels ~4 eV below E_F, and the Ba, Y, and La states above E_F. Chemisorption of oxygen produces a structure ~1.6 eV above E_F, and high energy electron bombardment eliminates this structure for $La_{1.85}Sr_{0.15}CuO_4$ but not for $YBa_2Cu_3O_{6.9}$. The deposition of Fe overlayers on $La_{1.85}Sr_{0.15}CuO_4$ reduces the emission within ~3 eV of E_F, indicating disruption of the hybrid Cu $3d_{x^2-y^2}$ -- O $2p_{x,y}$ states responsible for superconductivity. Fe adatoms also selectively alter the chemical states of Cu, quenching the Cu^{2+} configuration and leaving the surface region in a Cu^{1+} configuration. The growth of a thicker film results in the burial of the region containing Fe-O and the disrupted superconductor, although there continues to be evidence for O in solution in the Fe film.

The observation of high temperature superconductivity[1,2] has prompted intense research on the structure and electronic properties of these materials[3]. It has been found that the replacement of Y by La or many other rare earth elements does not significantly change the transition temperature, T_c, and this suggests that the CuO_{2+x} layers hold the key to superconductivity[4]. This is consistent with band structure calculations[5-7] for the model perovskite structure La_2CuO_4 which show antibonding hybrids derived from Cu $3d_{x^2-y^2}$ and O $2p_{x,y}$ states near the Fermi level.

Any comprehensive understanding of superconductivity in these materials will rely heavily on insight into the occupied and unoccupied electronic band states. There have been several photoemission studies of the occupied states of these materials[8-15]

0097-6156/87/0351-0212$06.00/0
© 1987 American Chemical Society

which have revealed Cu-O bonding states ~4 eV below E_F. In contrast, the unoccupied states of these materials, their surfaces, and the interface have not previously been studied.

In this paper we report the results of inverse photoemission (IPES) and high resolution x-ray photoemission (XPS) studies of cleaved, polycrystalline $La_{1.85}Sr_{0.15}CuO_4$ and $YBa_2Cu_3O_{6.9}$. These studies enable us to observe the low density Cu $3d_{x^2-y^2}$ - O $2p_{x,y}$ state from 0.5 eV below to 3.5 eV above E_F and to specify the atomic orbital origin of spectral features from the bottom of the valence band to 25 eV above E_F. We have investigated the stability of the $La_{1.85}Sr_{0.15}CuO_4$ surface under exposure to the ambient vacuum, to prolonged electron bombardment at low and high energy, to oxygen exposure, and to x-ray exposure.

With this information base for the clean surfaces, we have undertaken a variety of studies of interface formation. These latter studies are particularly important because chemical issues will control the application of the oxide-based superconductors in new technologies and their integration with existing technologies. A major concern is whether contacts will disrupt the crystal structure and atomic configuration required for superconductivity at high temperatures([16,17]). In this paper we discuss $Fe/La_{1.85}Sr_{0.15}CuO_4$ and show that Fe disrupts the Cu $3d_{x^2-y^2}$ - O $2p_{x,y}$ states at E_F, leaches oxygen from the surface to form Fe-O, and selectively alters the Cu^{2+} configuration. We show that reaction is kinetically limited at room temperature to the outermost 10-20 Å of the superconductor.

Experimental

In our inverse photoemission experiments, we directed a collimated, monoenergetic beam of electrons at clean or Fe-covered surfaces. The distribution of emitted photons of energy hν (the photon distribution curve or PDC) was measured. These photons resulted from the radiative decay of incoming electrons from initial states E_i of the solid to lower-lying final states E_f with hν = E_i - E_f. Such inverse photoemission experiments can be done at both ultra-violet (12-48 eV) and x-ray (1486.6 eV) energies. The latter is generally termed Bremsstrahlung Isochromat Spectroscopy (BIS)([18]) while the former is momentum- or k-resolved inverse photoemission (KRIPES)([19]). These are the time-reversed analogs of uv and x-ray photoemission.

The low energy inverse spectrometer consists of an f/3.5 grating and a position sensitive detector which collects photons dispersed by the grating in a wavelength window determined by the grating setting. The x-ray spectrometer is a 0.5 m Rowland circle mono-chromator with a quartz grating and a microchannel plate detector. Only photons of energy 1486.6 eV are transmitted by this mono-chromator, and discrimination against stray electrons is enhanced by a pinhole free, self-supporting Al window. Final state energies were scanned by ramping the accelerating voltage of the electron gun. The electron gun is a custom-designed Pierce-type gun from Kimball Physics which produces a collimated 1 mm x 5 mm electron beam([20]). The energy resolution (electrons plus photons) was determined by analysis of the Fermi level cutoff of a Au standard. It was 0.7 eV for BIS and 0.3-0.6 eV for KRIPES, depending on the photon energy.

The combination of uv and x-ray spectroscopies enabled us to vary the probe depth of the measurements via the electron mean-free-path. It also made it possible to identify the orbital character of the empty state structures through variations in their relative emission intensities(21).

In our x-ray photoemission studies, a monochromatized photon beam (Al K_α, hv = 1486.6 eV) was focused onto the sample surface, and the emitted electrons were energy analyzed with a Surface Sciences Instruments hemispherical analyzer. The take-off angle of the photoelectrons relative to the surface normal was 60° unless otherwise specified. A position-sensitive detector with 128 channels was used with a dedicated HP9836C computer to facilitate data acquisition.(22).

The $La_{1.85}Sr_{0.15}CuO_4$ and $YBa_2Cu_3O_{6.9}$ samples were high density, polycrystalline, and single phase with superconducting transition temperatures of ~35 K and ~92 K, respectively(23). The samples were cleaved *in situ* at pressures of 8 x 10^{-11} Torr, and all measurements were performed at room temperature. Typical beam currents were 15 μA for KRIPES and 100 μA for BIS. Immediately after each cleave, we acquired surface sensitive KRIPES spectra with incident electron energies of 20.25 and 26.25 eV, and these spectra were repeated at regular intervals to assess surface degradation. The typical data acquisition time for each KRIPES spectrum was ~1 hour. In XPS, the x-ray beam was focused to 300 μm and the surface degradation was monitored through the O 1s emission.

In our studies of interface reaction, Fe was evaporated from resistively heated W boats, and the deposition rates were monitored with Inficon quartz oscillators situated near the sample. The typical evaporation rate was ~1 Å/min, and the pressure during evaporation was ~1 x 10^{-10} Torr in both IPES and XPS systems.

Results and Discussion

$\underline{La}_{1.85}\underline{Sr}_{0.15}\underline{CuO}_4$. In Figure 1 we show KRIPES and BIS spectra for freshly cleaved surfaces. The incident electron energies, E_i, referenced to E_F, are given alongside each PDC. The uppermost spectrum shows the BIS yield within 25 eV of E_F. As can be seen, each spectral feature appears at the same position in all PDCs, indicating structure in the density of states. The consistency between KRIPES and BIS shows that changes in mean free path does not alter the features of the PDC's ($\lambda \approx$ 4-20 Å with probe depth of 3λ). This is important because it indicates that the bulk electronic properties can be studied reliably with surface sensitive techniques, even though the unit cell is ~12 Å in the c-direction.

The results of Figure 1 show a very low density of states at E_F and a monotonic increase until ~3.5 eV above E_F. This is consistent with photoemission spectra for the occupied states since they indicate a very low state density at E_F (see, for example, the bottom-most curve of Figure 5 of this paper). Band structure calculations for La_2CuO_4 (5-7) show that a single band breaks away from the high density Cu-O levels 2-6 eV below E_F and passes through the Fermi level. This band is formed by Cu $3d_{x^2-y^2}$ and O $2p_{x,y}$ antibonding orbitals and, according to the calculations, forms a nearly perfect nesting of the Fermi surface. It has been proposed that this nesting leads to a two dimensional charge density wave instability which is suppressed by the replacement of Sr for La(5,6).

The results of Figure 1 show a sharp peak 8.7 eV above E_F and a shoulder on the low energy side. From the band calculations(5-7), the peak can be associated with La 4f-derived levels and the shoulder can be associated with the La 5d states. It is important to note that the 4f states of metallic La lie 5.3 eV above E_F and that the 5d states are broad and ill-defined(25). The higher 4f energy in $La_{1.85}Sr_{0.15}CuO_4$ than in metallic La reflects differences in electron distribution. Significantly, the sharp, empty La 4f level provides us with the empty state equivalent of a core level, and chemical information can be inferred from changes in its energy, as will be discussed shortly.

Lineshape decompositions provide additional support for the association of 5d and 4f character with the empty state structures. To obtain the results shown in the inset of Figure 1, we subtracted a smooth background and fitted the resulting curve with two Gaussians. The fittings for the 4f contribution for E_i = 30.25 eV and hv = 1486.6 eV shown in Figure 1 were obtained with energy centers of 8.7 eV and full widths at half maximum (FWHM) of 2.15 and 2.0 eV, respectively. For the 5d contribution, the energy center was 5.8 eV and the full widths were 1.76 eV and 2.87 eV, respectively. Analysis indicates that the intensity ratio I_{4f}/I_{5d} increases from ~1.3 at E_i = 30.25 eV to ~5.9 at x-ray energies, consistent with the enhanced 4f cross section at high energy(21).

To monitor surface stability, we collected KRIPES spectra repeatedly with E_i = 20.25 and 26.25 eV. The results are summarized in Figure 2 where the PDCs shown at the right emphasize the energy window within 4 eV of E_F. Difference curves, enhanced by a factor of 5, are shown at the left, and each is referenced to the clean surface spectrum. These very surface sensitive measurements showed very little change in ~7 hours under low energy electron bombardment in an ambient vacuum of ~1 x 10^{-10} Torr (bottom panel) but that a structure of FWHM ~2 eV eventually appeared ~1.6 eV above E_F. To demonstrate that this feature was due to oxygen-induced states, we exposed the clean surface to O_2 with the electron gun retracted and no hot filaments within line of sight of the sample. After 10 L of exposure (1 L = 1 x 10^{-6} Torr-sec), the structure appeared with even higher intensity (Figure 2, center panel). Additional O_2 exposure to 100 L had little further effect. We conclude that oxygen chemisorbs on $La_{1.85}Sr_{0.15}CuO_4$ with energies comparable to those observed for Ni(26) and GaAs(27) surfaces (O $2p_{x,y}$ states 1.4 and 2 eV above E_F).

To investigate possible oxygen desorption from the surface and near surface region due to a high energy electron beam, we irradiated the surface with a ~1500 V, 0.1 mA beam (1x5 mm beam size). The results are shown in the top panel of Figure 2 where zero corresponds to a cleave exposed to ambient vacuum and low energy irradiation for 20 hours. After this surface had been irradiated for another 20 hours with the high energy beam, the oxygen-induced feature diminished and, indeed, the difference curve shows a *loss* in intensity relative to the clean surface.

This investigation of surface degradation on the $La_{1.85}Sr_{0.15}CuO_4$ suggests that the surface is reasonably stable under low energy electron bombardment and 1 x 10^{-10} Torr ambient vacuum but that

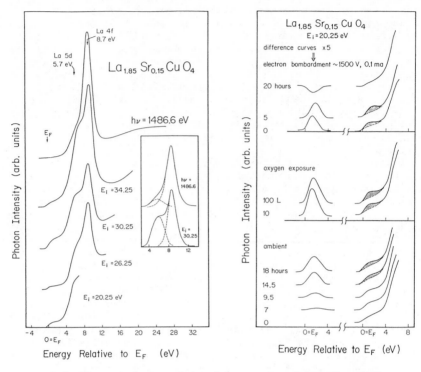

Left, Figure 1. Inverse photoemission spectra of $La_{1.85}Sr_{0.15}CuO_4$ showing the distribution of empty states within 25 eV of the Fermi level. The density of states at E_F is very low, and the prominent features at 5.8 and 8.7 eV are related to La 5d and La 4f states. The low incident electron energies provide maximum surface sensitivity. Comparison with the bulk-sensitive, high-energy results shows the viability of surface studies of these high-temperature superconductors. The inset shows the lineshape decomposition to highlight the La 4f and La 5d cross-section variation with electron energy.

Right, Figure 2. Inverse photoemission results for $La_{1.85}Sr_{0.15}CuO_4$ which show the slow growth of oxygen-induced emission under low-energy electron irradiation at ambient vacuum of $\sim 1 \times 10^{-10}$ Torr (bottom panel) and under controlled O_2 exposure (center panel). High-energy electron bombardment dissociates the chemisorbed oxygen and also reduces the oxygen content of the surface region, as indicated by the overall loss in intensity (top panel).

there is a gradual increase ~1.6 eV above E_F due to chemisorption of oxygen. High energy electron bombardment (~1500 eV) dissociates this oxygen, diminishing the oxygen content of the near surface, as indicated by the reduced emission of the Cu $3d_{x^2-y^2}$ - O $2p_{x,y}$ states. Given the dependence of the superconducting transition temperature on oxygen content, it is intriguing to speculate that the effect of high energy electron irradiation would suppress T_c for the near surface region and produce a nonsuperconducting skin.

$YBa_2Cu_3O_{6.9}$. Photon distribution curves for $YBa_2Cu_3O_{6.9}$ are displayed in Figure 3, scaled to electron incident current and data acquisition time. The uppermost spectrum shows the photon yield up to 25 eV above E_F. These KRIPES and BIS spectra show density of states (DOS) features as for $La_{1.85}Sr_{0.15}CuO_4$, but there is also a dispersive band that can be seen ~4 to ~6 eV, as indicated by the tic marks. Again, we see a very low density of unoccupied states from E_F to 3.7 eV above E_F. The structural similarities for the two ceramic oxides suggest that the states near E_F for $YBa_2Cu_3O_{6.9}$ are also due to the Cu $3d_{x^2-y^2}$ - O $2p_{x,y}$ antibonding orbitals.

The results of Figure 3 show a well-defined peak 14.0 eV above E_F (FWHM 2.1 eV). This feature exhibits a ~5 fold increase in relative intensity on going from E_i = 32.25 eV to E_i = 1500 eV when compared to the peaks lying between 4 eV and 9 eV. This large energy cross section dependence parallels that observed by Fauster and Himpsel(21) for the Gd 4f/5d intensity ratio. For $La_{1.85}Sr_{0.15}CuO_4$, we conclude that this high lying structure is due to the core-like empty Ba 4f levels.

The features ~4 eV to ~9 eV above E_F reflect emission from Y 4d and Ba 5d derived levels and are directly comparable to the La 5d derived levels for $La_{1.85}Sr_{0.15}CuO_4$. The environments of the Ba atoms in $YBa_2Cu_3O_{6.9}$ and the La atoms in La_2CuO_4 are very similar, suggesting that qualitative similarities should be observed in their electronic states as well. For La_2CuO_4, the empty La 5d levels are relatively flat except for one band which shows a ~2 eV dispersion downward. This would suggest that the feature at 6.5 eV is due to the Ba 5d levels because of the band which disperses ~2 eV from it (tic marks, Figure 3). In contrast, the Y atoms in the $YBa_2Cu_3O_{6.9}$ structure are sandwiched between CuO_2 layers. Since the latter form two-dimensional bands, we expect very weak interaction with Y, suggesting that the Y 4d levels are rather localized in real space and non-dispersive in k-space. From the results of Figure 3, we associate the Y 4d levels with the nondispersive peak 8.6 eV above E_F.

Support for these peak assignments comes from comparison between our results for $YBa_2Cu_3O_{6.9}$ with those for metallic Ba(28) and Y(29). The dashed line shown in the inset of Figure 3 results from shifting the Ba BIS spectrum 4.7 eV away from E_F. This chemical shift can be attributed to the change in environment of the Ba atoms, and an analogous shift has been observed for La (4f energy 5.3 eV for metallic La, 8.7 eV for the oxide superconductor). The shapes of the spectra are similar except for the extra emission 4 to 12 eV above E_F for $YBa_2Cu_3O_{6.9}$, which we attribute to the Y 4d levels. Furthermore, the BIS results for Y metal show structure 2.0 eV and 5.5 eV above E_F due to the d orbitals(29). Since the atom separation for Y atoms parallel at the basal plane is nearly the same as in Y, but the number of nearest neighbors is 4 rather than 12, we would expect the d band width to be substantially reduced in the ceramic. It then seems reasonable to associate the structure in $YBa_2Cu_3O_{6.9}$ at 8.6 eV with a 4-eV-shifted Y d-band manifold. Final support comes from results for the $Fe/YBa_2Cu_3O_{6.9}$ [30] interface where we found that the peaks at 6.5 eV and 14.0 eV persisted due to Ba outdiffusion while that at 8.6 eV attenuated rapidly.

Surface degradation of $YBa_2Cu_3O_{6.9}$ due to ambient vacuum exposure and high energy electron bombardment has also been monitored. The growth of the feature centered at ~1.8 eV was negligible until after ~4 hours of exposure to both the vacuum and the low energy electron beam used in KRIPES. We attribute this new state to chemisorbed oxygen since it is identical that shown in Figure 2 for $La_{1.85}Sr_{0.15}CuO_4$. The effects of the high energy beam, however, seemed to increase and broaden the chemisorbed oxygen peak formed by the ambient, in contrast to the situation for $La_{1.85}Sr_{0.15}CuO_4$ where a loss in intensity relative to the clean surface was observed. We speculate that the difference reflects oxygen transport to the surface of $YBa_2Cu_3O_{6.9}$. This ease in moving oxygen atoms into and out of $RBa_2Cu_3O_{7-x}$ structures has also been reported elsewhere [31].

<u>$Fe/La_{1.85}Sr_{0.15}CuO_4$</u>. The integration of high temperature ceramic superconductors into tomorrow's technology will rely in large measure on surface passivation and the ability to make stable contacts. As a first study of interface formation, we investigated Fe overlayers on $La_{1.85}Sr_{0.15}CuO_4$, with an emphasis on chemical phenomena. Iron was chosen because of its strong affinity for oxygen, which is a major constituent in all of the new high T_c materials.

In Figure 4 we show KRIPES spectra for Fe coverages, θ, between 0.25 to 30 Å. As shown, the effects of Fe adatoms are profound. For the La states, there is a broadening and a shift to lower energy. Comparison of the final La 4f position at 7.2 eV with the corresponding energy position in La metal at 5.3 eV provides clear evidence that the local bonds of La in the perovskite crystal have been altered by the addition of Fe. Analysis shows that the 4f feature for the reacting interface for θ = 2 Å appears as a doublet with a small shoulder on the higher energy side of the dominant peak. Lineshape decomposition shows two components, a weaker one at the unreacted position 8.7 eV and the dominant reacted component at 7.7 eV. Comparison of the emission intensity for the unreacted La in $La_{1.85}Sr_{0.15}CuO_4$ indicates that it is ~7% of that of the clean surface. The relative intensities of the two components are consistent with the ~4 Å mean free path of incident electrons of energy ~26 eV relative to E_F. We conclude that the outermost ~8 Å of the nominal surface is modified by the deposition of 2 Å of Fe as it disrupts the $La_{1.85}Sr_{0.15}CuO_4$ structure and breaks Cu-O bonds.

Additional insight into the destruction of the $La_{1.85}Sr_{0.15}CuO_4$ surface by the Fe overlayer, and the production of a nonsuperconducting skin, can be gained by examining changes in the PDCs near E_F. The results of Figure 4 demonstrate that the Fe 3d states appear ~2 eV above E_F whereas we would have expected them to grow at E_F if metallic Fe were forming. They then shift toward E_F and stabilize by 16 Å coverage. Further metallic Fe with a Fermi level cutoff was forming by ~8 Å. The convergence to metallic Fe

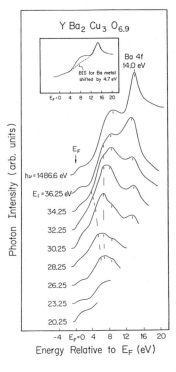

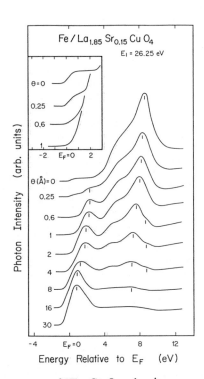

Left, Figure 3. Inverse photoemission spectra of $YBa_2Cu_3O_{6.9}$ showing the unoccupied states within 25 eV of the Fermi level. The density of states near E_F is very low, but strong emission at 6.5 eV, 8.6 eV, and 14.0 eV corresponds to the Ba 5d-, Y 4d-, and Ba 4f-derived states, respectively. The inset compares the BIS spectrum for $YBa_2Cu_3O_{6.9}$ with that of Ba metal, with the latter shifted by 4.7 eV.

Right, Figure 4. Inverse photoemission spectra for the evolving $Fe/La_{1.85}Sr_{0.15}CuO_4$ interface showing the shift to lower energy of the La_{4f} and La_{5d} states. Fe 3d states initially appear well above E_F but shift toward E_F as they grow; a metallic Fermi level cutoff becomes obvious after 8 Å of Fe. In the inset, we expand fivefold the portion of the PDCs for low coverage to show the loss of emission from the hybrid Cu–O states within 2 eV of E_F.

demonstrates that the chemical destruction of $La_{1.85}Sr_{0.15}CuO_4$ is limited at room temperature. The higher energy position of the Fe 3d states at the lowest coverage indicates the formation of a large band gap form of Fe-O(32) at the $Fe/La_{1.85}Sr_{0.15}CuO_4$ interface. At the same time, the results of Figure 4 indicate a dynamic evolution of Fe-O bonding configurations as the number of Fe atoms increases.

Another extremely important observation that can be made from Figure 4 is that the reacting Fe overlayer directly modifies the Cu $3d_{x^2-y^2}$ -- O $2p_{x,y}$ states near E_F. This is significant because the Cu $3d_{x^2-y^2}$ -- O $2p_{x,y}$ states appear to play an important role in superconductivity, even though the density of these states is very low near E_F because of the single band character,(5-7) As shown in the inset of Figure 4, the density of these states is decreased by the deposition of 0.25 Å of Fe. (The inset highlights the changes near E_F by enhancing the low coverage spectra five-fold.) Deposition of Fe to 0.6 Å diminishes these states further. This paucity of emission continues until the Fe 3d states of the growing metallic overlayer disperse to E_F and dominate. This provides further evidence that the highly reactive metal Fe depletes the near surface oxygen content as it forms an Fe-O phase. The loss of Cu-O antibonding states, the chemical shift of La, and the energy of the Fe 3d states all indicate the formation of a complex, multicomponent interface region. It appears likely that superconductivity in this region will be suppressed and that the scale of interaction will be enhanced by thermal activation. Indeed, the limited extent of the Fe-O interaction indicates kinetic limitations such that the interface is metastable.

To determine whether further intermixing follows the initial strong chemical interaction, we conducted BIS studies for Fe coverages 30-80 Å. For these measurements, the probing depth was substantially larger ($\lambda \sim 20$ Å) and the sensitivity to the La 4f states was much greater because of cross section effects(21). These BIS spectra showed a saturated 4f chemical shift, as discussed above. Moreover, the BIS spectra showed the attenuation of the reacted La 4f peaks, indicating the termination of chemical reaction by the time the nominal Fe coverage had reached ~8 Å. We estimate that this corresponds to disruption of ~20 Å of the superconductor.

XPS investigations of $Fe/La_{1.85}Sr_{0.15}CuO_4$ provide additional insight into this interface. In Figure 5 we show the Cu-O derived valence band states at binding energies of 4-6 eV and the low density of states at the the Fermi level for the clean interface. To highlight the Fe-induced changes near E_F, we show difference curves that were obtained by subtracting the clean surface results from those at the coverage indicated. The deposition of 0.25 Å Fe causes significant changes in the density of states at, and near, the Fermi level. The results are more remarkable at low coverage where they show the *loss* of emission within ~3 eV of E_F for the first ~1 Å of Fe coverage. Indeed, it is not until ~4 Å that there is a net gain in emission at E_F, analogous to that seen in the empty states (Figure 4). At higher coverage, the valence band EDCs of Figure 5 (and the empty state spectra of Figure 4) show the growth of the Fe d bands and the establishment of metallic d-band emission.

$Fe/La_{1.85}Sr_{0.15}CuO_4$ interface formation can be further explored by studying the core levels of different species. In the top panel of Figure 6 we show the O 1s core level for the freshly cleaved surface, the same surface after exposure to ambient vacuum for 14 hours, and for various coverages of Fe. The freshly cleaved surface shows a single peak (FWHM 1.45 eV) with a slight tail on the higher binding energy side. Comparison of the fresh and "aged" surfaces shows the growth of a shoulder at 2.2 eV higher binding energy, but neither the freshly cleaved nor the aged surface showed any detectable C or N contamination. This structure was first observable in XPS ~5 hours after cleaving and saturated after ~12 hours. The spectra for Cu, La, and Sr showed no changes over the same time period. Measurement of the O 1s emission from several places on the surface showed that the shoulder was present everywhere and not just in the region where the original x-ray spot was located. This eliminates the possibility of x-ray induced dissociation as the cause of the shifted component. Measurements at grazing emission, which changed the effective probe depth from ~18 to ~9 Å, enhanced the new feature, indicating that it is a surface phenomenon consistent. We conclude that the time dependent feature was due to ambient oxygen and was not related to instabilities of the chemical bonds of the superconductor surface. An analogous O 1s feature has been observed by Steiner *et al.*($\underline{9}$) for a $La_{1.85}Sr_{0.15}CuO_4$ surface immediately after scraping with a diamond file, suggesting that cleaving provides surfaces more suitable for studies of interface chemistry.

The results shown in the top panel of Figure 6 demonstrate that 0.25 Å of Fe broadens the O 1s peak by 0.33 eV and shifts it 0.5 eV toward higher energy. A broad shoulder also appears due to Fe reaction with adsorbed surface O, shifted 2.75 eV but is unrelated to interactions with the substrate. The measurement time for each coverage was ~5 hours, corresponding to 1.8 Langmuirs of ambient exposure. Since the position of the peak was independent of emission angle, whereas the main O 1s peak showed a shift of -0.3 eV upon changing the emission angle from 60° to 0°, we conclude that this feature is due to adsorbed O reacting on the surface with the Fe overlayer. The strength of the feature is due to the surface sensitivity of our technique. The main peak continued to shift toward higher energies with subsequent deposition until a maximum shift of 1.5 eV was obtained for an Fe coverage of 50 Å. Likewise, the peak width decreased after 2 Å to a FWHM of 1.55 eV at 40 Å. The integrated O 1s emission did not attenuate uniformly as a function of Fe coverage, but remained strong to 50 Å coverage, indicating outdiffusion of oxygen into the overlayer (see Figure 6 inset).

The lineshape of the Cu $2p_{3/2}$ emission has attracted considerable attention because of questions related to the valence(s) of Cu, and hence the bonding configurations, in these ceramic oxides. The spectrum for the clean surface shown at the bottom of the center panel of Figure 6 has been observed before, but its interpretation has remained controversial. In particular, the satellite structure is similar to that of CuO($\underline{33}$) where it is a final state effect related to the empty 3d levels($\underline{34}$). Previous XPS measurements of Cu, CuO and Cu_2O have shown that CuO has a binding energy 1.3 eV higher than Cu_2O and that of Cu_2O is degenerate with Cu metal($\underline{33}$). Our results demonstrate existence of two inequivalent states in the superconductor, namely Cu^{1+} which corresponds to the main peak and Cu^{2+} which accounts for the shaded regions.

Deposition of 0.25 Å of Fe sharply attenuates the Cu^{2+} emission and, by 1 Å coverage, only Cu^{1+} can be seen. Measurements at normal emission, which increased the probe depth from ~15 to ~30 Å showed no trace of the Cu^{2+} state by 1 Å coverage. The FWHM of the Cu^{1+} peak was 1.9 eV for θ = 1 Å and diminished to 1.2 eV at 8 Å as the homogeneity of the probed region increased. Although the results of Figure 6 are drawn to emphasize changes in lineshape, the emission intensity of the overall signal decreased rapidly with coverage. Quantitative analysis of the Cu emission showed exponential decay with a 1/e length of 10 Å, indicating layer by layer growth of Fe after reaction ceased. By θ = 30 Å, the Cu emission from the substrate was no longer detectable.

The growth of the Fe $2p_{3/2}$ core level emission provides additional insight into the chemical stability of the $Fe/La_{1.85}Sr_{0.15}CuO_4$ interface. The results shown in the bottom panel of Figure 6 reveal a broad, featureless peak shifted ~4 eV toward higher binding energy relative to that for metallic Fe. The onset of the metallic Fe $2p_{3/2}$ emission can be seen at 2 Å Fe coverage. This metallic feature grows with coverage and is sufficiently strong by 16 Å coverage that the Fe-O emission cannot be seen. This is consistent with the formation of a thin Fe-O layer and the subsequent burial of this layer as thicker films are grown. It is also consistent with the rapid approach to a final configuration for the O 1s emission. At the same time, the impact of these first few Angstroms of Fe are profound, as indicated above, since they modify *all* of the Cu^{2+} states within the probe depth by 1 Å coverage and the substrate O1s emission appears to be fully converted by 8 Å.

From this XPS and IPES study of $Fe/La_{1.85}Sr_{0.15}CuO_4$, we can develop a coherent picture of interface development. Adatoms of Fe initially destroy the two-dimensional character of the super-conductor, withdrawing oxygen to form more energetically favorable Fe-O bonds. This process continues until a critical coverage corresponding to the formation of Fe metal, and this coverage is related to the kinetics of oxygen transport to the region where Fe adatoms impinge. At high coverage, the Fe layer grows in thickness in a layer-by-layer fashion. At the same time, our results for high coverage show the presence of small amounts of O in the overlayer, probably in the form of a solid solution rather than phase separated regions of Fe-O.

Conclusion

These inverse photoemission studies offer the first clear picture of the empty states of high temperature superconductors and the occupied states are revealed by XPS. These studies represent the starting point for investigations of surface and interface properties, and we have reported the first work in this challenging area. Based on studies of Fe overlayers, it seems reasonable to predict that reaction metal overlayers on these ceramic super-conductors will deplete oxygen from the near surface region, thereby modifying the properties of the material. Studies of metal over-layers which are less prone to oxidize (Au) are underway for comparison to the results presented here. Likewise, preliminary analyses of Cu, Al, Ge, and Si overlayers are in progress and will be discussed elsewhere.

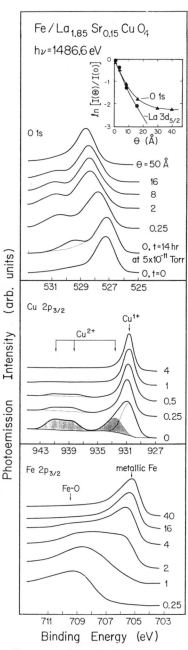

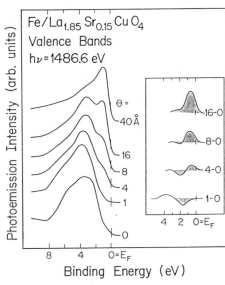

Left, Figure 5. Photoemission intensity from the $La_{1.85}Sr_{0.15}CuO_4$ valence bands for the clean surface, the aged surface, and for various coverages of Fe. The spectra are normalized for clarity. The inset shows that Fe deposition initially destroys Cu–O hybrid states near E_F.

Right, Figure 6. Photoemission intensity from the $La_{1.85}Sr_{0.15}CuO_4$ clean surface, the aged surface, and for various coverages of Fe for O 1s (top), Cu $2p_{3/2}$ (middle), and Fe $2p_{3/2}$ (bottom). The peak heights are normalized for clarity. The inset shows the rate of decay of O and La.

Acknowledgments

The work at the University of Minnesota was supported by the Office of Naval Research under ONR N00014-87-K-0029 and that at the Argonne National Laboratory was supported by Department of Energy under W-31-109-Eng-38.

Literature Cited

1. Bednorz, J.G.; Müller, K.A. Z. Phys. B 1986, 64, 189.
2. Chu, C.W.; Hor, P.H.; Meng, R.L.; Gao, L.; Huang, Z.J.; Wang, Y.Q. Phys. Rev. Lett. 1987, 58, 405.
3. A listing of recent papers related to high temperature superconductors appears in Phys. Rev. Lett. 1987, 58, No. 20 and Jpn. J. Appl. Phys. 1987, 26.
4. Hor, P.H.; Meng, R.L.; Wang, Y.Q.; Gao, L.; Huang, Z.J.; Bechtold, J.; Forster, K.; Chu, C.W. Phys. Rev. Lett. 1987, 58, 1891.
5. Mattheiss, L.J. Phys. Rev. Lett., 1987, 58, 1028.
6. Yu, Jaejun; Freeman, A.J.; Xu, J.-H. Phys. Rev. Lett. 1987, 58, 1035.
7. Pickett, W.E.; Krakauer, H.; Papaconstantopoulos, D.A.; Boyer, L.L. Phys. Rev. B 1987, 35, 7252.
8. Moss, S.C.; Forster, K.; Axe, J.D.; You, H.; Hohlwein, D.; Cox, D.E.; Hor, P.H.; Meng, R.L.; Chu, C.W. Phys. Rev. B 1987, 35, 7195.
9. Steiner, P.; Albers, J.; Klinsinger, V.; Sander, I.; Siegward, B.; Hüfner, S.; Politis, C. Z. Phys. B 1987, 66, 275.
10. Fujimori, A.; Takayama-Muromachi, E.; Uchida, E.; Okai, B. Phys. Rev. B (in press).
11. Onellion, M.; Chang, Y.; Niles, D.W.; Joynt, Robert; Margaritondo, G; Stoffel, N.G.; Tarascon, J.M. Phys. Rev. Lett. (submitted).
12. Johnson, P.D.; Qiu, S.L.; Jinag, L.; Ruckman, M.W.; Strongin, Myron; Hulbert, S.L.; Garrett, R.F.; Sinkovic, B.; Smith, N.V.; Cava, R.J.; Jee, C.S.; Nichols, D.; Kaczanowicz, E.; Salomon, R.E.; Crow, J.E. Phys. Rev. B 1987, 35, xxx - June 1.
13. Kutz, Richard L.; Stockbauer, Roger L.; Mueller, Donald; Shih, Arnold; Toth, Lowis E.; Osofsky, Michael; Wolf, Stuart A. Phys. Rev. B 1987, 35, xxx - June 15.
14. Fujimori, A.; Takayama-Muromachi, E.; Uchida, Y. Solid State Commun. (submitted).
15. Chang, Y.; Onellion, M.; Niles, D.W.; Joynt, Robert; Margaritondo, G.; Stoffel, N.G.; Tarascon, J.M. Solid State Commun. (submitted).
16. Cava, R.J.; Santoro, A.; Johnson, D.W.; Rhodes, W.W. Phys. Rev. B 1987, 35, 6716.
17. Geiser, U.; Beno, M.A.; Schultz, A.J.; Wang, H.H.; Allen, T.J.; Monaghan, M.R.; Williams, J.M. Phys. Rev. B 1987, 35, 6721.
18. Lang, J.J.; Baer, Y. Rev. Sci. Instrum. 1979, 50, 221.
19. Himpsel, F.J.; Fauster, Th. J. Vac. Sci. Technol. A, 1984, 2, 815; Johnson, P.D.; Smith, N.V. Phys. Rev. B 1983, 27, 2527.
20. Grioni, M.; Gao, Y.; Smandek, B.; Weaver, J.H.; Tyrie, T.; unpublished.
21. Fauster, Th.; Himpsel, F.J. Phys. Rev. B 1984, 30, 1874.

22. Chambers, S.A.; Hill, D.M.; Xu, F.; Weaver, J.H. Phys. Rev. B 1987, 35, 634.
23. Jorgensen, J.D.; Schuttler, H.B.; Hinks, D.G.; Capone, D.W.; Zhang, K.; Brodsky, M.B.; Scalapino, D.J. Phys. Rev. Lett., 1987, 58, 1024.
24. Herbst, J.F.; Lowy, D.N.; Watson, R.E. Phys. Rev. B 1972, 6, 1913.
25. Lang, J.K.; Baer, Y.; Cox, P.A. J. Phys. F 1981, 11, 121.
26. Himpsel, F.J.; Fauster, Th. Phys. Rev. Lett. 1982, 49, 1583.
27. Straub, D.; Skibowski, M.; Himpsel, F. Phys. Rev. B, 1985, 32, 5237.
28. Knaski, J.; Nilsson, P.O. J. Phys. F 1981, 11, 1859.
29. Spicer, W.; Fuggle, J.C.; Zeller, R.; Ackermann, B.; Szot, K.; Hillebrecht, F.U.; Campagna, M. Phys. Rev. B 1984, 30, 6921.
30. Gao, Y.; Wagener, T.J.; Weaver, J.H.; Arko, A.J.; Flandermeyer, B.; Capone, D.W.; inverse photoemission results for Fe/YBa$_2$Cu$_3$O$_{6.9}$ (in preparation).
31. Robinson, A.L. Science 1987, 236, 1063.
32. Mattheiss, L.F. Phys. Rev. B 1972, 5, 290.
33. McIntyre, N.S.; Cook, M.G. Anal. Chem. 1975, 47, 2208.
34. Wallbank, B.; Johnson, C.E.; Main, I.G. J. Phys. C 1973, 6, L493.

RECEIVED July 17, 1987

PROCESSING AND FABRICATION

Chapter 22

Processing of High-Temperature Ceramic Superconductors: Structure and Properties

L. E. Toth[1,4], M. Osofsky[1,5], S. A. Wolf[1], E. F. Skelton[1], S. B. Qadri[2], W. W. Fuller[1], D. U. Gubser[1], J. Wallace[1], C. S. Pande[1], A. K. Singh[3], S. Lawrence[3], W. T. Elam[1], B. Bender[1], and J. R. Spann[3]

[1]U.S. Naval Research Laboratory, Washington, DC 20375-5000
[2]Sachs-Freeman Associates, Landover, MD 20785
[3]Crystal Growth & Material Testing Association, Lanham, MD 20706

The processing of high T_c ceramic superconductors by traditional ceramic techniques is reviewed. All high T_c ceramic superconductors are layered Cu-O compounds that are closely related to each other. In each, Cu-O coordination polyhedra are typical of Cu^{+2} compounds. A critical step in processing these compounds is the intercalation of oxygen which changes the coordination polyhedra at a few atomic sites and causes a dramatic effect on the superconducting properties.

Following Bednorz and Müller's publication (1) on high transition temperature, T_c, ceramic superconductors, there has been a deluge of papers on the subject. Momentum increased with the announcement by Wu et al.(2) of a T_c breakthrough in excess of 90K. Since then a great deal has been learned about superconducting properties, crystal structures and phase relationships in the Y-Ba-Cu-O system. Knowledge of processing these materials has improved but not to the same extent as some other areas. Each research group seems to report a different set of processing procedures. It is widely acknowledged that it is difficult to transfer processing procedures from one group to another and achieve comparable superconducting properties. In this paper we attempt to review some of the processing procedures. We pay particular attention to the underlying structure: crystal and microstructure. At this time, relationships between structure and superconducting properties are better defined than are relationships between processing and properties. Therefore, one can use the evolution of the desired structure as a guide to processing.

[4]On sabbatical from the National Science Foundation, Washington, DC 20550
[5]Postdoctoral fellow, Office of Naval Technology, Arlington, VA 22217-5000

Our group has been mainly concerned with traditional ceramic processing techniques involving solid state particulate reactions such as sintering and hot pressing. A number of other techniques, such as co-precipitation and organometallics, have been successfully used by others, but will not be discussed here. Regardless of the method of preparation, it is believed that all materials must end up with the same structural modifications to ensure good superconducting properties.

General

Structure: Some 14 different ceramic compounds are reported to superconduct with T_c's in excess of 35K. These are listed in Table I. There are 5 distinct classes of materials with 4 different crystal structures.

Table I. High Temperature Ceramic Superconductors

Formula	A	Reference
La_2CuO_4		(3)
$La_{2-x}A_xCuO_{4-x/2+\delta}$	Ba,Sr,Ca	(4-6)
$ABa_2Cu_3O_{6.5+\delta}$	Y,Lu,Nd,Sm,Eu,Gd Er,Ho,Yb	(2,7,8)
$La_3Ba_3Cu_6O_{14+\delta}$		(9)
$Y_2BaCu_2O_{6-x/2+\delta}$		(10)

There are several chemical and crystal structure similarities in the high T_c ceramic superconductors. A general feature in their crystal chemistry is the occurrence of a small number of Cu-O coordination polyhedra. As discussed by Wells(11), these are the same polyhedra that one finds in Cu-O compounds in which Cu has a +2 valence. Copper is located in one of three configurations: (1) at the center of a square array of coplanar oxygen atoms (square planar); (2) at the center of the square base of a pyramid with oxygen at the vertices, (pyramidal (4+1)); and (3) at the center of a distorted octahedra with oxygen at the vertices, (distorted octahedra (4+2)), (see Fig 1). In square planar, the Cu-O spacing is about 1.95Å. In the pyramidal and distorted octahedra, there are 4 surrounding oxygen at short distances (1.95Å) comparable to those found in square planar, and 1 or 2 distances significantly longer (2.3Å). The notation (4+1) and (4+2) is used to denote the fact that 4 of the Cu-O distances are short and 1 or 2 are longer.

Table II lists the coordination of oxygen about the central copper atom in each of the high T_c ceramic superconductors and it lists typical Cu-O distances. In the crystal structures, the coordination polyhedra are arranged so that the square planar configurations are perpendicular to the c-axis of the unit cell and the long axis of the pyramids and octahedra are parallel to it. Within planes of Cu-O atoms, which are perpendicular to the unit

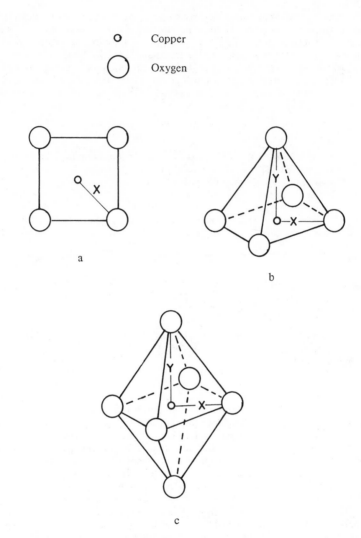

Figure 1. Coordination polyhedra of copper and oxygen atoms in high T_c superconductors: (a) square planar, (b) pyramidal (4+1) and (c) distorted octahedron (4+2). The distance x is about 1.95Å and the distance y about 2.3Å.

cell's c axis, the typical distance is 1.95Å. Parallel to the c axis, the Cu-O distances are either shorter or longer. Thus all structures can be viewed as having parallel planes of Cu-O sheets in which the Cu-O spacing is 1.95Å. The rare earth and oxygen atoms are also arranged in planes perpendicular to the c-axis. In $YBa_2Cu_3O_7$, there is a square planar ribbon at the basal planes of the unit cell which is parallel to the c axis and in which the Cu-O distance is only 1.85Å. The longest distance in this structure is also parallel to the c axis and is about 2.3Å, existing in both the distorted octahedra and 4+1 pyramidal structure. Figure 2 shows crystal structures of the four classes of superconducting materials, showing some elements of the Cu-O coordination polyhedra.

Table II.   Comparison of Coordination Polyhedra for Cu-O in Superconducting Ceramic Oxide Phases

Compounds	Coordination	Distances Å		Refs
$La_{2-x}A_xCuO_{4-x/2+\delta}$	distorted octahedra (4+2)	$Cu-O_1$ (x4)	1.937	(12)
		$Cu-O_2$ (x2)	2.27	
$La_3Ba_3Cu_6O_{14+\delta}$	square planar distorted octahedra (4+2) pyramidal(4+1)	Cu_2-O_1 (x4)	1.954	(13)
		Cu_2-O_1 (x4)	1.954	
		Cu_2-O_3 (x1)	1.723	
		Cu_4-O_2 (x4)	1.959	
		Cu_4-O_3 (x1)	2.333	
$La_{2-x}A_{1+x}Cu_2O_{6-x/2}$	pyramidal(4+1)	$Cu-O_1$ (x4)	1.94	(14)
		$Cu-O_2$ (x1)	2.27	
$YBa_2Cu_3O_{6.5+\delta}$	square planar	Cu_1-O_1 (x4)	1.85	(15,16)
		Cu_1-O_4 (x4)	1.94	
	pyramidal(4+1)	Cu_2-O_2 (x2)	1.94	
		Cu_3-O_3 (x2)	1.96	
		Cu_3-O_1 (x2)	2.3	

The superconducting and electronic transport properties of these materials are very sensitive to their oxygen content, thus it is important to understand where oxygen is added or subtracted in the unit cell. Table III lists positions where oxygen is added and its effect on T_c. For $YBa_2Cu_3O_x$, oxygen is added or subtracted from basal planes in the 0 1/2 0 and 1/2 0 0 positions. Also, oxygen can be ordered on the 0 1/2 0 sites leaving the 1/2 0 0 site vacant. This results in an orthorhombic distortion with b/a ≈ 1.7%, a unique one dimensional character to the structure and excellent superconducting properties (17). For $La_3Ba_3Cu_6O_{14}$ added oxygen fills the site 1/2 1/2 1/2 between two (4+1) pyramids, thus forming two distorted octahedra along the c axis. One octahedron is elongated and the other is compressed. Several groups (9,18) have found T_c onsets of 70K+ in $La_{3-x}Ba_{3+x}Cu_6O_{15}$ in samples annealed in a few atmospheres of oxygen.

If simple valences are considered, the addition of oxygen has the effect of raising a portion of the cations to a higher valence

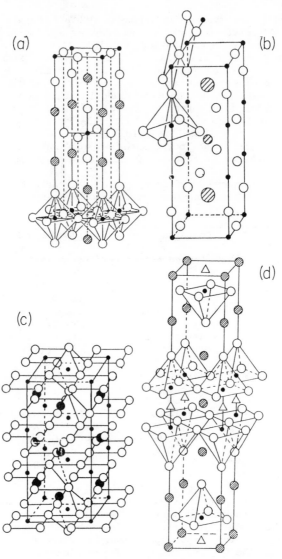

Figure 2. Crystal structures of ceramic high T_c superconductors. (a) K_2NiF_4 structure showing layers of distorted octahedra (4+2). (b) $YBa_2Cu_3O_7$ structure with nearly square planar and pyramidal (4+1) Cu-O coordination polyhedra, (c) Er-Rakho structure with square planar configurations perpendicular to, and distorted octahedra with the longer axis parallel to the c axis of the unit cell, (d) the $La_{2-x}Sr_{1+x}Cu_2O_{6-x/2+\delta}$ with (4+1) Cu-O pyramidal coordinations. The small solid circles are copper atoms, the large open circles are oxygen atoms, and the shaded circles are the rare earth, Y, Ba, or Sr atoms.

state, or more likely, hybridizing Cu-O bonds producing metallic hole-type conduction. It is convenient, however, to refer to these materials as if they have some Cu^{+3}, even though in a formal sense this ion may not exist.

Table III. Location of δ Oxygen Atoms and Effects on
Superconductivity

Compound	Location	Effect
$La_{2-x}A_xCu_{4-x/2+\delta}$	vacant octahedral corners	increases Cu^{+3}/Cu^{+2} ratio raises T_c
$La_3Ba_3Cu_6O_{14+\delta}$	vacant corner of octahedron at unit cell center	changes Cu coordination, pyramidal to octahedral increases $Cu^{+3/+2}$ ratio, semiconductor to 90K T_c
$YBa_2Cu_3O_{6.5+\delta}$	vacant site at 0 1/2 0	perfects square planar sites, increases $Cu^{+3/+2}$, orthorhombic distortion, raises T_c, 55 to 90K

<u>Processing</u>: All these materials can be processed in the same general manner. One starts with powders of the rare earth oxide (M_2O_3) or Y_2O_3, copper oxide CuO, and the carbonate of barium, strontium or calcium. The carbonates and rare earth oxides can be used in their as-received condition, but we have found that CuO requires additional milling to break up coarse particles. In addition, the powders should be predried to remove any adsorbed moisture prior to weighing. When drying, care should be taken to avoid agglomerate formation. The weighed powders are then thoroughly mixed in a mill or mortar and pestle. The powders are then calcined in open, flat crucibles. In the calcining step, carbonates are decomposed to the oxides and CO_2, and a multicomponent oxide is formed, i.e.

$$1/2 \ Y_2O_3 + 2 \ BaCO_3 + 3CuO \longrightarrow YBa_2Cu_3O_{6.5} + 2CO_2$$

The above formula assumes no addition or depletion of oxygen from the ambient during the calcining step, an assumption not always valid. The powders are not pelletized prior to calcining largely due to a large molar volume change (~30%) between reactants and products. One problem in calcining is that the carbonates remain stable and do not always decompose. Furthermore, at the reaction temperatures particle sintering and grain growth occur. Thus, the calcined materials must be remilled. By monitoring the calcining step with x-ray diffraction, we have found that 3-4 hr at 925C is sufficient time for calcining, provided all starting powders are 10μ or less.

Once calcining is complete, the major molar volume changes have taken place and the milled powders can be pelletized for sintering. Here a multiple of traditional ceramic processing steps can be used to form the material to its desired shape. The powders can be cold-pressed, with or without a binder, isostatically pressed, hot-pressed, mixed with binders and extruded into sheets, tubes, etc. Some members of our group, in collaboration with researchers at Brookhaven, have consolidated powders by plasma and flame spraying (19).

Once powders are consolidated and fired, the individual particles sinter together. This step is aided by a fine powder size (1-10μ) and a uniform dispersion of powders (no large voids). Typically 6-12 hours at 900-950C will sinter the particles to 80% density.

The final step in processing several of the high T_c ceramic superconductors is critical for good properties. The sintered samples are slowly cooled in flowing O_2 and held at some lower temperature to increase the oxygen content of the compound. For example,

$$ABa_2Cu_3O_{6.5} \text{ (from calcining)} + \delta/2\ O_2 \longrightarrow ABa_2Cu_3O_{6.5+\delta}$$

The effects of this step on superconductivity can be dramatic. For $YBa_2Cu_3O_7$, the fully oxygenated samples show a complete Meissner effect and R = 0 at temperatures in excess of 92K. Without the additional oxygen, the transitions are broad and the flux expulsion only partial (17). Table IV gives the composition ranges of intercalation of oxygen (20). The highest values of δ are obtained with long term anneals at about 500C under one atm of oxygen; lower values of δ are obtained with lower partial pressures. Table IV also shows the large effect intercalation has on the room temperature electrical conductivity, with samples becoming more metallic as oxygen is added.

Table IV. Known Ranges of Oxygen Intercalation in High T_c Ceramic
Superconductors

Compound	Intercalation Range δ	Log Cond. Range (300K)
$La_{2-x}A_xCu_{4-x/2+\delta}$	0-0.3	1-1.8
$La_{2-x}A_{1+x}Cu_2O_{6-x/2+\delta}$	0-0.2	1-2.5
$La_3Ba_3Cu_6O_{14+\delta}$	0-0.4 0.4-1.0 (3 atm)	0.5-1.9
$YBa_2Cu_3O_{6.5+\delta}$	0-0.5	- - - -

Individual Compounds

$YBa_2Cu_3O_x$:   Processing this compound follows the general procedure outlined above. At temperatures above 700C, $YBa_2Cu_3O_x$ has a tetragonal structure (21). Its oxygen content is believed to be $YBa_2Cu_3O_{6.5}$. Below 700C, the unit cell is orthorhombic and the composition adjusts towards $YBa_2Cu_3O_{7.0}$. The critical step in processing is to add this 0.5 oxygen atom per unit cell. This is critical because these oxygen atoms and their proper ordering on 0 1/2 0 sites dramatically improve superconducting properties. In such samples the T_c onsets are 93K, the transition widths 1K, and a full Meissner effect observed, i.e. the sample levitates in a strong magnetic field. The presence of the ordered intercalated oxygen atoms causes a distinct orthorhombic distortion in the unit cell of about 1.7% (17,21). Its presence can be readily observed from a splitting or shouldering the main X-ray diffraction peak at 32-33°. Normally the extra oxygen can enter during slow cooling (1°/min) from 900C in flowing oxygen. In some instances, furnace cooling is sufficient or alternatively one can anneal the sample at about 500-600C. We have observed, however, that some off-stoichiometric samples transform to the orthorhombic in a very sluggish manner. At this time, the reasons for this sluggishness in reaction are unclear. As another word of caution, we and others have observed that $YBa_2Cu_3O_{6.5}$ gradually decomposes when heated in air above 950C (22,23). By annealing a sample in air at 975C for 12 hours, we have nearly completely decomposed the phase into Y_2BaCuO_5, CuO and probably $BaCuO_2$. Reheating this same sample at 975C in flowing O_2 gradually reforms the $YBa_2Cu_3O_{6.5}$ phase.

The individual grains of samples with a pronounced orthorhombic distortion show striations or bands when viewed in an electron microscope (see figure 3). This is believed to be due to domain formation; in one band the "b" axis of the unit cell is oriented 90° to the "b" axis of the adjacent band. Electron microscopy also shows that small deviations from stoichiometry result in an amorphous second phase forming in the grain boundaries. Because most of the phases in equilibrium with $YBa_2Cu_3O_7$ are insulators, a grain boundary phase is probably an insulator.

We have measured a large number of superconducting properties on well characterized $YBa_2Cu_3O_7$ with a 1.7% orthorhombic distortion. These are listed in Table V.

$La_{3-x}Ba_{3+x}Cu_6O_{14+\delta}$: When processed by the general procedure, this sample is not superconducting. Mitzi et al. (9) showed that samples annealed for 24 hours at 450C in 3.5 atm of oxygen are superconductors, if x > 0.75, with T_c onsets of about 90K. We have verified these experiments (19). We find that the resistance behavior of $La_3Ba_3Cu_6O_{14+\delta}$ also becomes much more metallic after annealing in 3 atm of O_2. $La_2Ba_4Cu_6O_{14+\delta}$ becomes superconducting with T_c onset greater than 75K. The latter material shows a distinct second phase, however, so we cannot rule out the possibility that it is contributing to superconductivity. Also with x-ray analysis it is difficult to unambiguously decide if the crystal structure of this material is the Er-Rakho (13) or $YBa_2Cu_3O_7$ type. In both structures the cations have nearly identical positions, only the oxygen positions are significantly different. Both are very similar layered structures. The x-ray diffraction signatures, which are not sensitive to oxygen, are nearly identical. We see evidence for the very weak line at $2\theta=16$ as required for the (100) diffraction in the Er-Rakho tetragonal structure but also some evidence for peak shouldering which may indicate it is an orthorhombic structure and not tetragonal. It is also possible that

Figure 3. Transmission electron micrograph showing planar defects in $YBa_2Cu_3O_7$ uniformly spaced about 2000Å and dispersed in the specimens. These have been identified as twins formed because of the slight difference in the a and b axes.

Table V. Physical and Superconducting Properties of $YBa_2Cu_3O_7$
(NRL Samples)

Lattice Parameters (A)	
a	3.822 ± 0.002
b	3.888 ± 0.002
c	11.672 ± 0.005
Volume (A^3)	173.4
% Distortion	1.7
Oxygens per unit cell	6.94
T_c onset (K)	93
R=0 (K)	91
% flux expulsion	100
ρ (94K) ($\mu\Omega$-cm)	200
$(dH_c/dT)_{Tc}$ (kG/K)	22-36*
$H_{c2}(0)$ (kG)	1470-2370*
$H_{c1}(4.2K)$ (kG)	0.8
$H_c(0)$ (kG)	20-26*
$J_c(4K)$ (A/cm^2)	10^{5+}

*Calculated from critical field measurements
+Estimated from magnetization studies

both tetragonal and orthorhombic phases exist as suggested by Lee et. al. (24). Because of the importance of this material in understanding superconductivity in high T_c ceramics, its structure should be refined by neutron diffraction.

$La_{2-x}A_xCuO_{4-x/2+\delta}$:   A=Sr,Ba,Ca.

The optimum compositions for high T_c superconducting properties are for x=0.1 to 0.2. Samples are prepared according to the general procedure although processing temperatures of about 1100C are reported. The best samples are prepared in flowing O_2. Slow cooling and annealing at a lower temperature (500C) increase T_c by 1 to 2 degrees and also sharpen T_c. Likewise, annealing in vacuum destroys superconductivity. Reannealing in O_2 restores superconductivity. The degree of oxygen intercalation is small, δ=0 to 0.2 (20).

If x is small, less than 0.075, annealing in O_2 lowers T_c (25). If x=0, samples are not superconducting. Air-quenching samples annealed at 800-1000C cause a very small part of the sample to superconduct (27). This is probably grain boundary superconductivity because while R tends to zero, susceptibility measurements show only a trace of superconductivity. At very low temperatures, pure La_2CuO_4 transforms from an orthorhombic to an unknown structure (28). It has been speculated that this transformation may somehow inhibit superconductivity (29).

$Y_{2-x}Ba_{1+x}CuO_{6-x/2+\delta}$:   It has been suggested that this phase is a high T_c superconductor (10). This observation has never been confirmed, and in fact, it is doubtful that this particular phase even exists. Nevertheless, the crystal structure of this family of compounds, $La_{2-x}A_{1+x}Cu_2O_{6-x/2+\delta}$, is a layered one with coordination polyhedra, like those found in the high T_c superconductors and the compounds' electrical conductivities show the same sensitivity to oxygen intercalation. Thus the compounds are obvious candidates for superconductivity.

Summary: High T_c ceramic superconductors have layered crystal structures with Cu-O coordination polyhedra typical of Cu+2. Intercalation of additional oxygen is critical to the superconducting properties. All these materials can be processed in a similar manner by traditional ceramic processing techniques.

Acknowledgments

The authors acknowledge the help of W. Lechter and R. Rayne in sample preparation and useful discussions with D. Schrodt, M. Kahn, D. Lewis and W. Lechter.   B. Wood and R. Carpenter helped in manuscript preparation.

Literature Cited
1. J.G. Bednorz and K.A. Muller, Z. Phys. Rev. Lett, B 64, 189 (1986).
2. M.K. Wu, J.R. Ashburn, C.T. Torng, P.H. Hor, R.L. Meng, L. Gao, Z. Huang, Y.Q. Wang and C.W. Chu, Phys Rev Lett, 58, 908 (1987).
3. P.M. Grant, S.S.P. Parkin, V.Y. Lee, E.M. Engler, M.L. Ramirez, J.E. Vazquez, G. Lim, R.D. Jacowitz and R.L. Greene, Phys Rev Lett, June 1987 (to be published).
4. S. Uchida, H. Takagi, K. Kitazawa and S. Tanaka, Jpn J Appl Phys Lett 26,L1,(1987).
5. C.W. Chu, P.H. Hor, R.L. Meng, L. Gao, Z.J Huang, and Y.Q. Wang, Phys Rev Lett, 58, 405 (1987).
6. D.U. Gubser, R.A. Hein, S.H. Lawrence, M.S. Osofsky, D.J. Schrodt, L.E. Toth and S.A. Wolf, Phys Rev Lett, B 35, 5350 (1987) and references therein.
7. P.H. Hor, R.L. Meng, Y.Q. Wang, L. Gao, Z.J. Huang, J. Bechtold, K. Forster and C.W. Chu, Phys Rev Lett, 58, 1891 (1987)
8. D.W. Murphy, S. Sunshine, R.B. VanDover, R.J. Cava, B.Batlogg, S.M. Zahurak and L.F. Schneemeyer, Phys Rev Lett, 58, 1888 (1987).
9. D.B. Mitzi, A.F. Marshall, J.Z. Sun, D.J. Webb, M.R. Beasley, T.H. Geballe and A. Kapitulnik (submitted to Phys Rev).
10. Shiou-Jyh Hwu, S.M. Song, J. Thiel, K.R. Poeppelmeier, J.B. Ketterson and A.J. Freeman, Phys Rev Lett, B 35, 7119 (1987).
11. A. F. Wells, Structural Inorganic Chemistry 5th Ed. Claredon Press: Oxford, 1984.
12 R. J. Cava, A. Santoro and D.W. Johnson Jr., (to be published).
13. L. Er-Rakho, C. Michel, J. Provost and B. Raveau, J. Solid State Chem, 37, 151 (1981).
14. N. Nguyen, L. Er-Rakho, C. Michel, J. Choisnet and B. Raveau, Mate Res Bull, 15, 891 (1980).
15. M.A. Beno, L. Soderholm, D.W. Capone II, D.G. Hink, J.D. Jorgensen, Ivan K. Schuller C.U. Segre, K. Zhang and J.D. Grace, Appl Phys Lett (to be published).

16. F. Beech, S. Miraglia, A. Santoro and R.S. Roth, (to be published).
17. L.E. Toth, E.F. Skelton, S.A. Wolf, S.B. Qadri, M.S. Osofsky, B.A. Bender, S.H. Lawrence and D.U. Gubser, (submitted for publication to Phys Rev) and also E.F. Skelton, S.B. Qadri, B.A. Bender, A.S. Edelstein, W.T. Elam, T.L. Francavilla, D.U. Gubser, R.L. Holtz, S.H. Lawrence, M.S. Osofsky, L.E. Toth and S.A. Wolf, MRS Proceedings, California 1987 (to be published).
18. Unpublished result, NRL.
19. J.P. Kirkland, R.A. Neiser, H. Herman and W.T. Elam, (unpublished).
20. C. Michel and B. Raveau, Rev Chim Minerals, 21, 407 (1987).
21. Ivan K. Schuller, D.G. Hinks, M.A. Beno, D.W.Capone II, L. Soderholm, J.-P. Locquet, Y. Bruynseraede, C.U.Segre, and K. Zhang, (submitted for publication, Solid State Commun.)
22. Unpublished result, NRL.
23. T.L. Aselage, B.C. Bunker, D.H. Doughty, M.O. Eatough, W.F. Hammetter, K.D. Keefer, R.E. Loehman, B. Morosin, E.L. Venturini and J.A. Voigt, MRS Proceedings, California 1987 (to be published).
24. Sung-Ik Lee, John P. Golben, Sang Y. Lee, Xuao-Dong Chen, Yi-Song, Tae W. Noh, R.D. McMichael, Yve Cao, Joe Testa, Fulin Zuo, J. R. Gaines, A.J. Epstein, D.L Cox, J.C. Garland, T.R. Lemberger, R. Sooryakumar, Bruce R. Patton and Rodney T. Tettenhorst, MRS Proceedings, California 1987 (to be published).
25. J.M. Tarascon, L.H. Greene, W.R. Mckinnon, G.W. Hull and T.H. Geballe, Science, 235, 1373 (1987).
26. P.M. Grant, S.S.P. Parkin, V.Y. Lee, E.M. Engler, M.L. Ramirez, J.E. Vazquez, G. Lim, R.D. Jacowitz and R.L. Greene, Phys. Rev. Lett. (to be published).
27. E.F. Skelton, W.T. Elam, D.U. Gubser, V. Letourneau, M.S. Osofsky, S.B. Qadri, L.E. Toth and S.A. Wolf, (submitted Phy. Rev. Lett.).
28. R.V. Kasowski, W.Y. Hsu and F. Herman, Solid State Commun., (to be published).

RECEIVED July 6, 1987

Chapter 23

Relationship of Electrical, Magnetic, and Mechanical Properties to Processing in High-Temperature Superconductors

J. E. Blendell, C. K. Chiang, D. C. Cranmer, S. W. Freiman, E. R. Fuller, Jr.,
E. Drescher-Krasicka, Ward L. Johnson, H. M. Ledbetter, L. H. Bennett,
L. J. Swartzendruber, R. B. Marinenko, R. L. Myklebust, D. S. Bright,
and D. E. Newbury

National Bureau of Standards, Gaithersburg, MD 20899

The interrelation between processing, microstructure, and properties is an important factor in understanding the behavior of ceramic materials. This type of understanding will be particularly important in the development of the new high T_c superconducting ceramic oxides of the type $Ba_2YCu_3O_{7-x}$. As an initial effort in understanding these relations, a number of properties have been measured for these superconducting ceramics and related to their microstructure and processing sequence. The $Ba_2YCu_3O_{7-x}$ ceramics were prepared by powder processing techniques, followed by dry pressing and sintering in both air and flowing oxygen at various temperatures. The sintered bodies were annealed at various temperatures and environments. Superconducting properties, such as the transition temperature and the width of the transition, were measured by both electrical conductivity and AC magnetic susceptibility; both of these properties show a strong sensitivity to annealing temperature and atmosphere. The microstructure and density were also strongly dependent on processing conditions. In this regard, compositional mapping proved to be an important technique for quantifying microstructural variations. Mechanical properties, such as elastic modulus, hardness, and fracture toughness, which will be important for the reliable use of these materials in large scale structures, were also determined.

The discovery of the new high T_c, superconducting ceramics has generated excitement among physicists and material scientists alike. It is important to recognize that these materials are very similar to other advanced ceramics in that they are prepared from oxide powders which are pressed and sintered. They are polycrystalline materials whose electrical, magnetic, and mechanical properties will be critically dependent on their chemistry and microstructure.   Grain boundary chemistry and impurity phases may be the dominant factor in determining their current carrying capability as well as their resistance to fracture.   The objective of this work has been to determine the relationship between processing conditions, e.g., sintering and annealing temperatures and atmospheres, and the properties of $Ba_2YCu_3O_{7-x}$ ceramics through a careful characterization of their composition and microstructure.   The position and width of the superconducting transition temperature has been used as a primary measure of the effect of processing and microstructure variations. However, we also recognize that if these ceramics are to be incorporated into large scale structures, their mechanical properties, e.g., strength, are an important consideration.

Experimental Procedure

<u>Sample Preparation</u>.   Starting powders were mixtures of $BaCO_3$, Y_2O_3 and CuO.   The CuO was produced by heating basic copper carbonate at 500°C for at least 12 hours.   The powders were mechanically mixed, pressed into   large pellets, and   reacted in   air at 900°C for 24 hours.   During this firing, the pellets were placed on setters of the same composition in order to eliminate reactions with crucible materials.   (The Ba-Y-Cu mixtures were observed to react with platinum, gold, fused quartz, alumina, zirconia and magnesia crucibles.)   After heating, the pellets were ground by hand in an alumina mortar and pestle, repressed into new pellets and heated in air at 900°C for at least 24 hours.   After the above heat treatment, only single phase $Ba_2YCu_3O_{7-x}$ was observed by X-ray diffraction (Wong-Ng, W.; McMurdie, H. F.; Paretzkin, B.; Zhang, Y.; Davis K. L.; Hubbard, C. R.; Robbins, C. R.; Dragoo, A. L.; Stewart, J. M <u>Powd. Diff</u>. <u>2</u>, No.2, in press.).   This of course does not mean that other crystalline phases in trace quantities, or even amorphous phases, were not present.

The ground powders were milled in a vibratory rod mill using agate rods and acetone in a polyethylene jar.   The powder was dried in air at 200°C overnight.   The particle size after milling was observed to be a strong function of the calcination conditions. The powder was pressed into disks   at a pressure of 13 MPa and sintered under a variety of conditions.

Samples were sintered in air, flowing O_2, or flowing argon at temperatures between 880°C and 950°C (±10°C).   These conditions were chosen because after a sinter at 880°C, samples were strong enough to be cut and handled, while temperatures above 950°C were not used to avoid the incongruent melting point for $Ba_2YCu_3O_{7-x}$ of 1010°C (Roth, R. S; Davis, K. L.; Dennis, J. R. <u>Adv. Ceram. Matls</u>., in press.).   For the sintering and microstructural studies, the general procedure was to sinter a large pellet (prepared from a single batch of powder) at 880°C for 24 hours in air, cut it into bars, and subsequently sinter the bars at different temperatures, followed by an anneal at 600°C in air or oxygen and slowly cooling to room temperature (3 hours).

Shrinkage of the samples during sintering was measured using a dual-push-rod dilatometer with alumina as a reference material. The thermal expansion coefficient was determined from the slope of the length change versus temperature during cooling. The dilatometer samples were subsequently cut up and used for measurement of the strength and toughness. No variability in shrinkage was observed from batch to batch.

Microstructural Analysis. After sintering, the samples were examined by SEM. Both fired and fracture surfaces were examined uncoated. Samples were also mounted in epoxy and polished with diamond paste to 1μm. These samples were then used for compositional mapping or coated with Au and examined in the SEM.

Compositional Mapping. A new technique of compositional mapping by electron probe microanalysis (1) has been applied to characterize the compositional microstructure of the samples. Electron probe compositional mapping involves the use of computer-aided microscopy for the construction of images in which the displayed gray scale is related to the true composition of the specimen and not merely to X-ray intensity.

In this analysis, three wavelength-dispersive X-ray spectrometers were used to simultaneously measure the characteristic X-ray intensities for copper, barium, and yttrium, at each point in the scan, producing two-dimensional X-ray intensity arrays. Complete quantitative analysis corrections, using the NBS theoretical matrix correction procedure FRAME (2), were performed at each picture element (pixel) in the image scan.

In order to utilize the information available in the compositional maps, extensive use is made of computer image processing to modify the displayed intensity scale in order to improve the visibility of structures. Image processing techniques employed in the present work include contrast enhancement by modification of the input-display intensity relationship, progressive thresholding, Lapacian two-dimensional spatial derivatives, and segmentation of the image using pixel connectivity.

Magnetic Susceptibility. A small piece of each sample ($\approx$ 10mm by 2mm by 4mm) was mounted on a probe in an orientation to give the minimum possible demagnetizing factor, and its AC susceptibility measured as a function of temperature in a Hartshorn type bridge circuit at a frequency of 1.68 kHz. Temperature was measured to $\pm$2 K using a copper-constantan thermocouple. The thermocouple was in direct contact with the sample and gave no interfering susceptibility signal above 50 K. The AC signal amplitude was approximately 0.5 gauss. Both the real and imaginary parts of the susceptibility were measured. Only the real part of the signal, which arises from the diamagnetism of the sample due to the superconductivity (Meissner effect), is reported here. The susceptibility bridge was calibrated using manganese fluoride (NBS Standard Reference Material 766). The susceptibility plots have been arbitrarily normalized such that a factor of -1 represents a measured susceptibility of -0.023 emu/g. (With the actual porosities and shapes, this value of susceptibility would correspond roughly to a complete Meissner effect.) A completely superconducting sample with no demagnetizing factor and a theoretical density of 6.38 g/cc would give a susceptibility of 0.013 emu/g, with any demagnetizing factor or sample porosity tending to make this number more negative.

<u>Electrical Resistivity</u>.   The electrical resistivity of these materials was measured from room temperature to 15 K at 5 K intervals on bars, 6-10 mm long and 1 mm square, using a DC four-probe method.   The electrical contacts were made with fine platinum wire and silver paste.   A constant current of ≤10 mA was used.   The voltage difference was measured to 0.1 μV.   The temperature of the specimen was maintained by a closed-cycle cryogenic system and measured using a calibrated silicon diode.

<u>Mechanical Properties</u>.   There are a number of mechanical properties that will be important for the use of these superconducting ceramics in large scale structures.   These properties include hardness, elastic modulus, fracture toughness, strength, and susceptibility to environmentally enhanced fracture.   While critical fracture toughness, K_{IC}, gives a measure of the resistance of the material to large crack extension, the strength, which is determined by the growth of small flaws, 20 to 100 microns in size, will have a greater impact on engineering design.   Critical fracture toughness can be determined by two different procedures.

In the first procedure (3), a Vickers hardness indenter is used to place a controlled flaw into the polished surface of the ceramic.   The size of the cracks emanating from the hardness impression are measured on the surface for a particular indentation load, P. K_{IC} is given by the expression:

$$K_{IC} = \oint (E/H)^{\frac{1}{2}} (P/C_0^{3/2}) \qquad (1)$$

where E is Young's modulus, H is the hardness, C_0 is the crack size, and $\oint$ is an empirical constant established experimentally to be 0.016 (3).   The primary advantage of this procedure is the minimal amount of material required to perform the test.   In addition, the relative effects of external environments in promoting crack growth can be determined by indenting through the particular liquid.

The second indentation procedure involves placing an indentation in a bar or plate of material and subsequently fracturing the specimen in flexure (4).   The cracks extending from the hardness impression perpendicular to the applied tensile stress will grow to failure.   One of the advantages of the indentation-strength technique compared to the procedure described in the previous paragraph is that it is not necessary to measure a crack length, nor extensively polish the specimen surface, making specimen preparation much easier and reducing many of the uncertainties associated with direct crack measurement.   However, one must be sure that there are no residual surface stresses in the specimen, introduced during firing or machining, since these will add to the driving force for crack growth.   The governing equation for fracture is:

$$K_{IC} = \eta (E/H)^{1/8} (\sigma P^{1/3})^{3/4} \qquad (2)$$

where σ is the fracture stress, η is an empirical constant (= 0.59)([4]), and the other parameters are defined as before. Based on Equation 2, if K_{IC} is a constant for a material, then plotting the fracture stress versus the indentation load on a logarithmic scale should yield a straight line of slope -1/3. Deviations from this relationship at small indentation loads give information regarding the direct effect of the microstructure on the fracture process. Values of K_{IC} determined by this technique in the large indentation load regime agree well with those obtained by other fracture mechanics procedures ([4]).

The Vickers hardness of the ceramic was measured in air on a polished section at indentation loads of 5 and 10 Newtons. Indentation crack length measurements were made in air and water on the same section. The crack lengths introduced in air were used to calculate K_{IC} by Equation (1).

In addition, flexural bars, 1 mm x 1 mm x 19 mm, were diamond sawn from a billet. The edges of each bar were rounded to reduce the incidence of edge failures. Each bar was indented using a Vickers diamond at a load of 10 N. The indented bars were placed on a 3 point bend fixture (span = 10 mm.) with the indent on the tensile side of the bar, directly under the loading pin. The bars were fractured using a universal test machine at either ambient temperature in oil or immersed in liquid nitrogen, at a crosshead speed of 5 mm/min. K_{IC} was calculated from the fracture strength using Equation (2) per the method reported by Chantikul et al ([4]).

Unindented bars, sintered and annealed under various conditions, were also fractured in 3 point bending (span = 8 mm). A drop of oil was placed on the tensile side of the bar and the load ramped rapidly to failure in order to minimize environmental effects.

Polycrystalline, adiabatic, elastic moduli for the $Ba_2YCu_3O_{7-x}$ superconducting ceramics were evaluated from measurements of the ultrasonic wave speeds in these materials. The ultrasonic technique used for these measurements is described in detail elsewhere ([5]); specifics of the current procedure are briefly described here. Ultrasonic elastic moduli, C, are determined from the general relation:

$$C = \rho W^2 \qquad (3)$$

where ρ is the mass density and W is the ultrasonic wave speed. The wave speed is determined from the time required for an ultrasonic pulse to traverse the specimen, reflect from the back surface and return, the so-called pulse-echo technique. For the current studies the ceramic samples were cylindrical with flat and parallel opposing faces. Two specimens of differing mass densities were examined: 5.56 and 4.30 g/cm^3. An ultrasonic pulse of nominally 3 to 10 MHz frequency and 1 μs in duration was launched into the sample by means of an ultrasonic transducer bonded onto the flat surface. Two wave polarizations were used: longitudinal and shear. For the longitudinal waves with the higher density sample, a 10 MHz PZT transducer operating at 9.0 MHz was used. The transducer was bonded to the sample with stop-cock grease

("NoNaq"). The shear wave transducer for this sample was a 10 MHz lithium niobate transducer bonded with a cyanoacrylate polyester. For the lower density sample both PZT and quartz transducers operating at nominally 3 to 4 MHz were used for both longitudinal and shear wave propagation. They were bonded with either an alcohol-glycerine-based couplent or phenyl salicylate. The detected ultrasonic echoes were rectified and displayed on an oscilloscope equipped with time-delay circuitry for the transit time measurements.

Results and Discussion

Densification. For structural applications in particular, it will be extremely important that these superconducting ceramics be prepared to have as close to theoretical density (6.38 g/cc) as possible. Results to date show that samples sintered at 950°C in air are more dense than those sintered at the same temperature in oxygen, even though the grain sizes are the same. Densities determined by either Archimedes method or by geometrical measurements were lower than those determined by point counts on polished surfaces. For instance, a sample that was sintered in air at 950°C was determined to be 92.5% (±1%) dense by point counting, 87% (±1%) dense by geometrical measurement and 90% (±2%) by Archimedes method. The large cracks that can be seen in Figure 1 are the cause of some of these differences, and preclude accurate measurement of the density except by point counting. Point counting will tend to give a high estimate of the density if there are large porous regions, as these will be difficult to polish. Density measurements have not been made on all samples, but the highest densities, obtained on samples sintered at 950°C in air, are estimated to be above 95% of theoretical. Other samples sintered under nearly identical conditions were only ≈ 80% dense. The highest densities were obtained from powders that were initially sub-micron in size after grinding. Other powders had a much larger initial particle size, which resulted in a low sintered sample density. This result makes it clear that it is necessary to start with a very fine powder in order to produce high density parts.

Figures 2 and 3 show that very little densification occurs at 900°C in either air or oxygen. However, in air more densification occurs (Figure 2) than in oxygen (Figure 3), and there is a greater extent of grain growth as well, with the grains being at least two times larger than the initial particle size. Pores trapped inside grains, pores on two and three grain junctions and large pores due to poor packing can be seen in Figures 1c and 1d. Also, some large pores that are being entrapped have been identified (A in Figure 1c). The trapped pores tend to be large and about the same size as the pores due to packing defects and also tend to be located inside large grains. The pores on boundaries are much smaller and are mostly present at triple junctions.

The results of the shrinkage measurements, shown in Figures 4 and 5, also show the retardation of sintering in oxygen. Up to 300°C there is an apparent expansion of the samples due to the thermal expansion difference between $Ba_2YCu_3O_{7-x}$ and alumina. The

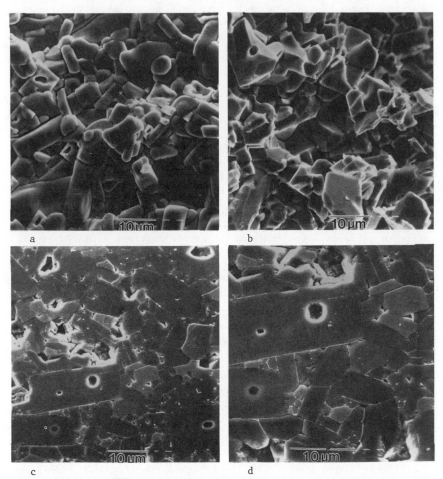

Figure 1.   Sample sintered in air at 950°C: a) as-sintered
surface: b) fracture surface; c) and d) polished surface. The
as-sintered surface shows extensive faceting. Elongated grains
can be seen along with large trapped pores on the polished
surfaces. A second phase along the interfaces can also be seen.

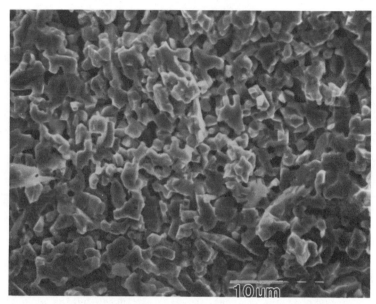

Figure 2. Fracture surface of a sample sintered in air at
900°C. Little densification has occurred, but the particles have
grown to about two times the initial powder particle size and
have become more rounded.

Figure 3. Fracture surface of a sample sintered in oxygen at
900°C. Very little densification or grain growth has occurred.

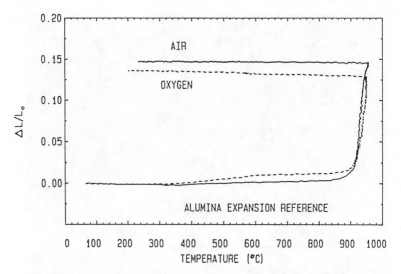

Figure 4.   Relative shrinkage ($\Delta l/l_o$) as a function of
temperature.   Alumina was used as a reference to compensate for
the expansion of the system.   In air, shrinkage begins at a lower
temperature and is more rapid than in oxygen.

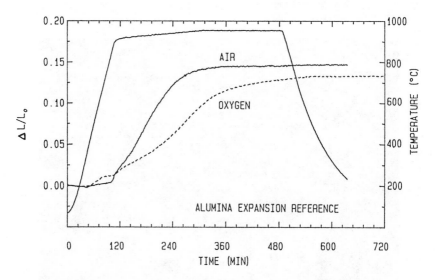

Figure 5.   Relative shrinkage ($\Delta l/l_o$) as a function of time.
Also shown is the temperature cycle used.

shrinkage of both samples between 300°C and 900°C does not represent densification. Shrinkage could be caused by rearrangement of grains under the push rod load (20 gm), due to coarsening by evaporation/condensation or to the presence of a low melting phase. It has been observed that $Ba_2YCu_3O_{7-x}$ gains weight in this temperature range during heating (Gallagher, P. K.; O'Bryan, H. M.; Sunshine, S. A.; Murphy, D. W. Matl. Res. Bull., in press.)(Chiang, C. K.; Cook, L. P.; Chang, S. S.; Blendell, J. E.; Roth, R. S. Adv. Ceram. Matls., in press.) (Cook, L. P.; Chiang, C. K.; Wong-Ng, W.; Blendell, J. E. Adv. Ceram. Matls., in press.), but differential scanning calorimetry results (Chiang, C. K.; Cook, L. P.; Chang, S. S.; Blendell, J. E.; Roth, R. S. Adv. Ceram. Matls., in press.) offer no evidence of a phase transformation below 900°C. We do not feel that there is any liquid present at these temperatures to account for this low temperature shrinkage. It seems more likely that the observed shrinkage occurs by surface transport followed by rearrangement to fill some of the void space. Densification is more rapid and starts at a slightly lower temperature (850°C) in air than in oxygen (900°C). Shrinkage had stopped after 3 hours at 950°C in air, while in oxygen shrinkage was still occurring, suggesting that the optimum sintering temperature in oxygen may lie higher than 950°C.

Thermal Expansion Coefficient.   The thermal expansion coefficient, as determined from the curves shown in Figures 4 and 5, is different for samples sintered in oxygen and air.   In air, $\alpha = 11.8 \times 10^{-6}/°C$, while in oxygen, $\alpha = 19.7 \times 10^{-6}/°C$.   This difference is not fully understood, but might be related to a greater extent of microcracking which occurs during cooling in air annealed samples.   The grain sizes of the samples sintered in air and oxygen were nominally the same, so this difference in microcracking is not due to a difference in the thermal expansion anisotropy stresses (6).   The measured expansion coefficients are relatively high for a ceramic, and thus thermal shock may be a real problem.   We have observed catastrophic failure of samples while removing them from liquid nitrogen and dipping them into alcohol at room temperature.

Microstructural Evolution.   Typical microstructures obtained for these materials sintered under a variety of conditions are shown in Figures 1-3 and 6-7.   Note in general that the fired surfaces of the grains are very faceted, indicating that there is a very strong orientation dependence of the surface energy.   On a large number of samples a grain boundary film was observed.   Even when this film was not observed, the presence of elongated grains with apparent 180° grain boundary dihedral angles indicates that a grain boundary film is present (7).   It has been suggested (Ekin, J. W.; Panson, A. J.; Braginski, A. I.; Janocko, M. A.; Hong, M.; Kwo, J.; Liou, S. H.; Capone, D. W.; and Flandermeyer, B. Materials Res. Soc., in press) that it is the presence of such films which limits the critical current density in these superconducting ceramics.

As shown in Figure 7, sintering in argon at 900°C produced a substantial quantity of liquid phase.   These samples were not superconducting at temperatures above 50 K, in agreement with

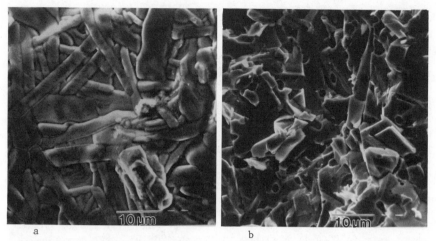

a   b

Figure 6.   Sample sintered in oxygen at 950°C: a) as-sintered
surface, b) fracture surface. Elongated grains and a second
phase can be seen.

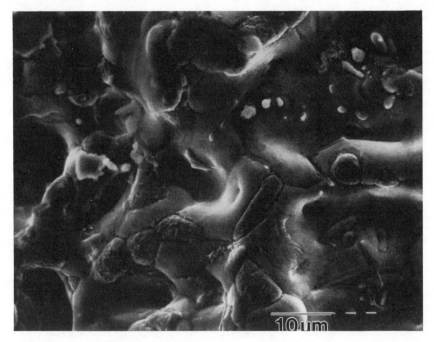

Figure 7.   As-sintered surface of a sample sintered in argon at
900°C. Extensive melting has occurred.

Gallagher et al (Gallagher, P. K.; O'Bryan, H. M.; Sunshine, S. A.; Murphy, D. W. Matl. Res. Bull., in press.).

When a grain becomes significantly larger than the average size of the matrix grains, it develops a much higher driving force for migration, thus increasing its rate of growth. The ability to eliminate pores from the material will be sensitive to the growth of these grains. Since large pores are less mobile, they will become separated from the grain boundaries while the smaller, more mobile pores will remain attached. Once trapped, pores shrink at a much lower rate than the ones that remain on grain boundaries. In these samples, the trapped pores represent only about half of the total porosity. The rest are due to packing defects that can be eliminated by better processing conditions such as a more uniform particle size distribution.

Compositional Maps. The compositional maps and corresponding SEM images shown in Figure 8 indicate that these materials are not homogeneous. In general, the composition appears to be uniform within the grains, but there are regions that are quite different from the average composition. Copper-rich and barium-rich inclusions were found. However, since the grain size is five times larger than the initial particle size, these do not represent unreacted material, but rather are due to segregation during sintering. Also, the phase diagram for this system (Roth, R. S; Davis, K. L.; Dennis, J. R. Adv. Ceram. Matls., in press.) indicates that the $Ba_2YCu_3O_{7-x}$ composition lies very near the liquidus at 950°C. Slight variations in composition, atmosphere or temperature could cause liquid formation. This compositional mapping technique allows the detection of much smaller amounts of a second phase than X-ray diffraction because of the spatial resolution of less than $4 \mu m^2$.

The most notable feature in these maps is the loss of yttrium in the regions near the pores. There are regions that show a yttrium loss, but no changes in the barium or copper concentrations (A in Figure 8). Either an impurity is present or else there is a mixture of phases in this location such that the Ba/Cu ratio is not changed. Based on the phase diagram, this ratio would correspond to the presence of CuO and $BaCuO_2$. To compensate, the yttrium would be found in a region containing the BaY_2CuO_5 phase (the "green" phase).

A barium rich region corresponding to a yttrium poor region can also be seen (B in Figure 8). This does not correspond to any feature in the SEM image. This could be an indication of a small melted region. There is another region surrounding a pore (C in Figure 8) that is barium rich, but there is no change in either the copper or yttrium concentrations. Compositional maps in conjunction with SEM images can yield important information about the chemistry and nature of the grain boundary films. In these samples it can be seen that the films appear to be yttrium poor, but more detailed examinations of the grain boundary regions, including TEM are needed to be more certain.

Compositional mapping procedures were also used to give an upper limit on the density. The threshold level on a composite image was set to a low level, and the area of the sample with

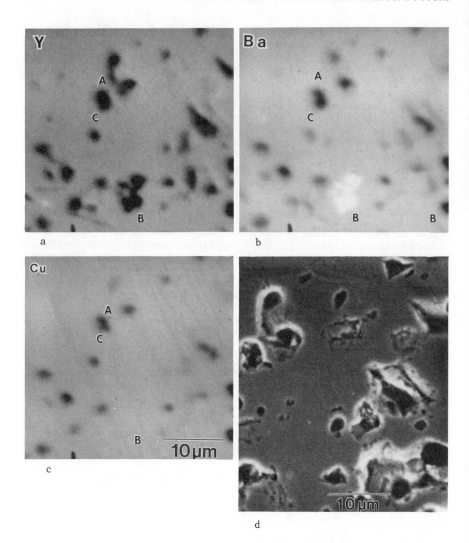

Figure 8.   Electron microprobe compositional maps and corresponding SEM image: a) yttrium, b) barium, c) copper, d) SEM image.   Region A shows a yttrium loss but no copper or barium enhancement.   Region B shows a yttrium poor region corresponding to a barium rich region with the copper concentration unchanged. Region C shows an enhancement of barium around a pore with no changes in copper or yttrium levels.

intensities above this level was determined. This gave a density of 97%. Because this method is not sensitive to small pores (which would appear as slight depressions in the surface by SEM) this is an upper limit. Nevertheless, this density is in good agreement with the point counting data.

Magnetic Susceptibility and Electrical Resistivity. The importance of annealing in the temperature range of 450 to 600°C in order to achieve the desirable superconducting properties is demonstrated in Figure 9. It was shown by Chiang et al (Chiang, C. K.; Cook, L. P.; Chang, S. S.; Blendell, J. E.; Roth, R. S. Adv. Ceram. Matls., in press.) through thermogravimetric analysis that maximum oxygen content is achieved in this temperature range. It is clearly seen in Figure 10 that annealing at low temperatures is necessary to increase the superconducting fraction in the sample.

As seen in Figure 10 the electrical resistivity of samples processed the same as those used for the susceptibility measurements shows a similar dependence on annealing conditions. It can be seen that the sample sintered in air at 950°C (which was not initially superconducting), upon annealing in air at 750°C became superconducting (resistivity = 0) below 50 K. However, the magnetic measurements show that less than 20% of the sample is superconducting at 50 K. Subsequent annealing at 600°C in oxygen increased the superconducting fraction at 50 K to approximately 80%, and also sharpened the transition.

A compilation of susceptibility data as a function of sintering and annealing conditions is presented in Figure 11. This data show that samples sintered and annealed in air do not contain as large a superconducting fraction as samples sintered in oxygen. It is possible to make samples sintered in air more superconducting by anneals in oxygen at low temperatures, but the superconducting transition remains broad despite the anneal. This is probably due to incomplete oxidation, but without longer anneals we can not determine whether this is a kinetic or a thermodynamic effect. The weight change measurements (Chiang, C. K.; Cook, L. P.; Chang, S. S.; Blendell, J. E.; Roth, R. S. Adv. Ceram. Matls., in press.)(Cook, L. P.; Chiang, C. K.; Wong-Ng, W.; Blendell, J. E. Adv. Ceram. Matls., in press.) indicate that oxygen diffusion is rapid in this temperature range and should not be a limiting factor. The fact that samples sintered in oxygen at 950°C do not contain as large a superconducting fraction as those sintered at 900°C also in oxygen may be an indication of the presence of a thermodynamic effect which inhibits reoxidation at the lower temperatures.

Figure 12 shows the resistivity as a function of temperature for a series of samples that were sintered in air at 950°C and then annealed in oxygen at various temperatures. It can be seen that both the room temperature resistivity decreases and T_c increases as the annealing temperature decreases.

Mechanical Properties. The hardness of the $Ba_2YCu_3O_{7-x}$ ceramic and its room temperature fracture toughness as determined in air by the indentation crack length procedure are given in Table I. The value of Young's modulus used in both this calculation as well as the

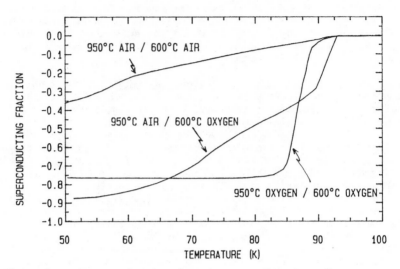

Figure 9.   Superconducting fraction as a function of temperature
as determined by AC magnetic susceptibility for the same sample
sintered in air at 950°C, annealed at 750°C in air and then
reannealed at 600°C in oxygen.

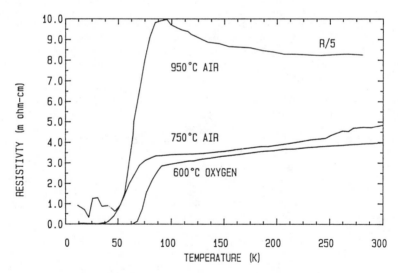

Figure 10.   DC electrical resistivity as a function of
temperature for samples identical to the one used in Figure 9.
The curve for the 950°C data is plotted as R/5 so as to fit on
the same scale.

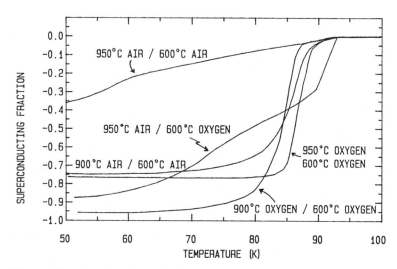

Figure 11. Superconducting fraction as a function of temperature as determined by AC magnetic susceptibility for samples sintered at various temperatures and annealed in air and oxygen.

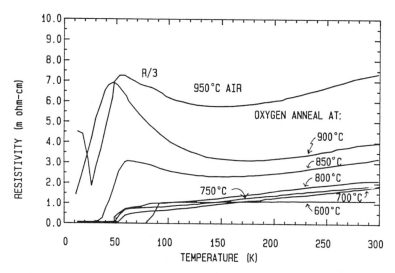

Figure 12. DC electrical resistivity as a function of temperature for samples sintered in air and annealed in oxygen at different temperatures. The curve for the sintered sample is plotted as R/3 so as to fit on the same scale.

subsequent indentation strength calculation was 130 GPa. As will
be seen later, there is a question regarding the absolute value of
E for these materials; we chose a value which would seem to apply
to a partially dense body. Different values of E would change the
value of K_{IC} determined through Equation 1 slightly, but would have
a negligible effect on the toughness determined by the indentation-
strength method (Equation 2). The relatively low value of hardness
is partially ascribed to the fact that this material has a low
density, $\approx 90\%$ of theoretical.

Both the strength and the fracture toughness of the indented
flexural specimens are also given in Table I. There is no
difference in fracture toughness between specimens tested in oil at
room temperature and those fractured in liquid nitrogen. The
values of fracture toughness determined by both indentation
procedures is quite low (comparable to barium titanate, a
polycrystalline ceramic having a similar crystal structure)
suggesting that the brittle failure of large components of this
material will be a definite problem. An additional consideration
is the fact that crack lengths obtained when the material was
indented in water were significantly longer than those produced by
indentation in air, indicating that moisture enhanced crack growth
leading to the delayed failure of components must be taken into
account.

Table I. Hardness and Fracture Toughness of $Ba_2YCu_3O_{7-x}$

Method	Environment	H (GPa)	K_{IC} (MPam$^{1/2}$)
Indentation	Air	3.7 ± .5	1.2
Ind.-Strength	Oil	-	1.0
Ind.-Strength	Liquid N_2	-	1.1

Three-point bend strengths were measured on samples from
another batch of this material, which had been sintered at 960°C in
either air or oxygen, and subsequently annealed at 600°C in either
air or oxygen. Results are given in Table II. The density of this
material was anomalously low for the sintering conditions, i.e., 75
to 80 percent of theoretical, which may affect the measured values
of strength. This low density probably resulted from a coarser
starting powder. On a relative basis, however, the data in Table
II suggest that materials sintered in oxygen are stronger than
those sintered in air. The data also strongly suggest that
materials annealed in oxygen after an air sinter will be
significantly weakened. The fact that the K_{IC} values obtained by
the indentation crack length procedure on polished surfaces of
these same materials showed the same trends (Table II) indicates
that the intrinsic properties of the material and not just the
generation of flaws is affected by the processing. Clearly,
however, this whole area of the relationship of mechanical
properties to processing needs much further study.

Table II.   Effects of Processing on Strength and K_{IC}

<u>Sintered in Air - 960° C</u>

As Sintered, σ = 115 ± 17 MPa (2)
K_{IC} = 1.8 MPam$^{\frac{1}{2}}$

<u>Annealed - 600° C</u>

| <u>Air</u> | <u>O$_2$</u> |

σ = 100 ± 25 MPa (6) σ = 59 ± 15 MPa (6)
K_{IC} = 1.5 MPam$^{\frac{1}{2}}$

<u>Sintered in O$_2$ - 960° C</u>

As sintered, σ = 160 ± 1 MPa (2)
K_{IC} = 2.5 MPam$^{\frac{1}{2}}$

<u>Annealed - 600° C</u>

<u>Air</u>   <u>O$_2$</u>

σ = 147 ± 14 MPa (4) σ = 127 ± 16 MPA (4)
K_{IC} = 1.8 MPam$^{\frac{1}{2}}$

(The number in parentheses gives the number of specimens tested.)

The ultrasonic longitudinal and shear wave speeds were measured at room temperature for two fully oxidized samples of $Ba_2YCu_3O_7$, as described above. From these values and the measured mass density, a longitudinal modulus, C_ℓ, and a shear modulus, C_s, can be calculated from Equation 3. The ordinary isotropic elastic moduli are related to these two moduli by:

Shear Modulus:	$G = C_s$		(4)
Poisson's Ratio:	$\nu = (C_\ell - 2C_s)/(2C_\ell - 2C_s)$		(5)
Young's Modulus:	$E = 2G(1 + \nu)$		(6)
Bulk Modulus:	$B = C_\ell - 4C_s/3$		(7)

Since heat does not have a chance to flow during the period of a high-frequency ultrasonic wave, these are adiabatic elastic moduli. Results are tabulated in Table III for both samples.

Porosity is known to decrease the elastic moduli of ceramic materials (8-9). To estimate the magnitude of this effect for the current elastic constant measurements, we have used theories for the elastic modulus decrement due to spherical pores by Mackenzie (10) and Ledbetter and Datta (11) to calculate the "zero-porosity" elastic moduli. Mackenzie relates the bulk modulus, B, and shear modulus, G, to the zero-porosity values, B_o and G_o, respectively, by

$$(B_o - B)/B = p(3B_o + 4G_o)/[4G_o(1-p)] + order(p^3) \qquad (8)$$

and

$$(G_o - G)/G_o = 5p(3B_o + 4G_o)/(9B_o + 8G_o) + order(p^2) \qquad (9)$$

where p is the porosity. The expressions from Ledbetter and Datta are more complicated and are not repeated here, but are more accurate for porosities greater than approximately 10 percent. Hence, the Mackenzie correction is not computed for the lower density sample. Using an X-ray density of 6.38 g/cm^3 for the $Ba_2YCu_3O_7$ superconductor' the calculated porosities for the two samples are 12.9 percent and 32.6 percent, respectively. Assuming that this porosity is in spherical voids, Equations 8 and 9 can be solved in an iterative fashion to give B_o and G_o, from which the other elastic moduli can be calculated from Equations 4-7.   A similar analysis is performed for the theory of Ledbetter and Datta. The results are given in Table III.

Table III. Adiabatic Elastic Properties of $Ba_2YCu_3O_{7-x}$

Property	Measured Value	Corrected for Porosity (10)	Corrected for Porosity (11)
POROSITY: 12.9%			
Longitudinal wave speed	4.87 km/s	5.27 km/s	5.22 km/s
Shear wave speed	2.76 km/s	2.98 km/s	2.92 km/s
Shear Modulus	42.4 GPa	56.6 GPa	54.6 GPa
Poisson's ratio	0.264	0.265	0.271
Young's Modulus	107 GPa	143 GPa	139 GPa
Bulk Modulus	75.5 GPa	102 GPa	101 GPa
POROSITY: 32.6%			
Longitudinal wave speed	3.63 km/s		4.32 km/s
Shear wave speed	2.06 km/s		2.35 km/s
Shear Modulus	18.3 GPa		35.2 GPa
Poisson's ratio	0.263		0.290
Young's Modulus	46.1 GPa		90.8 GPa
Bulk Modulus	32.4 GPa		72.2 GPa

Even after correcting for the porosity, the elastic properties for the two samples do not completely agree, indicating some further dependence on processing conditions. Furthermore, the values for bulk modulus are much smaller than that reported by

Block et al (Block, S.; Piermarini, G. J.; Munro, R. G.; Wong-Ng, W. Adv. Ceram. Matls., in press.) of 196 ± 17 GPa for $Ba_2YCu_3O_{7-x}$. Block et al measured the change in X-ray density with pressure to give the single-crystal isothermal bulk modulus. The correction between isothermal and adiabatic elastic modulus is only on the order of one or two percent and cannot account for the large discrepancy. A possible explanation for the differences might be a further elastic modulus decrement which would result from the presence of microcracks. Such microcracks, while contributing only insignificantly to the pore volume, could have a larger effect on elastic moduli. This possibility is a topic for further investigation.

Conclusions

It is clear that the superconducting properties of $Ba_2YCu_3O_{7-x}$ depend critically on both the sintering and the annealing conditions. Sintering in air led to a greater degree of densification than sintering in oxygen at the same temperature, but also produced samples in which subsequent annealing at 600°C led to a significantly smaller superconducting fraction than the oxygen sintered samples, as determined by AC magnetic susceptibility. Annealing at low temperatures (600°C) in oxygen appears to be necessary to obtain the highest critical temperature (T_c), the sharpest transition, and the largest superconducting fraction. Samples both sintered and annealed in oxygen have the highest superconducting fraction. In addition, it appears that sintering at 950° C is needed to achieve high densities, but leads to poorer superconducting properties than lower temperature sintering conditions. Both the sintering and annealing atmosphere had a marked effect on the strength and fracture toughness of this material as well.

The microstructures of the various materials sintered under different conditions do not vary significantly. Grain sizes are relatively uniform and average about 10 μm. Most samples contain a liquid phase along some of the grain boundaries and are compositionally inhomogeneous. The amount of second phase liquid is small (less than 1%), and the chemical inhomogeneities are located at grain boundaries or pores. The grain boundary films would reduce the area of current flow, but not prevent a continuous path for the current as it is possible to get a zero resistance path with a very small superconducting fraction. Large lath-shaped grains were observed in all samples, and pores were generally trapped inside the large grains. Microcracks were observed on the fired surfaces, and suggested by the elasticity measurements. The thermal expansion coefficient was very high for a ceramic and was different in air and oxygen. This could be due to oxygen uptake during cooling or to microcracking due to thermal expansion anisotropy. Microcracking can also contribute to the low strength, toughness, and elastic moduli for these materials.

In summary, the picture of these superconducting ceramics at this time is one which shows an extremely complex relationship between the various properties and the processing conditions. It is clear that almost every property measured, i.e. magnetic, mechanical, etc., depends strongly on the starting powder, sintering conditions, and thermal history. Much further work will be needed to understand and bring to the manufacturing stage, these interesting and potentially valuable materials.

ACKNOWLEDGMENTS

We would like to acknowledge the support of DARPA (order 6146) and to thank Carol A. Handwerker for a critical review of the manuscript. We also thank Theresa Baker and Grady White for their help in obtaining some of the data.

Literature Cited

1. Marinenko, R. B;   Myklebust, R.   L.;   Bright, D. S.; Newbury, D. E.
   J. Microscopy, 1987, 145, 207-23.
2. Yakowitz, H.; Myklebust, R. L.; Heinrich, K. F. J.   National Bureau of Standards (U.S.) Technical Note 796, Washington, DC, 1973.
3. Anstis, G. R.; Chantikul, P.; Lawn, B. R.; Marshall, D. B. J. Am. Ceram. Soc., 1981, 64, 533-38.
4. Chantikul, P.; Anstis, G. R.; Lawn, B. R.; Marshall, D. B. J. Am. Ceram. Soc., 1981, 64, 539-43.
5. Fuller, Jr., E. R.; Granato, A. V.; Holder, J.; Naimon, E. R. In Methods of Experimental Physics, Vol. 11, Coleman, R.V., Ed.; Academic Press: New York, 1974; Chapter 7.
6. Blendell, J. E.; Coble, R. L.; J. Amer. Ceram. Soc., 1982, 65, 174.
7. Kooy, C. Science of Ceramics, Stewart, G. W., Ed.; Academic Press: New York 1962; Vol. 1, p 21.
8. Wang, J.C. J. Matl. Sci., 1984, 19, 801-14.
9. Phani, K. K.; Niyogi, S. K. J. Matl. Sci., 1987, 22, 257-63.
10. J. K. Mackenzie, Proc. Phys. Soc.(London), 1950, 63B, 2-11.
11. Ledbetter, H. M.; Datta, S. K.; J. Acoust. Soc. Amer., 1986, 79, 239-48.

RECEIVED July 6, 1987

Chapter 24

Fabrication of $YBa_2Cu_3O_7$ Superconducting Ceramics

R. B. Poeppel[1], B. K. Flandermeyer[1], J. T. Dusek[1], and I. D. Bloom[2]

[1]Materials and Components Technology Division, Argonne National Laboratory, Argonne, IL 60439
[2]Chemical Technology Division, Argonne National Laboratory, Argonne, IL 60439

Recent experience concerning the preparation of $YBa_2Cu_3O_7$ powder and its fabrication into a variety of shapes is detailed. The fabrication techniques used were dry pressing, tape casting, and extrusion. Parameters found to affect superconductor quality are noted.

From the beginning, Argonne National Laboratory (ANL) has been a major center for research into the new oxide superconducting ceramics owing largely to the sizable superconductor program funded for some time by DOE Basic Energy Sciences. The expertise developed from this program has allowed ANL to capitalize on these new discoveries and begin a multidisciplinary research effort that has included work on phase diagram development (1), crystal chemistry (Beno, et al., Appl. Phys. Lett., in press), and heat capacity measurements (2) for both the La-Sr-Cu-O and Y-Ba-Cu-O systems. In addition to this basic materials research capability, ANL also possesses the ceramic fabrication expertise necessary to fabricate these compounds. Accordingly, a program to prepare and fabricate relatively large quantities of these materials (approximately 1 kg) has been initiated. This paper will provide an overview of the efforts to fabricate $YBa_2Cu_3O_7$ (hereafter denoted "1-2-3") at ANL. The efforts to fabricate the La-Sr-Cu-O system have been summarized in a recent publication by Wang et al. (3) and will not be discussed here.

Very briefly, the 1-2-3 compound has a T_c of about 90 K and a large anisotropy in current conductivity. The oxygen-defective perovskite crystal structure has the oxygen atoms in the two copper oxide sheets at the center of the unit cell buckling toward the yttrium atom. Oxygen is missing from two sides of the squares at the end of the unit, producing one-dimensional chains of copper and oxygen atoms.

Powder Preparation

The main focus on this effort so far has been on developing reliable procedures for producing the basic 1-2-3 composition with

0097-6156/87/0351-0261$06.00/0

good superconductive characteristics. Although other reaction methods (such as coprecipitation) are possible, only the solid-state reaction method has been used to form the 1-2-3 compound in this laboratory. In this process, $BaCO_3$, Y_2O_3, and CuO were wet milled, and the excess fluid then evaporated off. Other precursor materials such as BaO and Cu_2O have also been used without significant effect on the final results. Next, the raw powder must undergo a calcination procedure, a step which has proved to be surprisingly critical in the production of good superconductive material. Apparently, the $BaCO_3$-CuO system can preferentially form a eutectic at about 875°C, causing a phase separation that inhibits the formation of the desired 1-2-3 compound. Also, there seems to be a difference in relative reaction rates. In binary tests, the reaction of $BaCO_3$ with CuO to form $BaCuO_2$ is much faster than the reaction of Y_2O_3 with either CuO (forming $Y_2Cu_2O_5$) or $BaCO_3$ (forming $Y_2Ba_4O_7$). The difference in reaction rates again leads to phase separation. Improperly calcined material is very hard and shows excessive grain growth. To overcome the problems observed, two empirical procedures have been developed.

The first procedure uses a long-term precalcine (>24 h) at 850°C to decompose the carbonates below the eutectic and form the 1-2-3 compound as the major phase. This material is then lightly ground and final-calcined at 950°C for about 2 h in a laboratory kiln which reaches a temperature in about 15 min.

A second method is a slight variation of the first. The raw powders are taken to a temperature of 950°C for 2 to 6 h in the same quick-firing kiln mentioned above, cooled and reground, and the procedure repeated for a total of three calcinations. The powders are then checked for phase composition by x-ray diffraction (XRD). Chemical analysis was performed using inductively coupled plasma-atomic absorption analysis and has confirmed that the composition of these powders is correct to within the accuracy of the device.

An interesting aspect of both of these procedures is that a fast heat-up rate on the final 950°C calcine appears to be crucial to success. A second puzzling aspect of this system is that if one tries to sinter powders at temperatures of 950 to 975°C which have only been calcined to 850°C, poor superconducting materials will result, even though XRD shows the powder to be pure 1-2-3. It appears that calcination temperatures must be approximately 950°C to ensure good properties even though such calcination temperatures lead to powders that are relatively coarse and difficult to fabricate into good, dense, flaw-free ceramics.

After the 1-2-3 compound has been formed, the particle size must be reduced in order to enhance sintering. The 1-2-3 compound must also be protected from excessive contact with moisture and CO_2 which have been observed to destroy its superconducting properties. One can hypothesize that water and CO_2 are reacting with the Cu(111) centers in the 1-2-3 compound. It has been reported in the literature that Cu(111) in $NaCuO_2$ reacts very quickly with water and to produce Cu(11) hydroxides. The reported half-life for this reaction is on the order of 25 s. The exact reaction of CO_2 is unknown.

Experience has shown that considerable casual contact with the atmosphere can occur before the powder is degraded noticeably and that reasonable precautions are sufficient to allow the material to be handled without need for continuous dry, inert atmospheres.

This moisture sensitivity must also be taken into account when selecting a milling method. Obviously, dry milling is best for minimizing moisture degradation; however, this method does not do an efficient job of reducing the particle size. We have found that moderate milling times (2 to 4 h) in methanol, isopropyl alcohol, or xylene are acceptable and do a far better job of giving a relatively fine, uniform, particle size distribution. Direct contact with water is immediately detrimental.

Finally, it appears that the Al$_2$O$_3$ impurity introduced from ball milling with alumina media seems to enhance the formation of the eutectic and is therefore detrimental. Accordingly, only zirconia media are used to mill these powders.

Sample Fabrication

After an acceptable powder has been produced, it must be formed into the desired shape and sintered into a useful, dense ceramic. To date, our approach has been to use typical ceramic fabrication techniques such as dry pressing, tape casting, and extrusion.

The most straightforward method is dry pressing. Here, 2 to 5 wt % of binder and/or plasticizer is added to the powder to enhance particle rearrangement (and thereby, increase the green or unfired density) when uniaxially pressed in a metal die-punch set. It is possible to omit these organic binders if the powder is especially amenable, but the 1-2-3 compound requires such organic pressing aids to achieve high densities. Tape casting and extrusion are similar in that significantly higher quantities of organic binders (15 to 25 wt %) are used to suspend the powder into green shapes that are flexible and easy to handle. Desired cutting and forming operations can then be performed, followed by the sintering step after which the typical, brittle ceramic pieces are obtained. In tape casting, a fluid slip composed of the powder and an appropriate plastic-solvent system is prepared, which is then precisely doctor-bladed out onto a surface. After drying, this film is stripped and stored. In extrusion, a plastic mass is prepared and extruded through a die. For both methods, the object is to intimately disperse the powder into the organics with the general goal of achieving an even, unagglomerated distribution at the lowest binder content possible while still maintaining the desired plasticity. Though other techniques are possible, this objective is usually accomplished by ball milling a slip and then removing the excess solvent if an extrusion body is required.

After forming by any of the above means, the final fabrication step is sintering to form the ceramic. This procedure is done under a flowing oxygen atmosphere to encourage the equilibrium oxygen stoichiometry. A wide variety of firing schedules have been tried, with the following given as a typical example:

```
100°C/h to 500°C
4 h hold
200°C/h to 875°C
4 h hold
200°C/h to 975°C
2 h hold
100°C/h to 400°C
Furnace off
```

Two important observations have been made: (1) the 875°C
pre-anneal has proven to be helpful and (2) a slow cooling period
to allow oxygen equilibration is necessary. It is thought that
the cooling rate in this firing schedule will need to be slowed if
sample size or quantity increases. As mentioned above, the 1-2-3
powder is sensitive to moisture and CO_2. This makes the addition
of large quantities of organics a problem, as during the sintering
step they will decompose to water and CO_2 and tend to degrade the
powder. Clearly, a reduction in their quantity or the substitution of a less pernicious material is a goal to be pursued. Tape
casting is of particular interest to superconductor development
because it provides thin sheets of ceramic that turn out to be a
favorable form from which to construct a superconducting power
transmission line. This laboratory has been successful in
producing tapes that fire to near theoretical density and carry
several hundred amperes per square centimeter at 77 K. Extrusion
of green wires has also proved to be quite successful. Wires as
thin as 150 microns in diameter have been extruded, fired, and
proven to be superconductive with critical currents approximately
equal to that found in the tapes. Of course, these wires are
ceramic and therefore brittle; however, sintered wires this
thickness can already be wrapped into a coil that is about
0.3 meter in diameter. It is hoped that more experience will lead
to continued improvements in wire flexibility that will enhance
their usefulness.
     In regard to the moisture sensitivity of the 1-2-3 material,
it appears that the factors of binder content, milling times, and
atmosphere contact all have their effect and it is the total
exposure which is a major factor in determining the quality of the
superconductor. If milling fluids are used, then binder content
should be minimized. Milling times of 16 h or more have always
given poor results regardless of the fluid used. Tapes allowed to
age in the green state for several days before firing are clearly
degraded in quality, showing that with the 1-2-3 powder, typical
tape casting formulations do not confirm long-term stability.

Testing and Characterization

After fabrication, the samples are tested for the superconductive
transition temperature and critical current density by conventional methods. X-ray diffraction is the technique most consistently used to roughly determine the quality of the powders
prepared. In general, the XRD analysis must be free of second
phases (in particular, unreacted $BaCO_3$) to work well. A program
to examine these materials by electron microscopy is being
pursued. Initial results show considerable unreacted material

even in samples prepared with powders that show no inpurities under XRD analysis. Of course, such micro-inhomogeneity may be one of the undefined variables that cause the superconductive properties to vary unpredictably.

Summary

Techniques for the preparation of 1-2-3 have been determined. Such powders have been used in a variety of fabrication techniques to prepare technically interesting ceramic shapes such as wires and thin sheets. We confirm the tendency of the precursor powders to form the BaCO$_3$-CuO eutectic leading to a phase separation that degrades the desired properties, and the tendency of the 1-2-3 compound to degrade under contact with moisture and CO$_2$.

Future efforts at ANL will be aimed at completing the specification of the necessary processing parameters and in measuring the mechanical properties of such superconducting ceramics.

Acknowledgments

The work was supported by the U. S. Department of Energy, BES-Materials Sciences, under Contract W-31-109-Eng-38.

Literature Cited

1. Hinks, D. G.; Soderholm, L.; Capone, D. W. II; Jorgensen, J. D.; Schuller, Ivan K.; Segre, C. U.; Zhang, K.; Grace, J. D. Appl. Phys. Lett. 1987, 50(23), 1688.
2. Dunlap, B. D.; Nevitt, M. V.; Slaski, M.; Klippert, T. E.; Sungaila, Z.; McKale, A. G.; Capone, D. W.; Poeppel, R. B.; Flandermeyer, B. K. Phys. Rev. B Rapid Commun. 1987, 35(13), 7210.
3. Wang, H. H.; Carlson, K. D.; Geiser, U.; Thorn, R. J.; Kao, K. M.; Beno, M. A.; Monaghan, M. R.; Allen, T. J.; Proksch, R. B.; Stupka, D. L.; Williams, J. M.; Flandermeyer, B. K.; Poeppel, R. B. Inorg. Chem., 1987, 26, 1474.

RECEIVED July 6, 1987

Chapter 25

Processing, Structure, and High-Temperature Superconductivity

E. M. Engler, R. B. Beyers, V. Y. Lee, A. I. Nazzal, G. Lim, S. S. P. Parkin,
P. M. Grant, J. E. Vazquez, M. L. Ramirez, and R. D. Jacowitz

Almaden Research Center, IBM, San Jose, CA 95120

The superconducting behavior in oxygen defect perovskite oxides is found to depend on the amount and order of oxygen in the structure. In the case of $Y_1Ba_2Cu_3O_{9-y}$, the highest and sharpest transitions are related to ordering of one-dimensional Cu-O ribbons in the structure which are in turn coupled to a network of adjacent 2-dimensional Cu-O sheets. The isostructural rare earth derivatives of $Y_1Ba_2Cu_3O_{9-y}$ are found to display similar behavior.

Since the initial discovery by Bednorz and Muller (1) of higher temperature superconductivity, new breakthroughs and advances in this field have continued at a furious pace (2). Within the span of a few months, the superconducting transition temperature (Tc) of 35 K in the layered perovskite $La_{2-x}Ba_xCuO_{4-y}$ jumped to near 100 K in the distorted cubic perovskite $Y_1Ba_2Cu_3O_{9-y}$. Quite early it was recognized that the superconducting properties of $Y_1Ba_2Cu_3O_{9-y}$ depended on the processing conditions (3,4). This can be understood by considering the idealized structure of $Y_1Ba_2Cu_3O_{9-y}$ which is shown in Figure 1. This structure involves an ordering of Ba-Y-Ba in triplets (5) and in principle, can accommodate nine oxygens. Experimentally (4,6,7), however, this value has been determined to be approximately seven. How the oxygens distribute themselves in the structure turns out to be very dependent on processing conditions and is the key to achieving the highest and sharpest superconducting transitions.

Numerous X-ray and neutron diffraction studies (1,8) have determined that the Y plane is vacant, accounting for one oxygen vacancy. The Cu basal plane at the end of the unit cell accounts for the other vacancy which is split between them. It is the ordering of oxygen in these basal planes that appears to be a critical structural feature for high Tc(9). This ordering gives rise to 1-dimensional CuO_2 ribbons along the b-axis. These 1-D ribbons are coupled to two adjacent 2-dimensional networks of Cu-O sheets which lie in the a-b plane of the unit cell (Figure 1).

The ordering of the oxygens in these 1-D ribbons is intimately dependant on the processing of $Y_1Ba_2Cu_3O_{9-y}$. Further, the amount of oxygen can also be varied by different processing conditions. These compounds belong to a more general class of oxygen defect copper perovskites which are well-known to accommodate variable oxygen depending on the temperature at which they are prepared (10).

0097–6156/87/0351–0266$06.00/0
© 1987 American Chemical Society

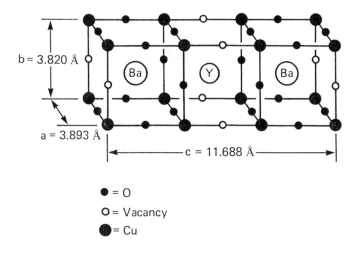

b = 3.820 Å

a = 3.893 Å

c = 11.688 Å

● = O
○ = Vacancy
◐ = Cu

Figure 1. Idealized crystal structure of $Y_1Ba_2Cu_3O_{9-y}$.

Thermal gravimetric analysis (TGA) reveals facile and reversible movement of oxygen in the 500-600 C range for $Y_1Ba_2Cu_3O_{9-y}$ (7,9,11) which decreases with increasing temperature. If samples are cooled too rapidly from high temperature to permit thermodynamic equilibrium to be obtained, then, lower amounts and disordering of oxygen in the structure can occur.

The sensitivity of the superconducting transition to processing is shown in Figure 2 where resistance vs. temperature plots are given for different preparative conditions. The highest and sharpest transition is obtained when the sample is heated in oxygen at 900 C and then allowed to cool slowly to room temperature over about 5 hours (2). More rapid cooling gives superconducting transitions which are considerably depressed and broadened. The partial pressure of oxygen is also important, since even with slow cooling, air annealed samples have lower and broader transitions. These changes in Tc appear to be intrinsic to the electronic structure of $Y_1Ba_2Cu_3O_{9-y}$ and not simply due to changes in the granular nature of the samples. This is seen in magnetic measurements which are plotted in Figure 3 on the same samples shown in Figure 2. The temperature of the diamagnetic shift moves to lower temperatures as would be expected if superconductivity is being changed on a microscopic scale (12). The magnitude of the diamagnetic shift is also considerably reduced.

With rapid cooling, both the amount and ordering of oxygen in the sample decreases in the basal planes. In the X-ray patterns shown in Figure 4, these changes are manifested by a decrease in the orthorhombic character of the unit cell (a≠b). By following diffraction lines near 46-48 °, which are sensitive to changes in the a and b lattice dimensions in the unit cell, one can follow the trend to a more tetragonal unit cell (a=b). When the sample is heated in argon, the unit cell appears tetragonal and an insulating material is obtained. This pattern of

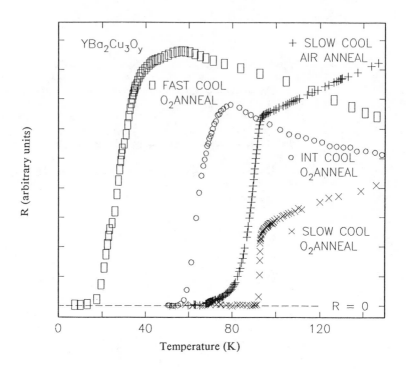

Figure 2. Resistance vs. temperature plots for different preparative conditions.

change in the unit cell is paralleled in high temperature X-ray analysis of $Y_1Ba_2Cu_3O_{9-y}$ which shows the same transformation to a tetragonal unit cell around 700 C (7,9). Further, derivatives of $Y_1Ba_2Cu_3O_{9-y}$, where Y is replaced by rare earth elements, also show similar behavior as illustrated in Figure 5. Again the trend to a more tetragonal unit cell leads to degraded and superconducting behavior. In fact, in the case of $Pr_1Ba_2Cu_3O_{9-y}$, the orthorhombic splittings seen in the Y compound are not evident (4) and this compound displays insulating electrical properties.

The fact that Tc is essentially unchanged by replacement of Y with the magnetic rare earth elements points to the Cu-O sublattice as the source of the superconducting properties of $Y_1Ba_2Cu_3O_{9-y}$. Mixtures of rare earths or alkaline earths (2,13) also do not affect Tc, again emphasizing the key role of the Cu-O sublattice.

The sensitivity of superconductivity to processing appears to be important in another newly discovered perovskite compound. Undoped La_2CuO_{4-y} has been extensively studied since its initial preparation in 1961. However, it is only recently that superconductivity has been found in this material (4-6). In this compound only a small fraction of the sample is actually going superconducting and the structure responsible for this behavior is unknown. The presence of small amounts of unreacted lanthanum oxide in our final product, suggests that some lanthanum deficient phases might be forming. This could give rise to Cu+3, a factor which has been recognized since Bednorz and Muller's initial discovery (1,3,14) as important to superconductivity in these copper oxide compounds.   On standing, samples of La_2CuO_{4-y} slowly lose oxygen, and eventually become non-superconducting which may account for this behavior not being discovered earlier (14).

The field of perovskite superconductors has witnessed an explosion of interest and activity over this past year.   While much remains to be done to better understand these compounds, some general structural and chemical features necessary for the superconducting behavior are beginning to emerge.   These can be summarized as:

1.   The Cu-O sublattice is the source of the superconducting behavior.   The ordering of this sublattice is very dependent on the processing.

2.   The low dimensional electronic structure of the Cu-O sublattice appears to be a general and important feature in all of the perovskite superconductors.

3.   The stabilization of the Cu+3/Cu+2 mixed valent states in the Cu-O sublattice appears to be a key reason why these materials exhibit high temperature superconductivity.

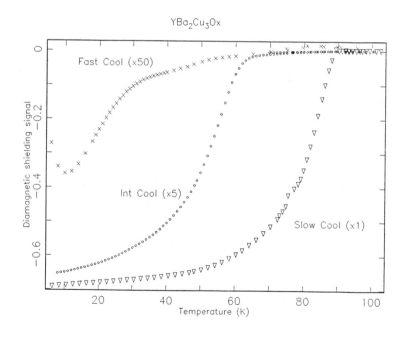

Figure 3.   Magnetic measurements vs. temperature plots for samples shown in Figure 2.

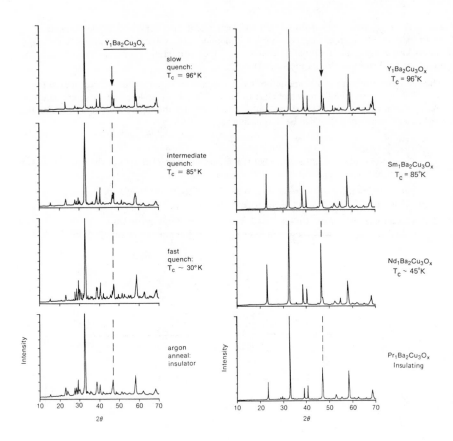

Figure 4. X-ray patterns showing
the effect of rapid cooling on
the amount and ordering of oxygen.

Figure 5. X-ray patterns showing
the effect of replacing Y with
rare earth elements.

Literature Cited

1. Bednorz, J.C. and Muller, K.A. **Z. Phys. B** 1986, 64, 189.
2. For a general overview of this progress see: Engler, E.M. **CHEMTECH** 1987, to be published.
3. Grant, P.M.; Beyers, R.B.; Engler, E.M.; Lim, G.; Parkin, S.S.P.; Ramirez, M.L.; Lee, V.Y.; Nazzal, A.; Vazquez, J.E.; and Savoy, R.J. **Phys. Rev. B1** 1987, 35, 7242.
4. Engler, E. M.; Lee, V.Y.; Nazzal, A.I., Beyers, R.B.; Lim, G.; Grant, P.M.; Parkin, S.S.P.; Ramirez, M.L.; Vazquez, J.E.; Savoy, R.J. **J. Amer. Chem. Soc,** 1987, 109, 2848.
5. Beyers, R.; Lim, G.; Engler, E.M.; Savoy, R.J.; Shaw, T.M.; Dinger, T.R.; Gallagher, W.J.; Sandstrom, R.L. **Appl. Phys Lett.** 1987, xx,xxx.
6. Cava, R.J.; Batlogg, B.; van Dover, R.B.; Murphy, D.W.; Sunshine, S.; Siegrist, T.; Remeika, J.P.; Reitman, E.A.; Zahurak, S.; and Espinosa, G.P. **Phys. Rev. Lett.** 1987, 58, 1676.
7. Gallagher, P.K.; O'Bryan, H.M.; Sunshine, S.A.; and Murphy, D.W. **Mater. Res. Bull,** 1987, xx, xxx.
8. Siegrist, T.; Sunshine, S.; Murphy D.W.; Cava, R.J.; and Zahurak, S.M. **Phys. Rev. B.** 1987, 35, 7137.
9. Beyers, R.; Lim, G.; Engler, E.M.; Lee, V.Y.; Ramirez, M.L.; Savoy, R.J.; Jacowitz, R.D.; Shaw, T.M.; LaPlaca, S.; Bochme, R.; Tsuei, C.C.; Park, S.I.; Gallagher, W.J. and Shafer, M.W. **Appl. Phys. Lett,** 1987, xx, xxx.
10. See for example Michael, C. and Raveau, B. **Revue de Chimie minerale** 1984, 21, 407.
11. Tarascon, J.M.; McKinnon, W.R.; Greene, L.H.; Hull, G.W.; and Vogel, E.M. **Phys.Rev. B** 1987, xx, xxx.
12. Parkin, S.S.P and Engler, E.M. to be published;
13. Willis, J.O.; Fisk, Z.; Thompson, J.D.; Cheong, S-W; Aikin, R.M.; Smith, J.L. and Zirngiebl, E. **J. Magn. Magn. Mat.,** 1987, 67, L139.
14. Grant, P.M.; Parkin, S.S.P.; Lee, V.Y.; Engler, E.M.; Ramirez, M.L.; Vazquez, J.E.; Lim, G.; Jacowitz, R.D.; and Greene, R.L. **Phys. Rev. Lett.** 1987, 58, 2482.
15. Beille, N.J.; Cabanal, R.; Chaillout, C.; Chevallier, B.; Demazeau, G.; Deslandes, F.; Etourneau, J.; Lejay, P.; Michael, C.; Provost, J.; Raveau, B.; Sulpice, A.; Tholence, J.L.; and Tournier, R. **Physique Mat. Condensee** 1987, xx, xxx.
16. Sekizawa, K.; Takano, Y; Takigami, H.; Tasaki, S.; and Inaba, T. **Jpn. J. Appl. Phys.** 1987, 26, L840.

RECEIVED July 6, 1987

Chapter 26

Correlation of Resistance and Thermogravimetric Measurements of the $Er_1Ba_2Cu_3O_{9-\delta}$ Superconductor to Sample Preparation Techniques

Sung-Ik Lee, John P. Golben, Yi Song, Xiao-Dong Chen, R. D. McMichael, and J. R. Gaines

Department of Physics, Ohio State University, Columbus, OH 43210

The resistance dependence and thermogravimetric analysis (TGA) of $Er_1Ba_2Cu_3O_9$ has been measured in the temperature range 27 C to 920 C. The heat treatments and oxygen flow rates simulated actual sintering and annealing processes used in sample preparation. Evidence of a phase transition in $Er_1Ba_2Cu_3O_{9-\delta}$ near 680 C is discussed, as well as the implications of the maximum oxygen uptake near 400 C. The impact of sample preparation procedures on sample features is also discussed.

Recently there has been an accelerated interest in copper oxide superconductors. Higher transition temperatures have been reported for the La-Ba-Cu-O system (1), the Y-Ba-Cu-O system (2), and the Y-Ba-Cu-O-F system (3), respectively. Key factors involved in the synthesis of these materials include the sintering temperature range, the incorporation of oxygen in the annealing or cooling process, and a sufficiently slow cooling rate. In spite of differences in the specific time periods and temperatures used in these synthesis procedures, similar superconducting successes have been reported. In this paper we have measured the resistance dependence and oxygen content of $Er_1Ba_2Cu_3O_9$ and $Y_1Ba_2Cu_3O_{9-\delta}$ samples above room temperature under the conditions of a simulated sintering and annealing process. The results provide insight into the importance of specific sample preparation procedures. We will start with a detailed discussion of the two most common preparation procedures for copper oxide superconductors at this moment; the solid state reaction method and the coprecipitation method. A discussion of the resistance and thermogravimetric analysis (TGA) data and how it correlates to the sample formation will follow.

0097–6156/87/0351–0272$06.00/0

Sample Preparation

The solid state reaction method begins with the oxide or carbonate forms of the individual constituents in stoichiometric ratios.
The mixed powder is pulverized with a mortar and pestle to a uniform
color, and "prebaked" to 900 C for 9 hours to remove water and lib-
erate CO_2. The pulverizing and prebaking procedures are repeated
at least one more time. The actual importance of the prebaking
procedures is as yet unclear. Powder x-ray diffraction data of the
prebaked powder demonstrates the presence of the 1-2-3 phase. It is
believed that the series of prebaking and pulverizing steps aids in
homogenizing the sample and extending the formation of the solid
solution. For 1-2-3 superconductors, the samples are pressed to
15000 p.s.i., sintered at a temperature around 900 C for 12 hrs, and
slowly cooled to 500 C. At this temperature, an oxygen flow of about
2 ft^3/hr is maintained for 12 hours. The furnace is then slowly
cooled to room temperature.

Many different versions of this method have been reported; the
main variable being the length of time used for the various steps.
Superconductors have been synthesized in just a few hours (4) and
certainly the convenience of this method is one reason why it is
predominantly used.

The coprecipitation method starts with the nitrate forms of the
constituents in solution and then precipitates them out in carbonate
form through the addition of Na_2CO_3. At least one group has also
suggested titrating KOH into the solution to suppress the formation
of bicarbonates, thereby retaining the delicate stoichiometric bal-
ence. (5,6) The slow incorporation of Na_2CO_3 results in a uniform
blue precipitate that is free from clumping. After the two solutions
have been mixed for about 45 minutes, the heat and the stirrer are
turned off and the precipitate is allowed to settle for about two
more hours. The precipitate is then dried overnight at a temperature
of about 140 C. The remaining steps for the coprecipitation method
are the same as those used for the solid reaction method.

The advantage of the coprecipitation method is that the constituents are mixed on an atomic scale which helps in the formation of
the desired phase. The nitrate version of this method also tends to
preserve the starting stoichiometric ratio throughout the process.
In contrast we have found that our solid state reacted samples often
have other phases present as observed by x-ray diffraction. These
other phases have appeared in the x-ray spectra presented by other
groups (4,7) and are likely the result of local inhomogeneities in
the sample mixture. The most common extra phases observed are $BaCuO_2$
and CuO.

Discussion of Resistance and TGA Data

What actually occurs during these final preparation procedures
can be inferred from resistance measurements taken on pellets within
the furnace. A DC technique with current reversal was used for these
measurements. Four platinum leads were attached to the pressed pellet sample with silver epoxy. The epoxy was covered with a layer
of protective ceramic paste. (8) To compensate for induced thermal
emf's, an average of the forward and reversed current directions was

used for each measurement. The temperature changes were made in small increments so as to obtain thermal equilibrium at each temperature.

Figure 1 shows the resistance data upon heating to 920 C for $Er_1Ba_2Cu_3O_{9-\delta}$, with data for $Y_1Ba_2Cu_3O_{9-\delta}$ shown for comparison. Initially the resistance is large, but it decreases in general as a function of temperture up to 500 C. A slight dip near 130 C is observed upon heating but is absent in the cooling curve. This dip is present in both curves and is likely due to the curing of the silver epoxy contacts. A very slight bump in the Er data occurs near 360 C as well as a gradual bump from 600 to 800 C. The resistance along the cooling curve (not shown) is considerably smaller than the initial resistance upon heating, however an indication of the resistive bump between 600 C and 800 C still remains.

The general resistive trend from room temperature to 500 C is most likely an effect of the sample grain formation. We mentioned previously that the superconducting phase is present in the prebaked powder. However after being pulverized and pressed, the unsintered pellet contains a large number of grain boundaries. As the temperature is increased, the grains grow by merging with other grains. This results in the formation of more conductive paths as well as reducing the number of high resistance grain boundaries. This explanation is supported by scanning electron microscope (SEM) pictures. After the extended grain connections are formed and the sample is subsequently cooled, the sample retains its high conductivity (comparable to a dirty metal).

The grain sizes present in the sample may also explain why the resistance of the $Er_1Ba_2Cu_3O_{9-\delta}$ sample is about an order of magnitude less than the $Y_1Ba_2Cu_3O_{9-\delta}$ sample. SEM photographs reveal that the $Er_1Ba_2Cu_3O_{9-\delta}$ grains are much larger than the $Y_1Ba_2Cu_3O_{9-\delta}$ grains present in our samples. Grain size may have also some influence on the oxygen access to the sample which is believed to be inherently important to the electrical properties of these materials. We will discuss the oxygen interaction next.

Oxygen deficiencies in perovskite compounds have been correlated with conducting and superconducting behavior. (9) Full oxygen occupancy generally results in insulators. For the 1-2-3 compounds, a full oxygen occupancy is unfavored due to the limits set by the Cu atom valences. On the other hand, a severe oxygen deficiency is also likely to result in an insulator. This is indirectly evidenced by the fact that many samples that are quickly sintered in air and cooled do not show superconducting behavior, while other samples prepared in a high oxygen environment are superconductors with high values for T_c (zero resistance). (10,11) The TGA results of Murphy et. al (12) suggest that the actual occupancy range for the oxygen constituent in these superconductors is rather limited. However due to the natural state set by the Cu valences, this occupancy range is commonly achieved. This may explain why it is relatively easy to synthesize these superconductors, and why high T_c superconductivity in the 90 K range was first observed in these compounds.

TGA data taken on a prebaked but unsintered piece of $Er_1Ba_2Cu_3O_{9-\delta}$, shows that a majority of the weight loss upon heating occurs above 750 C. Some of the weight loss can be attributed to the liberation of CO_2 in spite of the previous prebaking steps. However a large portion of the loss is due to the liberation of O from the Cu-O bonds in the sample. Though sintering at a temperature as high as possible without forming an incongruent liquid phase is important for the formation of large grains, the high temperatures involved are clearly unfavorable to the incorporation of oxygen into the sample. Therefore it is particularly important to slowly cool the sample in oxygen below this temperature, or to include an oxygen annealing step in the preparation.

The role of oxygen incorporation along with the TGA data can provide a qualitative explanation for the observed resistive features (Figure 2). On heating the sample in air, a major uptake of oxygen begins at about 340 C and maximizes at 410 C. The maximum uptake upon cooling is at around 360 C. This oxygen inhaling corresponds to the slight resitance bump at 360 C (Figure 1, inset). One of the possible locations for the extra oxygen incorporated would be on the Cu planes between the Ba layers of adjacent unit cells. In a recent neutron diffraction study on $Y_1Ba_2Cu_3O_{9-\delta}$ (13) it was concluded that there were ordered oxygen vacancies on these outer planes such that the remaining oxygen atoms formed "lines" with the Cu atoms. If a portion of the conductivity in the sample were associated with these lines, then the added incorporation of oxygen into these planes might have a disruptive effect. The slight resistive bump could be the result of this, or any other oxygen incorporation effect.

The extended resistive bump between 600 C and 800 C can be interpreted similarly. In this case though, the resistive anomaly is pronounced, but the TGA curves only show a slight deviation near 680 C. The resistance curve for $Y_1Ba_2Cu_3O_{9-\delta}$ demonstrates a bump at 750 C which correlates well with the reported tetragonal to ortho-rhombic phase transition in $Y_1Ba_2Cu_3O_{9-\delta}$ at this temperature (14,15). The corresponding resistance bump for $Er_1Ba_2Cu_3O_{9-\delta}$ indicates that this type of transition also occurs in this compound near this temperature. It is notable that the maximum oxygen uptake in $Er_1Ba_2Cu_3O_{9-\delta}$ at 410 C is below the reported maximum uptake for $Y_1Ba_2Cu_3O_{9-\delta}$ at 500 C. (16)

The temperature at which an incongruent phase is formed in $Y_1Ba_2Cu_3O_{9-\delta}$ is much higher than the $Er_1Ba_2Cu_3O_{9-\delta}$ compound. This fact and the difference in the grains sizes of these two compounds can be expected to alter their respective oxygen access and absorption. This may explain the differences in the resisitive behavior upon cooling of these compounds.

From the correlation between the observed oxygen uptake and resistive features for the compounds $Er_1Ba_2Cu_3O_{9-\delta}$ and $Y_1Ba_2Cu_3O_{9-\delta}$, we can understand why a quick solid state reaction procedure would be successful. This is in spite of "impurity" phases being present in the compounds. The quick sintering to 900+ C allows for the formation of a connecting path of grains, while cooling slowly from 700 C (especially in the vicinity of 500 C) in the presence of oxygen can possibly serve as a replacement for an annealing step. In seems possible that only a relatively short length of time is needed for the sample transitions to occur. However the extreme case of "quenching" a sample from the oven often does not produce a superconductor.

In summary, we have shown that the previously empirical sample preparation procedures of slow cooling and annealing in oxygen seem

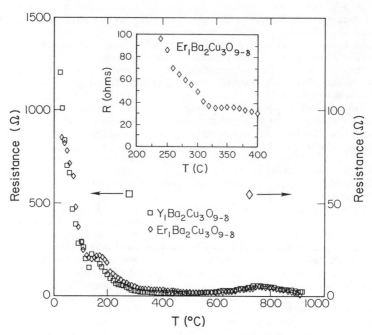

Figure 1. The resistance above room temperature for nominal
compositions of $Er_1Ba_2Cu_3O_{9-\delta}$ (triangles) and $Y_1Ba_2Cu_3O_{9-\delta}$
(squares) upon heating. The inset enhances the region between
200 C to 400 C for the $Er_1Ba_2Cu_3O_{9-\delta}$ sample.

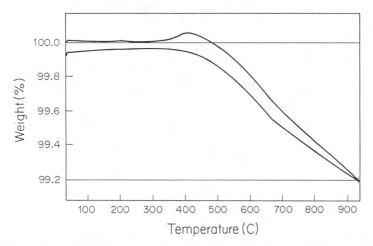

Figure 2. The TGA curves for a previously sintered $Er_1Ba_2Cu_3O_{9-\delta}$
sample upon warming in air (top) and cooling in oxygen (bottom).
The slight dip near 680 C and the maximum oxygen uptake near
400 C correlate with resistive features.

to be actually important towards insuring that the proper sample transitions take place. In $Er_1Ba_2Cu_3O_{9-\delta}$, a tetragonal to ortho-rhombic phase transition near $680^\circ C$ is implied when compared to the corresponding data on the $Y_1Ba_2Cu_3O_{9-\delta}$ compound. The maximum oxygen uptake of $Er_1Ba_2Cu_3O_{9-\delta}$ near $400^\circ C$ also correlates well to the reported uptake temperature for $Y_1Ba_2Cu_3O_{9-\delta}$. Since oxygen defi-ciencies and positions in the unit cell seem to be intricately related to the superconducting mechanism, the sample preparation procedure should be structured to maximize the oxygen interaction with the sample.

Acknowledgments

We wish to thank Roy S. Tucker and Sang Young Lee for help in sample preparation, Margarita Rohklin for her work on the SEM, and M. S. Wong for help with the TGA. The financial support of the National Science Foundation through a grant DMR 83-16989 to the Ohio State University Materials Research Laboratory and grant DMR 84-05403 is gratefully acknowledged.

Literature Cited

1.   J.G.Bednorz and K.A.Muller, Z. Phys. B 1986, 64, 189.
2.   M.K. Wu, J.R.Ashburn, C.J.Torng, P.H.Hor, R.L.Meng, L.Gao, Z.J.Huang, Y.Q.Wang, and C.W.Chu, Phys. Rev. Lett. 1987, 58, 908.
3.   S.R. Ovshinsky, R.T. Young, D.D. Allred, G. DeMaggio, and G.A. Van der Leeden, Phys. Rev. Lett. 1987, 58, 2579.
4.   M. Hirabayashi, H. Ihara, N. Terada, K. Senzaki, K. Hayashi, S. Waki, K. Murata, M. Tokumoto, and Y. Kimura, Jap. J. Appl. Phys. 1987, 26, 1454.
5.   Hau H. Wang, K. Douglas Carlson, Urs Geiser, Robert J. Thorn, Huey-Chuen I. Kao, Mark A. Beno, Marilyn R. Monaghan, Thomas J. Allen, Roger B. Proksch, Dan L. Stupka, Jack M. Williams, Brian K. Flandermeyer, and Roger B. Poeppel, (submitted to Inorg. Chem. [Commun.]).
6.   Aravinda M. Kini, Urs Geiser, Huey-Chuen I. Kao, K. Douglas Carlson, Hau H. Wang, Marilyn R. Monaghan, and Jack M. Williams, (submitted to Inorg. Chem. [Commun.]).
7.   E. Takayama-Muromachi, Y. Uchida, Y. Matsui, and K. Kato, Jap. J. Appl. Phys. 1987, 26, 1476.
8.   We use Acme E-Solder 3022 with hardener #18 available from Acme Chemicals & Insulation Company; a Division of Allied Products Corp.; New Haven Conn. 06505, and Alumina Cement available from L.H. Marshall Co.; 270 W. Lane Ave; Columbus Ohio 43201. We emphasize that no particular endorsement of products is implied.
9.   Francis S. Galasso, Structure, Properties, and Preparation of Perovskite-Type Compounds (Pergamon Press, New York, 1969).
10.   Sung-Ik Lee, John P. Golben, Sang Young Lee, Xiao-Dong Chen, Yi Song, Tae W. Noh, R.D. McMichael, Yue Cao, Joe Testa, Fulin Zuo, J.R. Gaines, A.J. Epstein, D.L. Cox, J.C. Garland, T.R. Lemberger, R. Sooryakumar, Bruce R. Patton, and Rodney T. Tettenhorst, Extended Abstract Materials Research Society, 1987.

11. Sung-Ik Lee, John P. Golben, Sang Young Lee, Xiao-Dong Chen, Yi Song, Tae W. Noh, R.D. McMichael, J.R. Gaines, D.L. Cox, and Bruce R. Patton, Phys. Rev. B 1987 (to be published).

12. D.W. Murphy, S. Sunshine, R.B. van Dover, R.J. Cava, B. Batlogg, S.M. Zahurak, and L.F. Schneemeyer, Phys. Rev. Lett. 1987, 58, 1888.

13. J.E. Greedan, A. O'Reilly and C.V. Stager, (submitted to Phys. Rev. Lett.).

14. Ivan K. Schuller, D.G. Hinks, M.A. Beno, D.W. Capone II, L. Soderholm, J.-P. Locquet, Y. Bruynseraede, C.U. Segre, and K. Zhang, (submitted to Sol. State Comm.).

15. P. Strobel, J.J. Capponi, C. Chaillout, M. Marezio, and J.L. Tholence, Nature 327, 306 (1987).

16. J.M. Tarascon, W.R. McKinnon, L.H. Greene, G.W. Hull and E.M. Vogel (submitted to Phys. Rev. Lett.).

RECEIVED July 6, 1987

APPLICATIONS

Chapter 27

Applications of High-Temperature Superconductivity

A. P. Malozemoff[1], W. J. Gallagher[1], and R. E. Schwall[2,3]

[1]Thomas J. Watson Research Center, IBM, Yorktown Heights, NY 10598-0218
[2]Francis Bitter National Magnet Laboratory, Massachusetts Institute of Technology, NW 17-060, Cambridge, MA 02139

The new high temperature superconductors open up possibilities for applications in magnets, power transmission, computer interconnections, Josephson devices and instrumentation, among many others. The success of these applications hinges on many interlocking factors, including critical current density, critical fields, allowable processing temperatures, mechanical properties and chemical stability. An analysis of some of these factors suggests which applications may be the easiest to realize and which may have the greatest potential.

In January 1986, J. G. Bednorz and K. A. Mueller of the IBM Zurich Research Laboratory discovered high temperature superconductivity in a certain class of copper oxides (1). Only a year later, superconducting transition temperatures in this class of materials had been pushed to almost 100 K (2-3), far beyond the previous record of 23 K that had remained unbroken since 1973, and well above the benchmark temperature 77 K of boiling liquid nitrogen. In addition to the scientific excitement of this extraordinary development, the possibilities for practical application of superconductivity clearly need to be re-examined, if only because of the vastly greater ease in cooling to 77 K compared to the liquid helium temperature range ($\simeq$4 K) used in all previous applications of superconductivity(4).

While the possibility of achieving yet higher temperatures in new compositions continues to be actively pursued, we focus in what follows on the most studied group of superconductors with transition temperatures in the range from 99-100 K, namely the $YBa_2Cu_3O_{7-\delta}$ group with the "123" layered perovskite superstructure (5), which we henceforth refer to as YBaCuO. Certainly the possibilities for application would be increased immeasurably if stable room temperature superconductors were found.

Previous applications of superconductivity have entailed about $20 million in yearly sales of finished superconducting material (mostly NbTi and

[3]IBM visiting scientist

0097–6156/87/0351–0280$07.75/0

Nb_3Sn wire and tape) used primarily in magnets. A quite small fraction of the total has gone into shielding or SQUIDs. By far the dominant (70-80%) magnet application has recently been in medical magnetic resonance imaging, where the roughly quarter million dollar cost of the superconducting magnet and the approximately 400 units sold in 1986 imply a market of order $100 million yearly. Most of the remainder involves high energy physics or laboratory magnets. While this is a substantial market for superconductivity, the new high-temperature superconductors have now raised vastly greater hopes.

In this paper we review some possible applications for the new high temperature superconductors. The scope is huge -- almost as broad as the entire field of electricity and magnetism -- so of necessity the treatment here of any single application will be brief and superficial. Furthermore, even though research has progressed at an unprecedented pace, the amount known about the technical properties of the new superconductors is still limited. Thus there is a great unpredictability hovering around our assumptions. Our conclusions should be regarded in this light.

The applications we discuss here, in magnets, power transmission, computer interconnections, Josephson devices and instrumentation, have almost all been studied before, during two decades of active work in applied superconductivity (4). Thought of simply as a standard superconductor with a higher transition temperature, YBaCuO does not by itself imply new kinds of applications, even though it may improve the commercial prospects for applications previously burdened by overhead costs associated with helium refrigeration. More novel applications may well emerge in the future.

Our analysis below reveals much promise in a variety of applications, but also many challenges. Clearly a long-term view will be essential to exploit the full potential of this new technology.

Material Parameters for Applications

The prospects for various applications depend, of course, on the materials properties of the superconductor. Many properties are important, including mechanical properties, interfacial interactions and contacts, high frequency loss, and so forth. Here we emphasize as an example the critical current density, which is one of the most vital properties for most applications.

First we review some typical materials parameters obtained from measurements on randomly oriented ceramics (6-7). Since the YBaCuO structure (5) and electronic properties (8) are highly anisotropic, the orientationally-averaged values obtained from studies of ceramics are only an initial indication until more complete experimental results on single crystals and oriented films and ceramics become available. For material with a resistivity just above the transition of 400 $\mu\Omega$cm, a Hall carrier density of 4×10^{21}cm^{-3}, and dH_{c2}/dT of 2 T/K (6-7), one deduces a BCS coherence length $\xi(0)$ of 1.4 nm, a London penetration depth $\lambda(0)$ of 200 nm, a mean free path ℓ of 1.2 nm, a thermodynamic critical field $H_c(0)$ of 1 T (10000 Oe) and an upper critical field $H_{c2}(0)$ of 120 T. The large ratio of λ to ξ implies that YBaCuO is a Type II superconductor, i.e, magnetic flux can penetrate the bulk material in a certain range of field strength.

The largest low-field 77 K critical current density (6) measured so far in randomly-oriented ceramic material is only 1000 A/cm^2, a disappointingly small value which would exclude all but a very few of the possible applications. More recent results on preferentially oriented epitaxial films on SrTiO$_3$ substrates (9-11) show values 100 times higher at 77 K, namely more than 10^5A/cm^2. Measurements on single crystals (8) as well as the epitaxial films show currents as large as 3×10^6 A/cm^2 at 4.2K, and the crystal measurements show that the critical current density has a large temperature and field dependent anistropy. The 10^5 A/cm^2 current level at 77 K would be adequate for many applications, although this value was only achieved in zero field.

The current density improvement is apparently related either to orienting the YBaCuO material so that current flows only along favorable directions or to the elimination of grain boundaries (12), which may form weak (Josephson-coupled) obstacles to current flow. Unfortunately single crystals and single-crystal substrates are not appropriate for many applications. Ways must be found, perhaps using oriented ceramics, to achieve higher current density in a polycrystalline environment. A recent report (13) of 3×10^4A/cm^2 in films on polycrystalline substrates indicates rapid progress in this direction.

Insight into the temperature dependence of the zero-field current density can be obtained by considering a theoretical upper limit to current density of a superconductor, namely, the "depairing current density", where the kinetic energy of the superconducting electrons equals the condensation energy $H_c^2/8\pi$. At low temperatures and zero applied field, one finds (14)

$$J_d = 10H_c/4\pi\lambda \tag{1}$$

where J_d is the critical current density in Amps/cm^2, H_c is the thermodynamic critical field in Oe and λ is the London penetration depth in cm. Using the parameters given above, one calculates 3×10^8 A/cm^2, a very large value indeed (7). Rather similar values are obtained for the conventional high field superconductors because their lower transition temperatures and hence lower thermodynamic critical fields are compensated by their higher electron density which lowers the penetration depth.

However, at higher temperatures J_d is reduced. Near T_c, an estimate can be made with the Ginzburg-Landau theory (14):

$$J_d = 10H_c(1 - t)^{3/2}/3\sqrt{6}\,\pi\lambda \tag{2}$$

where t is the reduced temperature T/T_c. There is a factor 1.84 reduction in the prefactor and an all-important temperature dependent factor $(1 - t)^{3/2}$, which can cause a drastic decrease in J_d near T_c. For example, the depairing critical current density of YBaCuO with T_c of 92 K is reduced to 1.2×10^7

A/cm^2 at 77 K, more than an order of magnitude lower than at 4.2 K. Thus YBaCuO at 77 K has a fundamental disadvantage in zero-field critical current density compared to, say, Nb$_3$Sn at 4.2 K. The same effect impacts the hope for practical room temperature superconductors: To maintain reasonable critical current density at 300 K or 27 C, the superconducting transition would have to be closer to 400 K or 127 C!

The 77 K, zero-field depairing current density estimated above would still be adequate for most applications. But another effect limits the actual critical current density J$_c$ to lower values: This is the depinning of flux or vortex lines which penetrate Type II superconductors like YBaCuO. Currents exert forces on flux lines, and when they depin above a threshold current, they generate a "flux flow resistance" in their normal cores (14). In high-field magnet materials, current densities are maximized by introducing structures that pin the flux lines, such as dislocation walls and α − Ti precipitates in NbTi (15). But even in this well-studied case, the maximum achieved current density is still over an order of magnitude less than the depairing limit.

Little is known about pinning in YBaCuO. But the experience with Nb$_3$Sn and NbTi suggests that even after extensive effort to optimize pinning, critical currents in YBaCuO are likely to remain at least an order of magnitude lower than the depairing limit. This implies a likely limit of 10^6A/cm^2 for YBaCuO at 77 K and zero field, a factor of over ten less than what has already been demonstrated in Nb$_3$Sn at 4.2 K and zero field. This is a serious disadvantage for YBaCuO, as will be seen below. Clearly even a modest increase in transition temperature would improve the limit substantially; T$_c$ of 120 K would give a factor of 3 improvement.

Critical current density also depends on magnetic field, decreasing monotonically to zero at the upper critical field H$_{c2}$ (16-17). In YBaCuO the slope of H$_{c2}$ with temperature is unusually large, of order 2 T/K when field is applied parallel to the predominant conduction planes of the structure (18-19). This implies record values (up to 200 T has been estimated) for the upper critical field at low temperatures and opens up the possibility of very high field magnets.

The ability to achieve ultra-high fields (over 20 T), however, will certainly be limited in practice by the strain induced from stresses connected with the containment of high magnetic fields. It is customary to have conductor tensile strains of several tenths of a percent in NbTi magnets (at fields ≤10 T), for example, and it is unlikely such strains can be supported by the brittle ceramic YBaCuO (20). For example, consider a solenoid of 20 cm inner diameter operating at 20 T. We assume the winding to contain 50% steel by area which carries the hoop stress. If the overall current density is 3×10^4A/cm^2 (which implies over 6×10^4 in the superconductor), then the hoop stress is 600 MPa (87 ksi) overall and 1200 MPa (174 ksi) in the steel. An elastic modulus of 207 GPa or 3×10^7 psi for steel yields strains of almost 0.6% in the winding.

At fields well below 20 T, there are also serious obstacles to overcome. For applications, it is the ability to carry substantial currents in high fields that is important, not merely the value of H_{c2} determined by the zero of resistance or, more optimisticly, by the breakaway from normal resistance behavior (21-22). High field values for zero resistance have only been observed so far in single crystals (18-19). Results so far achieved with potentially more useful ceramic wire configurations have been disappointing (12,23-24). The resistive transitions as a function of field are broad and show a "foot" near zero resistance, which causes superconductivity to be suppressed very rapidly with field (22). Presumably this effect arises either from grain boundaries or from strong anisotropy, and again techniques for orienting ceramics may be important.

Progress in this direction could help to open up applications for high field magnets operating at 4.2 K, where the critical current density of YBaCuO in high fields could be much higher than for Nb_3Sn because of the high intrinsic values of H_{c2}. (YBaCuO current densities of 1.7×10^6 were measured at 4 T in Ref. 8.) Even more interesting for applications would be magnets operating at 77 K. Taking $dH_{c2}/dT \simeq 2$ T/K for a 92 K superconductor, one can estimate an H_{c2} of 30 T, which is still larger than the 4.2 K value of about 20 T for Nb_3Sn and 10 T for NbTi. On the other hand, if one uses 0.4 T/K which characterizes dH_{c2}/dT for field perpendicular to the YBaCuO conduction plane (19), then H_{c2} at 77 K is only 6 T. In this case applications are only likely in a low field range where the refrigeration cost advantage predominates (see below). Clearly, higher transition temperatures or larger dH_{c2}/dT would increase the leverage of YBaCuO vis-a-vis Nb_3Sn .

Superconducting Magnet Applications

The impact of the critical materials parameters, especially the critical current density and the allowable operating strain, is seen quite clearly in magnet applications. Superconducting magnets are the dominant present commercial application of superconductivity and underlie many of the potential but presently unrealized applications (15). The existing commercial applications include:

1. Medical Magnetic Resonance Imaging (MRI): As noted above, MRI is currently the dominant market for superconductive materials and devices (25), although the first superconductive MRI magnet was only delivered in 1980. This application does not tax the superconducting properties of existing commercial NbTi wire. Peak fields are below 2.5 Tesla and the required critical current density is rather low. The primary concerns are field homogeneity, field stability and device reliability. This has led to very conservative designs with the superconductor operating at only 50% to 70% of its critical current. Stresses are rather low and large amounts of copper are used for stabilization. While there are other techniques for providing fields for MRI such as permanent magnets or resistive electromagnets, superconductive magnets have captured the bulk of the MRI market by providing higher fields and better field stability which translates into superior image quality.

2. High Energy Physics (HEP):   HEP was until recently the largest consumer of superconducting materials. Most development of high field materials and magnets was, in fact, performed in HEP labs or in response to HEP needs. The demand is quite cyclical however, peaking when a large machine such as the Tevatron is built and declining to almost zero in the years between major machines. The next major superconducting HEP project is likely to be the highly publicized Superconducting SuperCollider (SSC). Interestingly, the currently favored magnet design uses NbTi operating at 6 T and 2.75×10^5 A/cm^2, rather than Nb$_3$Sn which could be capable of higher field operation. In HEP, superconductivity has made possible devices such as the Tevatron and large bubble chambers which would be prohibitively expensive to operate if built with conventional resistive magnets.

3. High Field Magnets for Scientific Applications: High field magnets for scientific applications have been a small but important industry for over 20 years. With the increasing importance of analytical NMR in the chemical and biotechnology industries and the continued work on gyrotron tubes employing superconductive magnets, there has been a modest resurgence of growth over the past few years. In these applications superconductors are used to generate very high fields in small volumes making more widely available analytical techniques which would otherwise be restricted to a few large laboratories having multi-megawatt motor generators and high field copper magnets.

4. Magnetic Separation: Although true commercial applications of magnetic separation are rare, many pilot projects are underway. The devices are used to separate minerals and scrap and to purify polluted water and flue gas. The role of superconductivity is to increase the field and field gradients available and hence enhance the efficiency of the operations.

Another group of magnet applications, those not yet commercialized, is much larger. Primary candidates at the present time include:

1.   Magnetic Confinement Fusion
2.   Rotating Machinery (e.g. turbine or homopolar generators)
3.   Energy Storage
4.   Levitated Trains - MAGLEV
5.   Magnetic Launchers
6.   Magnetohydrodynamics

In most of these applications superconductive technology is not the factor limiting commercialization, but the cost advantages offered by higher temperature operation cannot be ignored. The degree to which YBaCuO magnets operating at 77 K impact any of these applications depends on the performance characteristics of the material, the competing technologies (in most cases conventional NbTi or Nb_3Sn superconductor operating at 4.2 K) and the economics of each application.

Critical current density considerations. As mentioned above, perhaps the most crucial materials parameter is the critical current density J_c. To illustrate the requirements on J_c, we consider a simple model system, the long thin-walled solenoid. The central field (in Oersteds) is given by

$$H = 4\pi Jt/10 \tag{3}$$

where J is the current density in Amps/cm² and t is the thickness of the superconducting winding in cm.

Most of the applications listed above require fields of 20,000 to 100,000 Oe (2 to 10 T). For reasonable thickness windings this implies a winding current density well in excess of 10^4 Amps/cm². Since the winding area must include stabilizing material (see below) as well as structure and insulation between turns, the minimum useful current density in the superconductor itself is likely to be about 5 x 10^4 A/cm². The requirement is likely to exceed 10^5 A/cm² in those applications requiring high fields or high current density to achieve precise field geometries such as multipole magnets for accelerators, high field magnets for scientific applications, high gradient magnetic separators and compact rotating machinery. As stated above, the achievement of such current densities in production quantities of YBaCuO material will require considerable additional materials and process development.

Materials cost per ampere meter. In those applications where the critical current density and strain characteristics make the magnet technically feasible, we must consider the cost. We focus here on the superconducting materials cost and the total refrigeration cost including the cryostat and capitalized refrigerant costs.

Using typical present commercial prices for 99.9% pure components, one finds a raw materials cost of $0.3/cm³ for YBaCuO. Assuming the finished superconductor is a factor of 4 more costly than the raw materials (as is roughly the case for Nb_3Sn wire), we can make a first very rough cost estimate of about $1.2/cm³. By comparison, for multifilament Nb_3Sn wire at $100 per pound the equivalent number is $2.0/cm³.

In many applications, the relevant materials cost is the per volume cost divided by the current density because less of a higher current density material need be used to achieve the same required current. The figure of merit can then be expressed in units such as dollars per ampere-meter, where meter refers to the length of the wire.

If at zero field the critical current of YBaCuO at 77 K remains a factor of 10 lower than that of Nb_3Sn at 4.2 K, as discussed in the previous section, the cost per ampere meter of Nb_3Sn would be about a factor of 6 below that of YBaCuO. At higher fields the cost per ampere-meter depends on the field dependence of J in each case (26-27). In Fig. 1 we compare some speculative

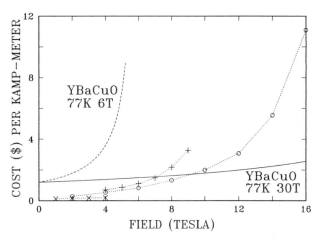

Fig. 1.   Cost per ampere meter for various superconducting magnet wires: circles - Nb_3Sn values for 10^6 lbs. quantity (26-27); crosses - NbTi values for present SSC spec (26); x's - 4.2 K YBaCuO value from $1.2/cm³ cost estimate and measured single crystal critical current densities at 4.2 K (8). For YBaCuO at 77 K (solid and dashed lines), a current density form $J(1-H/H_{c2})$ was used with $J=10^5 A/cm^2$ and H_{c2} either 30 T (single crystal value for H⊥c (19)) or 6 T (H∥c).

costs for YBaCuO with costs for current state-of-the-art commercial super-conductors. The NbTi data shown is for the material proposed in the reference design study for the Superconducting SuperCollider. This material has a critical current density at 5 T of over twice that of the material used to build the Tevatron. The Nb_3Sn data is for large quantities (26) of multifilamentary wire produced via the internal tin core process (27).

For YBaCuO at 4.2 K we use critical currents from single crystals (8) and the materials cost estimated above. This of course assumes that the single crystal critical currents can be realized in long lengths of conductor wound in a magnet geometry and also that the strain limits of the material are not ex-ceeded as discussed earlier. Because of the very high upper critical field of YBaCuO at 4.2 K, the critical current of this material dominates at very high fields reducing its cost-per-ampere-meter below that of its conventional com-petitors. This is an exciting opportunity for the new superconductor, although obviously it is hardly an immediate threat to the conventional materials since the arduous optimization process for YBaCuO wire fabrication has hardly be-gun and the strain limits are far from clear.

At 77 K, we estimate the YBaCuO cost per ampere-meter by taking a current density of the form $J(1 - H/H_{c2})$, with $J=10^5 A/cm^2$ and H_{c2} either 30 T or 6 T, as discussed in the previous section. Fig. 1 shows that 77 K YBaCuO crosses under Nb_3Sn above about 10 T if its critical field is suffi-ciently high (compare the 30 T and 6 T curves in Fig. 1), although this field range is still in jeopardy because of the mechanical yield strain limit. At low fields, the unfavorable cost-per-ampere-meter must be offset by reduced re-frigeration cost if YBaCuO is to compete with the conventional superconduc-tors. It should be emphasized that the specific numbers for YBaCuO given here are only illustrative of the kinds of crossovers which are likely to occur; meaningful engineering evaluation must await data on real YBaCuO wires. A further caveat is that our analysis ignores the cost of fabricating material into a magnet. Currently NbTi magnets are less costly than Nb_3Sn up to about 9 T partly because ductile NbTi is much easier to use than brittle Nb_3Sn .

Refrigeration costs. The cryostat of a magnet operating at helium temperature contains components operating at three different temperatures. The outermost shell is a radiation shield operating at or near 77 K which intercepts the bulk of the heat transferred from room temperature by radiation. This shield is normally cooled by a liquid nitrogen reservoir or a single stage refrigerator. The next shield operates at about 20 K and is cooled by cold gas from the he-lium boiloff or by a two stage refrigerator. This intercepts almost all of the re-maining radiation. The final stage is the magnet container which operates at 4.2 K and is cooled by helium boiloff or by a three stage refrigerator. The magnet weight and any operating forces must also be transferred to room temperature via support members. These support members are heat sunk to the intermediate heat shields and are in many cases the dominant heat loads for the 20 K and 4.2 K thermal stations.

If we substitute an YBaCuO magnet for that operating at 4.2 K, the shields at 77 K and 20 K are eliminated and the thermal mass previously at 4.2 K now operates at 77 K. This will reduce the capital cost of the cryostat by a factor of about 2. The heat leak to the 77 K mass should remain unchanged so the refrigeration cost for a YBaCuO magnet should be essentially the same as that for cooling the 77 K shield on an equivalent helium cooled magnet. For a commercial MRI magnet the liquid nitrogen consumption is about 1 liter/hr and the helium consumption is about 0.3 liter/hr. Assuming transferred cryogen costs of $0.25 and $8.00/liter for nitrogen and helium respectively yields a factor of 10 reduction in overall refrigeration operating costs. This reduction would probably be less for cryostats cooled by small closed cycle refrigerators.

Interplay of Materials and Refrigeration Costs. To estimate the effects of these materials and refrigeration costs on magnet costs we again consider the long thin solenoid. The refrigeration costs (both the cryostat and the capitalized costs of the cryogens) are divided into two parts corresponding to the two sources of heat leak into the system. The first term, corresponding to radiative heat transfer, is proportional to the cryostat area. For a very thin warm bore solenoid this is $4\pi RLC_{r1}$, where C_{r1} is the cost of refrigeration for the heat transmitted through one cm^2 of cryostat surface, R is the mean radius, L is the cryostat length, and the area of the inner room temperature bore is assumed roughly equal to that of the outer shell.

The second term, corresponding to the heat leak via conduction through the load bearing members, is assumed proportional to the conductor volume and can be written $2\pi RLtC_{r2}$, where C_{r2} is the cost of refrigeration for the heat transmitted through the support members for each cm^3 of conductor and t is the winding thickness. The materials cost is given by $2\pi RLtC_m$, where C_m is the cost per cm^3 of superconductor. Combining these terms, substituting for t from Eq. 3, and dividing by the area gives the total cost per area of cryostat as:

$$C/A = C_{r1} + (C_{r2} + C_m)5H/(4\pi J) \qquad (4)$$

While this simple analysis ignores costs such as coil forms and winding labor, it does illustrate some interesting scaling relationships. We will compare the systems Nb_3Sn operating at 4.2 K and YBaCuO operating at 77 K in some limiting cases. First consider the case of a large light magnet where the heat leak is primarily via radiation. One can then ignore C_{r2}. In the limit of low field, the refrigeration costs dominate, and hence YBaCuO is the lower cost system. At higher fields, the materials cost per ampere-meter will be the deciding factor, and, according to Fig. 1, Nb_3Sn could be the lower cost system up to about 10 T. A crossover may occur at some intermediate field, depending on the exact parameter values.

The other limiting case is that of a compact heavy magnet where the radiation heat leak is negligible. Then the cost ratio is independent of field and depends only on the ratio of the refrigeration costs and the superconducting materials costs each normalized to the volume of superconductor. Note that in both cases the costing must be done for a fixed time - usually the estimated lifetime of the device - since the C_r-terms include the capitalized cost of cryogens.

Although any meaningful engineering analysis is impossible given the lack of good cost and critical current data, one can make an order of magnitude estimate. We consider a model 100 cm bore MRI magnet and cryostat system providing 1.5 T with a cryostat area of about $2.1 \times 10^5 cm^2$ and a conductor volume of about $1.4 \times 10^5 cm^3$. The heat leak is assumed to be evenly divided between radiation and conduction (a reasonable assumption for an MRI magnet and intermediate between the two limiting cases) and cryostat costs of $100,000 at 4.2 K and $50,000 at 77 K are assumed. Helium cost is $30,000 per year for 10 years in the 4.2 K case and nitrogen is $3000 per year in the 77 K case. Using the cost per cm^3 for Nb_3Sn and YBaCuO given above and assuming field independent critical currents of $10^5 A/cm^2$ for Nb_3Sn and a conservative 10^4 for YBaCuO gives a crossover of about 1.7 T, i.e. the YBaCuO magnet is more costly above 1.7 Tesla. While the actual number obtained from this exercise is clearly highly sensitive to the assumed input parameters, it does indicate that the possibility that YBaCuO could be useful even at a J_c of 1×10^4 and that the advantage would increase rapidly with increasing J_c and decreasing materials cost.

Stabilization, and Conclusions on Magnets. Another consideration in the design of superconductors is "stabilization". Conventional superconductors are used in the form of multifilamentary wires which consist of 100 to 10,000 filaments 6 μm to 100μm in diameter distributed in a high conductivity matrix, usually copper. This fine subdivision of the superconductor is necessary to prevent thermal runaway or quenching of the conductor at currents below the critical current. The matrix material serves to provide an electrical shunt around small normal zones and to conduct away heat. Both the degree of subdivision and the amount of normal metal needed for proper stabilization depend on the specific heat of the superconductor and the slope of the J_c versus temperature curve. At this point it is not clear how one would stabilize an YBaCuO conductor, but the higher specific heat of the material at 77 K does significantly increase the maximum adiabatically stable strand size, thus easing this problem.

Some general conclusions which can be drawn from this analysis are:

- If YBaCuO can demonstrate the promise of usable current density at higher fields than Nb_3Sn, it could be useful for very high field magnets provided the attendant structural problems can be solved.

- If the cost per ampere-meter of YBaCuO at 77 K remains higher than conventional superconductors it will not be competitive in applications where the superconductor materials cost dominates, i.e. very large magnets at relatively high fields, unless there are very special demands on space and weight.

- For applications at relatively low fields where little superconductor is used, refrigeration becomes a significant fraction of the overall cost, and YBaCuO could enjoy a cost advantage. The crossover points, in field and device size, below which YBaCuO is attractive will depend strongly on the device configuration and application.

- Additional leverage emerges in those applications requiring minimal weight, particularly in space or in ship propulsion. Space offers the special advantage of a naturally low temperature environment so that all refrigeration could conceivably be eliminated.

- In addition to the challenge of achieving high current density at appropriate fields, manufacturing problems caused by contacts, brittleness of the material and chemical stability in the non-superconducting matrix are major obstacles yet to be solved.

Superconducting Power Transmission Lines (SPTL)

Another possible large scale application of superconductivity is in power transmission. This is inherently different from the magnet applications in several ways. First, high field is not a critical factor in most designs. Second, superconducting magnets are in many cases an **enabling** technology, that is they make possible applications which have no realistic non-superconducting competitor. In power transmission, by contrast, there exist a variety of highly developed technologies for transmitting electrical energy and the superconductor is valuable only when it brings a distinct cost advantage relative to current technology. Moreover this cost advantage must outweigh a perceived reliability disadvantage in an industry where unscheduled downtime must be infinitesimal.

The costs of a SPTL divide naturally into two parts which must be combined for overall costing purposes. The capital costs are the resources required to construct the line. The operating costs are primarily the cost of power consumed by the line with some minor additions for maintenance. For a SPTL the operating costs must include the efficiency of the refrigerators, i.e. the fact that removal of one watt of heat from a 4 K line will consume several hundred watts of power at room temperature. For overall costing purposes one either capitalizes the operating costs or depreciates the capital costs over the lifetime of the line. In either case one must make assumptions about the cost of funds (i.e. the interest rate) and the busbar cost of power over the life of the line. Since both of these factors vary considerably with time and geography, comparisons of various cost estimates must be made with great caution. It was largely changing views about the future busbar cost of power that led, in the late 1970's, to the abandonment of the SPTL development projects which were begun during the oil crisis of the early 70's. In view of this we will restrict ourselves here to an investigation of the scaling relationships inherent in the physics of the line using rough cost estimates only where necessary.

The early work of Garwin and Matisoo (28) focused on d.c. transmission because the hysteretic losses in superconductors of the late 1960's made a.c. lines uneconomical. Transmission at d.c. is most favorable, however, for long lines and high powers to amortize a.c.-to-d.c. converter costs. The prospects for practical introduction of SPTL's, however, are greatest for smaller a.c. systems, on which we focus below.

Capital Costs.   We consider in turn the superconducting materials costs, cryostat-related costs and refrigerator capital costs.

-*Superconductor Cost*: The superconductor is certain to be the most costly material in a line and hence an efficient design will minimize its volume. Its cross sectional area depends on the fault current the line is required to carry and its surge impedance. In the simplest approximation, the amount of superconductor required will be inversely proportional to its critical current density at the line operating temperature and low field. The total amount of superconductor can also be expected to scale directly with the power-length product of the line (usually expressed in MVA-miles). The arguments made earlier concerning relative costs of YBaCuO and Nb_3Sn remain valid here. The cost of conductor per MVA-mile for a 77 K YBaCuO line would exceed that of a 4 K Nb_3Sn line by a factor estimated roughly as 6. This increased conductor cost must be offset by reduced cryostat and operating costs for the line to be competitive. Typical values for the surface current density, σ of a coaxial line are 500 A/cm (where the length is measured around the circumference of the coaxial conductor). A J_c of $10^5 A/cm^2$ implies a 50 μm thick film. Even with a factor of 10 for fault current, the conductor thickness is not excessive.

-*Cryostat, Dielectric and Structural Costs*: To first order the cryostat, dielectric and structural costs can be scaled as the line capacity, i.e. per MVA-mile in the limit of large installations. The estimate of a factor of 2 reduction in cryostat costs made for the magnet case should also hold here. The cryostat cost probably scales somewhere intermediate between the cryostat area and its volume because a larger diameter cryostat will require heavier walls and support members.

-*Refrigerator Capital Costs*: In small and intermediate size systems the capital cost of a refrigeration system scales more slowly than the refrigeration power. In the limit of large systems, however, the system cost probably varies approximately as the refrigeration capacity. This is not exactly true since a long, low-loss line will require more circulation pumps and compressor stations than a short lossy line of the same total dissipation. For our purposes we assume the refrigerator capital cost to be proportional to the power input to the refrigerator. This is the refrigeration load at the operating temperature multiplied by the actual thermodynamic efficiency of the system, i. e.

$$W = (T_1 - T_2)H/T_2E \tag{5}$$

where T_1 is the heat discharge temperature, i. e. about 300 K, T_2 is the heat input temperature - 77 K for an YBaCuO line and about 7 K for an actual Nb_3Sn line (29-30), H is the heat input at T_2, E is the refrigerator efficiency, i. e. the fraction of ideal thermodynamic efficiency (about 0.5 at 77 K and 0.2 at 7 K (28)), and W is the power input at room temperature.

Operating Costs. The operating cost is primarily the power required to operate the refrigerators as given by Eq. 5. The heat input H has three major components, superconductor loss, dielectric loss and the cryostat heat leak.

–Superconductor Loss: In a d.c. line the superconductor can be considered to be lossless. In an a.c. line however, there is a hysteretic loss connected with the movement of flux through the material. Since the magnitude of this loss is determined by surface pinning and the penetration of flux into surface irregularities, it scales with the surface current density σ (amps/cm). This loss has been well characterized for Nb_3Sn but no measurements have been made to date for YBaCuO. Since the magnitude of the loss is very dependent on surface condition, one might be required to use different conductor fabrication techniques for an YBaCuO SPTL conductor than for a magnet conductor.

–Dielectric Loss: If an organic dielectric is incorporated into an a.c. line the heat generated by dielectric loss in this material must be included in the operating costs. Unfortunately most polymers are poor dielectrics (have relatively large loss tangents) at low temperatures so this term can be comparable to the superconductor loss for lines operating near 4.2 K.

–Cryostat heat leak: The cryostat heat leak includes radiation from room temperature and heat conducted through the structural supports. We assume that the line is long enough that the leak at the terminations is negligible. As in the magnet case, we would assume the heat leak for a 77 K line to be similar to that of the 77 K shield of a 4.2 K line.

The major contribution of a 77K conductor would be to reduce the capital and operating costs associated with refrigeration. The refrigerator efficiencies of 0.5 at 77 K and 0.2 at 4 K (28) yield thermodynamic efficiencies (watts in per watt of refrigeration delivered) of approximately 200 at 7 K for a Nb_3Sn line and 6 for a 77 K line. Actual operating experience at the Brookhaven demonstration line (29-30) was closer to 400 for 7 to 8 K operation. One could thus see a reduction of about 30 in the operating costs associated with refrigeration. Reductions in refrigerator capital costs might not be as dramatic but should still be significant.

Scaling and the Impact of 77 K Superconductor. There is a close analogy here to the earlier cost scaling discussion for magnets (see Eq. 4) if one considers capacity (the power-length product) instead of magnetic field as the operative variable. Again the cost scaling depends on whether the refrigeration and operating costs are relatively independent of capacity or whether they scale with capacity.

The former would be the case for a d.c. or low-loss a.c. line where the heat leak through the cryostat is comparable to or exceeds the superconductor a.c. loss. In this situation there will be a cost-advantage crossover as a function of capacity between 77 K YBaCuO and 4 K Nb$_3$Sn, with the former favorable at low capacity because of lower refrigeration cost and the latter favorable at high capacity because of lower net materials cost (which results from the higher low-field current density). There will also be a crossover with the conventional resistive line (usually insulated by SF$_6$ gas) because of its yet lower cooling cost balanced by higher power dissipation scaling with the line capacity.

These crossovers would leave a "window" for 77 K YBaCuO at intermediate capacity. Of greatest importance, the lower refrigeration cost of 77 K YBaCuO would shift the conventional-to-superconducting crossover to lower capacity than for the earlier superconducting systems. Given the large refrigeration cost advantage, such a downward shift could be substantial. Since hurdles to introduction of superconducting technology depend strongly on capacity, such a shift could be of great importance in enabling commercial SPTLs.

If the a.c. loss of YBaCuO is significant, the refrigeration costs scale more directly with capacity. The cost-crossovers are then much more sensitive to the absolute costs, and meaningful evaluation awaits more reliable data on real YBaCuO conductors, which would allow the development and analysis of realistic designs.

In summary, YBaCuO operation at 77 K could possibly reduce the capacity (power length product) at which the SPTL becomes attractive compared to conventional lines. While substitution for overhead a.c. lines will likely remain problematic given their low capital cost, substitution for underground compressed gas (SF$_6$) lines may prove practical. One must bear in mind, however, that the installed transmission capacity in the U.S. is growing at a rate of only a few percent a year and hence there are limited opportunities for new large capacity lines. Replacement of existing lines implies a much more stringent financial test since the savings must be sufficient to provide a rapid payback of the discretionary investment.

Interconnects in Computers

Another possible application for high temperature superconductors is as interconnects in computer systems with semiconducting devices (31). These could be called hybrid systems, since they involve both superconductors and semiconductors. In particular CMOS devices are well-known to have enhanced performance at 77 K and are thus potentially compatible with YBaCuO.

The competitor to YBaCuO in this application is copper, which has a resistivity at 77 K of 0.24 $\mu\Omega$cm, compared to a value at 300 K of 1.7 $\mu\Omega$cm. In fact at room temperature, copper alloys with even higher resisitivity are used to reduce electromigration, which is practically eliminated at the lower temperatures. Thus copper offers a reduction in resistance of more than a factor of 6 at 77 K.

For relatively short lines, as might be of interest in interconnections on a microcircuit chip, a lumped circuit analysis is appropriate, and the performance can be estimated from an RC time constant. Here it is not sufficient to consider only the interconnect line resistance in the total resistance R. The effective output impedance of the CMOS device also contributes and is a key feature which limits the leverage of the superconductor.

To estimate a typical CMOS device output impedance (31), one can use, for example, experimental results of Sun et al. (32) on a ring oscillator with a fan-in and fan-out of 3 in a NAND configuration. The devices have a 0.5 μm channel length and a 9 μm channel width. At a supply voltage of 2.5 volts and a load capacitance of C = 0.2 pf, the observed delay per stage is 450 psec, implying an effective output impedance of 2250 Ω. This is of course only a sample value since device characteristics are highly non-linear and so the value can change with loading.

The superconductor has significant leverage over 77 K copper only if the Cu line resistance is significant compared to the output impedance. But even a line with a very small cross-section of 0.5x0.5 μm^2 and a relatively long length of 5 mm has a resistance of only 400 Ω. Relative to 2250 Ω, this would give less than 20% contribution to the RC time constant. Larger gate widths reduce the output impedance and thus increase performance and the superconductor leverage, but at the expense of density.

Another problem concerns the current density. For a typical drain current (see Fig. 5 of Ref. 30) of a few mA, the current density for a 0.5x0.5 μm^2 line is over 10^6A/cm^2, a value approaching the 77 K limit for YBaCuO discussed in the section on material parameters. For larger cross-section lines, say 2.5x1 μm^2, the current density drops to 10^5A/cm^2, but the leverage over 77 K copper is reduced from 20% to a mere 2% for the numbers discussed above.

In summary, if superconducting lines could be fabricated with the same ease and reliability as copper, they would doubtless be used in computing systems with 77 K semiconducting devices. But the above estimates, while only a rough guide, indicate that for the standard CMOS devices, it may be difficult to achieve sufficient performance advantage to outweigh the likely processing problems. Clearly other device configurations need to be found in which lower output impedance offers a greater leverage for superconductivity. An area where superconductivity may be more useful is in packaging, where power losses and voltage drops of very long lines can be eliminated and current densities can be kept lower. Here again copper lines offer a viable alternative with much less risk. It is likely therefore that simple substitution of superconducting for normal metal lines will not be enough, and more creative designs will have to be devised to exploit the advantages of high temperature superconductivity.

Electronics Based on Superconducting Devices

Superconducting devices offer possibilities of higher speed, more sensitivity and greater precision in a broad range of electronic applications, including mm-wave detection, high speed digital and analog signal processing, ultra-low-noise low-frequency measuring devices and dc voltage standards. Commercial instruments operating at or near 4 K and utilizing Josephson devices for high sensitivity measurements are now available from several sources, and the first commercial instrument employing high speed Josephson circuitry is being introduced. Digital computer research utilizing Josephson devices has been pursued seriously by a number of laboratories. In addition there are a number of noncommercial applications, e.g. the maintenance of national voltage standards and SIS mixers in radiotelescopes, that are best performed with superconducting devices. The question naturally arises with each of these as to how well they could work at liquid nitrogen temperatures.

Digital Josephson Technology

Digital Josephson technology for high performance computer applications has been pursued over the past 15-20 years in a number of laboratories worldwide (33-37). While successful in the demonstration of working logic chips and partially functional memory, this technology has seen its projected performance advantage narrowed by the rapid progress of semiconductor-related technologies like VLSI, MLC packaging, grooved Si cooling and GaAs HEMT. Josephson technology is constrained by the two-terminal nature of the Josephson device and the resultant need to use threshold logic.

   For general purpose digital applications, the merits of Josephson-based technology must be considered against those of more conventional technologies that could be available at the same time as Josephson technology. It quickly becomes apparent that inductive circuit families, such as the magnetically coupled interferometer-based circuits have density potentials too low to be competitive. For logic, several circuit families that do not involve inductive loops are known and have been used to make complex LSI-level chips. For memory, there is at present no known alternative to inductive storage, and the magnetic cross-coupling between adjacent cells sets limits on achievable memory densities (36,38).

   Let us examine possible ways in which the new high temperature superconductors might affect digital Josephson electronics: (1) The new superconductors might allow additional performance to be achieved with Josephson circuits operating either at liquid helium or liquid nitrogen temperatures. (2) Convenience of liquid nitrogen cooling might ease the introduction of Josephson technology.

The principal means to higher performance with the new superconductors is through the increased size of the superconducting gap Δ. The increased gap promises higher band widths for superconducting lines ($\Delta/2\pi\hbar$), a faster fundamental limit for the switching time of a gate (also $\Delta/2\pi\hbar$), as well as increased voltage capability for driving transmission lines. In practice the switching speeds of devices, and hence the fastest pulses put onto superconducting lines, are limited not by the gap frequency of conventional superconductors, but rather by junction capacitances C_j to $Z_0 C_j$ where Z_0 is the transmission line impedance. Therefore the larger gap of the new superconductors is not significant as far as this aspect of high speed operation. The ability to better drive lines is one that could be usefully employed in parts of circuits, particularly Josephson memory, but there is no direct effect on memory density. Another way of making use of the larger gap would be to run at higher current levels, keeping the transmission-line impedance constant, so as to improve device noise margins. The limited cooling capability of liquid helium, however, would necessitate lower ultimate circuit densities at the higher power levels. Thus it does not appear that the larger gap of the high temperature superconductor can significantly improve the performance projections for Josephson technology at 4 K.

Using the new superconductors to permit higher temperature operation of digital Josephson circuitry is another possibility. Increased thermal noise in this case dictates higher operating current levels, but with the increased cooling capability available at liquid nitrogen temperature increased power could be tolerated. Thus the technology comparison for 77 K Josephson similar to that of 4.2 K Josephson. (Refrigeration is not a primary concern in either case.) One new option for a digital system at 77 K is the use of Josephson logic circuitry with dense semiconductor memory, the prospects for dense Josephson memory not being favorable. The challenge with this option is to develop the requisite high speed interface circuitry between the two technologies.

In addition to these circuit considerations, there is a considerable amount of materials and processing development needed to make Josesphson technology with YBaCuO a viable option. In particular, YBaCuO's short coherence length and its sensitivity to oxygen diffusion are hurdles to be overcome in the development of high quality tunnel junctions. Until reasonable quality junctions can be demonstrated, meaningful prospects for YBaCuO digital Josephson systems are very difficult to assess.

New Cryogenic Device Possibilities

There has been active interest in new transistor-like devices compatible with superconductivity for a number of years (39). Part of the interest derived from difficulties in using the two-terminal Josephson device in digital circuitry. With the discovery of superconductors with transition temperatures above 77 K, the device needs and opportunities need to be reassessed. MOSFET's, GaAs HEMT's, and heterojunction bipolar transistors (HBT's) work well at 77K, so the demand for a new three-terminal device at 77 K is certainly less than it was at 4.2K. Nevertheless, enhanced performance or lower power would be desirable for 77 K transistors. Enhancements are certainly possible within conventional semiconductor approaches: e.g. scaling, sharper doping profiles, etc., and there is much discussion of the possibilities in semiconductor device literature. Here we discuss new features which superconductivity might add to known devices (FET's) and new devices that might be possible with superconductors.

Superconductivity and FET's. FET's with channels that can be induced into the superconducting state were proposed some time ago, and first demonstrated in the last couple of years with both a Si-based device (40) and a InAs-based device (41). In these devices, the channel is induced into the superconducting state by a proximity effect from superconducting source and drain electrodes. The strength and decay length of the proximity effect distance are governed by the superconducting coherence length $\xi = (\hbar^3 \mu / 6\pi m e k_B T)^{1/2} (3\pi^2 n)^{1/3}$ induced in the channel, relative to the channel length. Here μ and n are respectively the mobility and the carrier density of the semiconducting channel. The higher mobility and lower carrier density in InAs resulted in devices with channels approaching 1 μm showing some effects, whereas the Si devices had channels of 0.1 to 0.2 μm. The output voltage range over which superconductivity makes some difference is restricted to voltages on the order of the superconducting gap voltage, ≤ 5 mV for conventional superconductors. On the other hand, the voltages required to significantly modify the carrier density are on the scale of a few tenths of a volt to a volt. For example, in the best demonstration of a superconducting FET (40), the Si-device had an output voltage swing of ~1 mV, but had an input gate voltage of ~120 mV.

With the new oxide superconductors, there are prospects for much larger output voltages, approaching perhaps 15 mV to 30 mV, as the gap voltage (or perhaps twice the gap voltage) is in this range. While it is conceivable that semiconducting thresholds could be lowered to ~30 mV, it is does not appear practical to expect superconducting channels to be a big part of digital FET's. Depending on their noise properties, superconducting FET's might have potential for analog applications.

One issue of importance for these superconducting FET's is the attainment of very low contact resistance from the source or drain to the channel. Obviously, poor contacts will inhibit the induction of superconductivity via the proximity effect into the channel. If low, or zero, contact resistance can be achieved, this could give a valuable contribution to the performance of an FET even without the channel becoming superconducting.

Over the years there has been consideration given to field-effect devices where an electric field acts directly on a superconductor (42,43). Only in the most favorable cases for superconducting semiconductors with very low carrier densities (e.g. $SrTiO_3$ with n = 10^{18} cm^{-3} and $T_c \leq 1K$) is such an effect expected to be of a significant size. Estimates place the carrier densities of the new oxide superconductors at least three orders of magnitude higher than in lowest carrier density superconducting semiconductors.

Low-Voltage Hot Electron Transistors. Hot electron transistors with significant gain were demonstrated for the first time in the last couple of years(44). These devices operate most favorably at low temperatures and might be suitable for liquid nitrogen operation. If the base could be made metallic and superconducting then there is the possibility of a nonballistic mode of operation, as proposed for the SUBSIT device due to Frank et al. (45). In this case, the long recombination time in the superconducting base relative to a short time for excitations to tunnel out of the superconductor leads to the possibility of gain. The limiting aspect for this mode of operation is the need for a very special low-barrier-height contact between the superconducting base and the collector.

Devices based on Nonequilibrium Superconductivity. Nonequilibrium superconductivity has been the source of a number of device proposals, though in most cases so far the three-terminal devices lack the important property of nonreciprocity, characteristic of a true transistor-like device. Nonreciprocity seems to be part of at least one proposal, originally due to Giaever (46), for a device in which there is gate-controlled tunneling through superconducting gap. Because of the high current density junctions required for its implementation, it has never been seriously pursued.

Generally the times characteristic of nonequilibrium effects get much faster at higher temperatures, which might be good for device performance, but the coupling of electronic modes and lattice and other modes, which have considerably more heat capacity at high temperature, will be much stronger. This may imply that only heating effects are possible. For devices that rely on the resistivity of the superconducting material when it is in the normal state, the high resistivity of the oxide superconductors might be advantageous. Also the somewhat lower carrier densities in the new materials may make them easier to drive out of equilibrium.

Superconducting Instruments

Electronic instruments utilizing superconductivity can be broadly separated into two classes according to whether or not they are based on the Josephson effects. The former include sensitive SQUID magnetometers, voltage standards, and high speed samplers. SIS mixers and superconducting bolometers are the best known among devices that do not involve Josephson effects. The Josephson-effect devices are available, or close to becoming available, in commercial instruments and it is these that appear to be the most likely candidates for the first commercial applications of high temperature superconductivity, aside from simple shielding applications. Our discussion will be limited to Josephson-effect-based devices.

SQUIDs. Superconducting QUantum Interference Devices (SQUIDs) are flux-to-voltage transducers and are the most sensitive detectors of magnetic flux. Because currents can be used to create magnetic flux, and voltages can drive currents, SQUIDs can be configured as sensitive current and voltage detectors. SQUIDs can be well matched for low impedence measurements. In the best of cases at low temperatures with extremely small input (inductive) impedance, SQUIDs can have white noise levels dominated by quantum mechanical zero-point fluctuations. Generally devices better suited for more general applications have a white noise level dominated by thermal fluctuations. SQUIDs that operate at 77 K will thus have ~20-fold worse noise sensitivity than their helium temperature counterparts. For the highest performance SQUIDs it is clearly desirable to operate at low temperatures and the new high temperature superconductors offer no particular performance advantage. The high performance arena includes neuromagnetic imaging, an application under active development because of its large growth potential.

Many applications of present commercial SQUIDs (for example those applications in which background noise levels are high) do not make use of their extreme sensitivity, and these applications could benefit from easier-to-use instrumentation. The energy sensitivity of the best currently available commercial dc SQUIDs is ~4000$\hbar$. The flux gate magnetometer is the most nearly competitive magnetic field sensing device, having an equivalent energy sensitivity of 10^4 to $10^5\hbar$. Coupled SQUIDs made with IBM-like Josephson technology typically have sensitivities of 100$\hbar$ to 1000$\hbar$. If these could be fabricated with the new superconducting material and their performance scaled as the ratio of operating temperatures, 77 K/4.2 K = 18, then their sensitivity would be in the range of 1800 to 18000$\hbar$. Thus it appears possible, at least on paper, to make 77 K SQUIDs that outperform all presently available commercial SQUIDs. A key implicit assumption in this calculation is that the the high temperature SQUIDs will have junctions with no greater capacitance than the C ~ 0.1 pF of present day Josephson technology.

An aspect of importance to some laboratory applications of dc SQUIDs that were developed and are used at IBM   is that the SQUIDs operate over a broad temperature range. For instance, small particle susceptometers must have samples in intimate contact with them such that the entire SQUID has to be warmed and cooled with the experiment (47). High T_c material could potentially extend the temperature range approximately 10 fold.

Simple SQUID devices in a few forms have already been demonstrated to operate in the liquid nitrogen temperature range. Koch et al. (48) successfully operated at 68 K a one-layer thin film dc SQUID with long, granular, weak-link bridges, while Iguchi et al. (49) and Zimmerman et al.(50) successfully operated rf- and dc-SQUIDs, respectively, based on bulk ceramic loops with break-junction or point contact Josephson elements. The rf-SQUID of Zimmerman et al. was successfully operated in a flux-locked loop at 77 K.

Technology issues of importance for developing high temperature, thin film SQUID technology will be patterning complex multilevel integrated circuits, the variable temperature properties of the Josephson element (if it is a constriction link or a short normal conducting link and not a tunnel junction), shunt impedence (capacitive or inductive), and film critical current density. If the SQUID is always to be operated at a fixed temperature that can be well regulated (say 77K), then the temperature dependence of non-junction Josephson elements may be less of an issue. The critical current density demand stems from the current concentration due to the step-down transformer that usually comprises the SQUID input. In magnetometer applications large currents may be involved; in some gradiometers designs currents can be kept small.

High Speed Josephson Electronics. High speed electronic sampling circuits have been developed with Josephson technology. The highest performance achieved for on-chip measurement of Josephson waveforms has 2.1 ps risetime and 3.7 ps pulse half-width (51). The first commercial superconducting time-domain reflectometer/sampler is just being introduced (52). Its time resolution should be in the range of a few picoseconds, which is about an order of magnitude better than the best available conventional sampling oscilloscope. Josephson samplers are also capable of one to two orders of magnitude better sensitivity than the 0.1 mV sensitivity of conventional samplers, though their dynamic range can be limited. There does not appear to be a semiconductor technology that can match the performance of Josephson technology in this application, though there is a potential for electro-optic instruments that could have comparable or even higher temporal resolution.

The intrinsic sampler resolution is essentially determined by the ratio of current density to specific capacitance of a Josephson junction pulser contained in the sampling circuit. For picosecond resolution with existing helium temperature superconducting technologies, junction current densities are approaching 10^4 to 10^5 A/cm^2. The upper end of this is at the limit that has presently been demonstrated for the YBaCuO superconductor itself at 77 K, so achieving such density in a junction will be a major challenge.

If a suitable junction technology can be developed to meet the sampler performance demands, then it is obviously true that operation at 77K would allow substantial cost reductions in the instrument manufacture and operation. The sensitivity would degrade some at 77 K, but this should not be much of a problem. One aspect that makes the sampling circuit application attractive is that it is a few gate-count circuit and the circuit vertical structure can be relatively simple.

Voltage Standard. The voltage standard in most countries is based on the Josephson relationship between the ac signal applied to a junction and the resultant dc voltage output of a junction. Until very recently, the output voltage has been at the relatively low level of a few millivolts, but recent advances in fabricating uniform series arrays of junctions and engineering the uniform application of microwave power to the whole array, have made possible Josephson standards with output at the one volt level (53). This achievement makes possible the development of relatively simple commercial calibration standards with voltage precisions of a few parts in 10^9.

Once modest quality tunnel junctions can be fabricated from the new superconductors, liquid nitrogen mV-level voltage standards can be readily made. When uniform arrays of junctions can be fabricated, a liquid nitrogen, one-volt level standards could be easily made. These could well be among the first applications for the new high temperature superconductors.

Conclusions

Space does not permit consideration of a vast number of other possible applications for high temperature superconductivity, including magnetic shielding, infrared sensors, telecommunications, and so forth. This review has revealed many design constraints set by the known or extrapolated properties of existing YBaCuO materials. The tremendous potential impact of the applications makes the development of these properties or the discovery of yet better materials of great importance. A more reliable assessment awaits data on these new high temperature superconductors fashioned into more realistic wire or patterned film configurations.

It seems safe to predict that there will be commercial applications of even the existing YBaCuO group of the new superconductors, with the simplest applications like magnetic shielding appearing in a short time frame. Next could come instrument applications. The more complex applications in large magnets or computing systems will probably take longer to develop. The promise is there, but in view of the many challenges described in this article, a long-term view will be essential in fully exploiting this promise. The coming years will surely be exciting ones in the history of electromagnetic technology.

The authors thank R. L. Garwin, D. Heidel, C. Tsuei, R. Koch, B. A. Zeitlin and many other colleagues for valuable discussions.

Literature Cited

1. J. G. Bednorz and K. A. Müller, Possible High T_c Superconductivity in the Ba-La-Cu-O System, Z. Phys. B **64**, 189 (1986).

2. M. K. Wu, J. R. Ashburn, C. J. Torng, P. H. Hor, R. L. Meng, L. Gao, Z. J. huang, Y. Q. Wang and C. W. Chu, Superconductivity at 93 K in a New Mixed Phase Y-Ba-Cu-O Compound at Ambient Pressure, Phys. Rev. Lett. **58**, 908 (1987).

3. Z. Zhao, L. Chen, Q. Yang, Y. Huang, G. Chen, R. Tang, G. Liu, C. Cui, L. Chen, L. Wang, S. Guo, S. Li and J. Bi, Superconductivity Above Liquid Nitrogen Temperature in Ba-Y-Cu Oxides, Kexue Tongbao **6** (1987), to be published.

4. **Superconductor Applications: SQUIDs and Machines**, ed. B. ᴿ Schwartz and S. Foner (Plenum Press, New York 1976).

5. F. Beech, S. Miraglia, A. Santoro and R. S. Roth, Neutron Study of the Crystal Structure and Vacancy Distribution of the Superconductor $Ba_2YCu_3O_{9-\delta}$, to be published, Phys. Rev. B Rapid Commun.

6. R. J. Cava, B. Batlogg, R. B. van Dover, D. W. Murphy, S. Sunshine, T. Siegrist, J. P. Remeika, E. A. Rietman, S. Zahurak and G. P. Espinosa, Bulk Superconductivity at 91 K in Single Phase Oxygen-Deficient Perovskite $Ba_2YCu_3O_{9-\delta}$, Phys. Rev. Lett. **58**, 408 (1987).

7. A. J. Panson, A. I. Braginski, J. R. Gavaler, J. K. Hulm, M. A. Janocko, H. C. Pohl, A. M. Stewart, J. Talvacchio and G. R. Wagner, The Effect of Compositional Variation and Annealing in Oxygen on Superconducting Properties of $Y_1Ba_2Cu_3O_{8-y}$, submitted to Phys. Rev. B.

8. T. R. Dinger, T. K. Worthington, W. J. Gallagher and R. L. Sandstrom, Direct Observation of Electronic Anisotropy in Single Crystal $Y_1Ba_2Cu_3O_{7-x}$, Phys. Rev. Lett. **58**, 2687 (1987).

9. P. Chaudhari, R. H. Koch, R. B. Laibowitz, T. R. McGuire and R. J. Gambino, Critical Current Measurements in Epitaxial Films of $YBa_2Cu_3O_{7-x}$ Compound, Phys. Rev. Lett. **58**, 2684 (1987).

10. M. Beasley, at International Workshop on Novel Mechanisms of Superconductivity, Berkeley CA, June 22-26, 1987.

11. R. Kwo, at International Workshop on Novel Mechanisms of Superconductivity, Berkeley CA, June 22-26, 1987.

12. R. W. McCallum, J. D. Verhoeven, M. A. Noack, E. D. Gibson, F. C. Laabs and D. K. Finnemore, Problems in the Production of $YBa_2Cu_3O_x$ Superconducting Wire, in **Ceramic Superconductors** , special issue of Advanced Ceramic Materials, ed. D. R. Clarke and D. W. Johnson, to be published July 1987.

13. K. Kitazawa, private communication on critical currents obtained in polycrystalline films at Sumitomo Electric Co., Osaka, Japan.

14. M. Tinkham, **Introduction to Superconductivity** (McGraw-Hill, New York 1975).

15. G. Bogner, **Large Scale Applications of Superconductivity**, in Ref. 4, p. 547.

16. D. C. Larbalestier, A. W. West, W. Starch, W. Warnes, P. Lee, W. K. McDonald, P. O'Larey, K. Hemachalam, B. Zeitlin, R. Scanlan and C. Taylor, High Critical Current Densities in Industrial Scale Composites Made from High Homogeneity Nb 46.5 Ti, IEEE Trans. **MAG-21** , 269 (1985).

17. R. E. Schwall, G. M. Ozeryansky, D. W. Hazelton, S. F. Cogan and R. M. Rose, Properties and Performance of High Current Density Sn-Core Process of MF Nb_3Sn, IEEE Trans. **MAG-19** , 1135 (1983).

18. Y. Iye, T. Tamegai, H. Takeya and H. Takei, The Anisotropic Upper Critical Field of Single Crystal $YBa_2Cu_3O_x$, submitted to Japan. J. Appl. Phys.

19. T. K. Worthington, W. J. Gallagher and T. R. Dinger, Upper Critical Field Measurements and Anisotropy in Single-Crystal $Y_1Ba_2Cu_3O_{7-x}$, submitted to Phys. Rev. Lett.

20. T. R. Dinger, R. F. Cook and D. R. Clarke, Fracture Toughness of $YBa_2Cu_3O_{7-x}$ Single Crystals, submitted to Applied Physics Letters.

21. T. P. Orlando, K. A. Delin, S. Foner, E. J. McNiff, Jr., J. M. Tarascon, L. H. Greene, W. R. McKinnon and G. W. Hull, Phys. Rev. B Rapid Comm., May 1987.

22. D. O. Welch, M. Suenaga and T. Asano, On the Anisotropy of H_{c2} and the Breadth of the Resistive Transition of Polycrystalline $YBa_2Cu_3O_{7-x}$ in a Magnetic Field, submitted to Phys. Rev. Lett.

23. S. Jin, R. C. Sherwood, R. B. van Dover, T. H. Tiefel and D. W. Johnson, Jr., High T_c Superconductors - Composite Wire and Coil Fabrication, Appl. Phys. Lett. **51**, July 1987, to be published.

24. G. W. Crabtree, J. W. Downey, B. K. Flandermeyer, J. D. Jorgensen, T. E. Klippert, D. S. Kupperman, W. K. Kwok, D. J. Lam, A. W. Mitchell, A. G. McKale, M. V. Nevitt, L. J. Nowicki, A. P. Paulikas, R. B. Poeppel, S. J. Rothman, J. L. Routbort, J. P. Singh, C. H. Sowers, A. Umezawa and B. W. Veal, in **Ceramic Superconductors**, special issue of Advanced Ceramic Materials, ed. D. R. Clarke and D. W. Johnson, to be published July 1987.

25. R. E. Schwall, MRI - Superconductivity in the Marketplace, IEEE Trans. **MAG-23** (1987), to be published.

26. B. A. Zeitlin, Intermagnetics General Corporation, private communication: NbTi J_c vs. H curve scaled to SSC spec with cost fixed at 5 T by the current SSC design cost document. Multifilamentary Nb_3Sn made by the internal tin core process, with costs for quantities of 10^6 pounds; critical current fixed at 10 T and scaled using Nb_3Sn tape magnet design curves at other fields. (See Fig. 1).

27. R. E. Schwall, G. M. Ozeryansky, D. W. Hazelton, S. F. Cogan and R. M. Rose, Properties and Performance of High Current Density Sn-Core Process Multifilamentary Nb_3Sn , IEEE Trans. **MAG-19**, 1135 (1983).

28. R. L. Garwin and J. Matisoo, Superconducting Lines for Transmission of Large Amounts of Electrical Power over Great Distances, Proc. IEEE **55**, 538 (1967).

29. E. B. Forsyth, M. Garber, J. E. Jensen, G. H. Morgan, R. B. Britton, J. R. Powell, J. P. Blewett, D. H. Gurinski and J. M. Hendrie, Factors Influencing the Choice of Superconductor in AC Power Transmission Applications, **Proceedings of the Applied Superconductivity Conference** , IEEE Pub. No. 72CH0682-5 TABSC, 202 (1972).

30. E. B. Forsyth and R. A. Thomas, Performance Summary of the Brookhaven Superconducting Power Transmission System, Cryogenics **26**, 599 (1986).
31. D. F. Heidel, private communication.
32. J. Y.-C. Sun, Y. Taur, R. H. Dennard and S. P. Klepner, Submicrometer-Channel CMOS for Low-Temperature Operation, IEEE Trans. Electron Devices **34**, 19 (1987).
33. J. Matisoo, The Superconducting Computer, Scientific American **242**, No. 5, 50 (May, 1980).
34. Special Issue on Josephson Computer Technology, IBM Journal of Research and Development **24** No. 2, pp. 107-252 (1980).
35. Special Issue: Josephson Computer Technology, Bulletin of the Electrotechnical Laboratory, **48**, No. 4, pp. 1-131 (1984).
36. W.H. Henkels, L.M. Geppert, J. Kadlec, P.W. Epperlein, H. Beha, W.H. Chang, and H. Jaeckel, Josephson 4 K-bit Cache Memory Design for a Prototype Signal Processor. I. General Overview, J. Appl. Phys. **56**, 2371 (1985).
37. H.H. Hayakawa, Josephson Computer Technology, Physics Today **39**, No. 3, 46 (March, 1986).
38. H. Jaeckel and D.F. Moore, Stray Magnetic Coupling between Neighboring Superconducting Memory Cells in Dense Arrays of Josephson Interferometers, J. Appl. Phys. **53**, 5936 (1982).
39. W.J. Gallagher, Three-Terminal Superconducting Devices, IEEE Trans. **MAG-21**, 709 (1985).
40. T. Nishino, M. Miyake, Y. Harada, and U. Kawabe, Three-Terminal Superconducting Device using a Silicon Single-Crystal Film, IEEE Electron Dev. Lett., **EDL6**, 297 (1985).
41. H. Takayanagi and T. Kawakami, Superconducting Proximity Effect in the Native Inversion Layer on InAs, Phys. Rev. Lett. **54**, 2449 (1985).
42. For early references see references 24-27 in cited by Gallagher in Ref. 39.
43. M. Gurvitch, H.L. Stormer, R.C. Dynes, J.M. Graybeal, and D.C. Jacobson, Field Effect on Superconducting Surface Layers of $SrTiO_3$, Extended Abstracts, Superconducting Materials, 1986 Fall Meeting of the Materials Research Society, 47 (1986).
44. M.I. Nathan and H. Heiblum, A Gallium Arsenide Ballistic Transistor, IEEE Spectrum **23**, 45 (1986).
45. D.J. Frank, M.J. Brady, and A. Davidson, A Superconducting Base Transistor, IEEE Trans. **MAG-21**, 721 (1985).
46. I. Giaever, U.S. Patent No. 3,116,427.
47. M. B. Ketchen, T. Kopley and H. Ling, Miniature SQUID Susceptometer, Appl. Phys. Lett. **44**, 1008 (1984).
48. R.H. Koch, C.P. Umbach, G.J. Clark, P. Chaudhari, and R.B. Laibowitz, Quantum Interference Devices (SQUIDs) Made from Superconducting Oxide Thin Films, Appl. Phys. Lett.,July 1987, to be published.

49. I. Iguchi, A. Sugishita, and M. Yanagisawa, DC SQUID Operation at 77 K Using All Ceramic Josephson Junctions, submitted to Japan. J. Appl. Phys.

50. J.E. Zimmerman, J.A. Beall, M.W. Cromar, and R.H. Ono, Operation of a Y-Ba-Cu-O rf SQUID at 81 K, submitted to Appl. Phys. Lett.

51. P. Wolf, B.J. van Zeghbroeck, and U. Deutch, A Josephson Sampler with 2.1 ps Resolution, IEEE Trans. MAG-21, 226 (1985).

52. Hypres Inc., Elmsford, NY.

53. J. Niemeyer, J.H. Hinken, and R.L. Kautz, Microwave-Induced Constant-Voltage Steps at One Volt from a Series Array of Josephson Junctions, Appl. Phys. Lett. 45, 478 (1984).

RECEIVED July 10, 1987

RESEARCH NEEDS AND OPPORTUNITIES

Chapter 28

Research Opportunities and New Directions in High-Temperature Superconducting Materials

David L. Nelson[1], M. Stanley Whittingham[2], and Thomas F. George[3]

[1]Chemistry Division, Office of Naval Research, Arlington, VA 22217-5000
[2]Schlumberger-Doll Research, Old Quarry Road, Ridgefield, CT 06877-4108
[3]Departments of Chemistry and Physics & Astronomy, 239 Fronczak Hall, State University of New York at Buffalo, Buffalo, NY 14260

Various directions of future research needs and opportunities are discussed. These directions include the development of theoretical descriptions of the superconducting behavior and formation of high T_c materials, the synthesis and characterization of such materials, and the fabrication of corresponding wires, fibers, thin films and device structures, with applications such as in microcircuitry. New types of chemistry associated with superconducting solids and surfaces are also suggested.

The recent breakthrough in high-temperature superconductivity of inorganic materials emphasizes the need for a better understanding of the phenomenon of superconductivity from both experimental and theoretical points of view. Inorganic superconductors are not new: a number of oxides and sulfides and their intercalated products were found to be superconducting below 10 K more than fifteen years ago (1), and superconducting structures containing paramagnetic ions were observed (2). In 1973, lithium titanium oxide was found with T_c at 13 K (3) followed by barium lead bismuthate in 1975 (4). However, essentially all research in the following years was aimed at optimizing the A15 compounds for commerical applications, with noted success in applications to superconducting NMR machines. The finding of superconductivity beyond 90 K in copper-containing oxides is not expected to be unique to this class of materials, and demands an understanding as to whether those characteristics can be extended to other systems, including organics.

Overall, there is a need to:
- Describe superconducting behavior in high T_c materials and the reaction dynamics leading to such materials (are these, by necessity, metastable?) using microscopic and statistical theories, and test predictions experimentally.
- Synthesize new high T_c superconducting materials.
- Address preparation of bulk materials, thin films and device structures containing high T_c materials.
- Characterize high T_c superconducting materials to establish structure-property relationships leading to predictive capabilities.
- Enhance environmental stability of high T_c superconductors.
- Explore new types of chemistry associated with superconducting solids, surfaces and interfaces.
- Develop processing science and fabrication approaches for controlling composition, crystal structure, microstructure and morphology in processing wires, fibers, tapes, thin films and device structures with optimized properties, for both small- and large-scale demonstration projects.
- Develop the bases for design techniques for low-power device structures and methods for their integration into microcircuits, as well as for high-power device structures, including concepts to assure their mechanical integrity and power-handling capacity above T_c.

We now expand briefly on the above points. Theoretical models for high T_c materials need to be developed. Work must be carried out on the basic general mechanisms to develop models that are predictive and capable of being tested. Such models would point the way to achieving desirable characteristics of materials, leading to a more scientific approach for developing even higher T_c materials. Ab initio electronic methods (e.g., generalized valence bond and configuration interaction) should be utilized to examine localized states, together with approaches based on highly-localized chemical interactions, the objective being to predict properties for new systems derived by atom substitutions. Density functional-based and coupled-cluster ab initio methods can be used to compute ideal single crystal properties such as the electronic structure of layered perovskites, to elucidate bonding mechanisms, valence electron fluctuations, vacancy and defect energies, pairing mechanisms, total energy calculation of structural stability (including strain behavior), and phonon frequencies and force constants. A theoretical treatment of the unique properties of thin films, grain boundaries, epitaxially-grown interfaces consisting of metals, semiconductors and insulators (including strain considerations), and reduced geometry structures is desirable. The information so generated can be used in molecular dynamical computer simulations of reactions leading to high T_c superconducting structures. Such simulations will involve, for example, Monte Carlo techniques and nonequilibrium statistical procedures to describe the time evolution of dynamical variables in a statistical ensemble, satisfying the principle of maximum entropy.   It would also be useful to explore the transition region from a cluster to bulk material and the onset of superconductivity.

In the absence of a universally accepted predictive theory describing high T_c superconductivity, the basic characteristics of known high T_c materials must be fully investigated and understood, since they are likely to be important factors in producing other classes of materials with even more spectacular properties. The chemist needs to determine why copper, and why in particular the relatively unstable Cu(III), is special, and why the defect-rich and/or non-isotropic crystalline (one- and two-dimensional) structure is apparently necessary for a high superconducting temperature. Are such materials by definition metastable, that is, must one form such compounds by low-temperature modifications (for example, oxygen intercalation) of high-temperature structures? For example, $YBa_2Cu_3O_{7-\delta}$ can apparently not be formed below 800°C, yet its structure can be modified by a low-temperature anneal (500-700°C) to incorporate excess oxygen to convert it into the high T_c phase. In addition, the preparation of new inorganic compounds carried out by novel techniques, such as the sol-gel route and the synthesis and use of new classes of organometallic precursors, need to be emphasized. Novel methods of preparing both metastable and stable compounds of superconducting oxides, including other cation-anion pairs such as nitrides, sulfides, chlorides, hydrides and fluorides, as well as chalcogenide materials and composites, should be considered. The structure and properties of multicomponent phases must be thoroughly investigated as a function of composition. By systematically varying structure and composition, theoretical predictions of superconductivity in these novel materials can be tested. Organic and polymeric compounds should also be explored as potential high T_c superconductors. Possible candidates are long molecules with side chains that can be polarized by electrons travelling down the spine of the molecule and thus might serve to attract other electrons to form Cooper pairs.

The preparation of well-characterized thin films is particularly challenging since the surface chemistry and physics related to nucleation and growth need to be examined within the context of the rather complicated ternary and quaternary component systems. Emphasis must be placed on optimizing the use of techniques such as reactive sputtering, e-beam or thermal-source atomic or molecular beams, chemical vapor deposition of novel organometallic precursors of high T_c materials, and plasma-assisted deposition (or combinations of these) for growing thin films. A special emphasis should be given to growing epitaxial thin films of the highest T_c superconducting materials that are sufficiently robust to meet the demands of each of the processing steps to fabricate demonstration devices, and that are environmentally stable within the context of various applications.

A close tie between the synthesis and processing of new superconducting compounds and the measurements of their properties is essential. The measurement of structural properties of bulk phases, thin films and microstructures utilizing techniques such as X-ray diffraction, electron microscopy, SEXAFS, STM, photoemission, and electron- and photon-induced desorption needs to be undertaken, and the relation of material processing to properties evaluated. Magnetic characterization, electronic transport, thermodynamic measurements such as critical magnetic fields, critical temperatures and specific heats, possibly optical properties, and anisotropy of critical currents aid in refining synthesis conditions and in optimizing phases, compositions and processing conditions. The correlation of microstructure, grain boundary character and single crystal behavior (e.g., critical transition temperature, critical current, coherence length and structural anisotropy) with critical current for polycrystalline materials must be examined.

It is especially important to address environmental degradation of high T_c superconducting materials, since reliable use of these materials in all applications is determined to a large extent by external interactions. Emphasis is placed on surface studies of both bulk and thin-film materials to examine chemical stability as a function of temperature, the effect of electron and photon bombardment, interfacial reactivity of atomic oxygen, more generally the effects of oxygen depletion or enrichment, and chemical corrosion by water and other chemically hostile environmental constituents. Since the known high T_c materials are prone to rapid environmental degradation by water, all degradative mechanisms should be addressed together with passivating and corrosion inhibiting chemistry for protection. Important to environmental degradation and to materials processing is a knowledge of the ionic mobility of the constituent ions, in particular oxygen and copper, as a function of composition, temperature and oxygen partial pressure. The ionic mobility not only determines the temperature and times of processing, but also how rapidly one might expect degradation of the processed material under operational conditions. These electronically-conducting, highly non-stoichiometric cuprates might also find use as electro- and oxidation catalysts, where this diffusion information would also be essential.

The recent development of high T_c superconductors opens a whole new realm of possible chemistry involving supercondcting solids and surfaces. It had always been assumed that a perfect conductor did not exist, but now comes a special kind of perfect conductor available at temperatures readily accessible to chemists. An example of new chemistry is laser-induced chemical vapor deposition, where a laser could be used to "write" an insulating layer or a "normal" conducting layer on top of a superconductor, or even to write a superconductor on a semiconducting substrate.

Identifying and addressing issues related to high T_c materials required to process components for both small- and large-scale demonstration projects should be undertaken. For example, the growth, pattern generation and transfer of specific material structures with fully optimized properties for VLSI devices must be developed. Examples include superconducting interconnects and thin-film transmission lines for superconductor-semiconductor integrated circuits, magnetic and infrared sensors, Josephson junctions, SQUIDS and material configurations for high-power devices. Fundamental issues arising from processing required to control the microstructure and morphology of wires, fibers, tapes, monoliths and other macroscopic bodies for large-scale demonstrations should be intensively addressed.

Even in the absence of a complete research-knowledge base, efforts should continue to address design and fabrication basics, focussing on device and components architectures and processes for these new materials, including related temperature stabilization and refrigeration approaches. However, in the drive to commercially exploit these new materials, the basic understanding must not be neglected, and chemistry has a key role to play alongside physics and materials and engineering science to ensure rapid progress toward full commercial implementation.

Literature Cited

1. Whittingham, M. S. and Ebert, L. B. In <u>Intercalated Layered Materials</u>; Levy, F. Ed.; D. Reidel Publ. Co.: Dordrecht, Holland, 1979; p 555.
2. Gamble, F. R. and Thompson, A. H., Solid State Commun. 1978, <u>27</u>, 379.
3. Johnston, D. C.; Prakash, H.; Zachariasen, W. H. and Viswanathan, R., Mater. Res. Bull. 1973, <u>8</u>, 777.
4. Sleight, A. W.; Gillson, J. L. and Bierstedt, P. E., Solid State Commun. 1975, <u>17</u>, 299.

RECEIVED July 7, 1987

Chapter 29

Synthesis, Chemistry, Electronic Properties, and Magnetism in the Y–Ba–Cu–O Superconductor Systems

Ruth Jones[1], M. F. Ashby[1], A. M. Campbell[2], Peter P. Edwards[1], M. R. Harrison[1], A. D. Hibbs[1], D. A. Jefferson[1], Angus I. Kirkland[1], Thitinant Thanyasiri[3], and Ekkehard Sinn[3]

[1]University Chemical Laboratory, University of Cambridge, Lensfield Road, Cambridge, CB2 1EW, Great Britain
[2]G.E.C. Research Laboratories, Hirst Research Centre, Wembley, Great Britain
[3]Department of Chemistry, University of Virginia, Charlottesville, VA 22901

The relation of sample preparation conditions, including oxygen concentration, annealing and hot isostatic pressing, to the superconducting critical temperature of the title materials is described. New synthetic materials are examined and characterized by a variety of techniques, including magnetic susceptibility measurements, ESR, electron microscopy and X-ray diffraction. The relationship between magnetization and critical current is examined.

In the spirit of this symposium, we outline a snapshot in time of our present work on inorganic superconducting materials. The aim is to illustrate how a wide range of physical techniques from various disciplines and laboratories have been brought to bear on this remarkable new class of material. Our ultimate aim is to pursue material development in parallel with the development of useful devices. The $YBa_2Cu_3O_{7-\delta}$ materials already available have the possibility of immediate application and it is essential that the development stage of these materials be carried out in parallel with ongoing research on new materials.

Sample Preparation and Superconducting Transitions

Y_2O_3, CuO and $BaCO_3$ were used as starting materials, each 99.999% pure (Aldrich). The purity and phasic nature of the starting materials were checked by X-ray powder diffraction. The powders were thoroughly mixed, ground (pestle and mortar) and pressed into pellets, which were fired at 935 °C for 16 h in air. The samples were then annealed for 24 h in O_2 at 1000 °C, and then for a further 6 h in O_2 with the furnace temperature reduced to 420 °C. Samples were characterized by powder Cu $K\alpha$ X-ray diffraction.

Resistivity measurements revealed characteristic transitions of the type shown in Figure 1 for both warming and cooling cycles. We have observed frequent, but not reliably repeatable sudden changes in resistance when a sample was cooled slowly. Figure 1 includes a resistive transformation in which the resistance appeared to fall to zero for about a second at 150K, before returning to its original value. These effects generally occur near this temperature although smaller effects have been seen up to 230K. During the excursion the total voltage across the sample may fall, but does not rise so we do not think cracking of the specimen can explain the results. However, more work is required before we could attribute the effect to superconductivity. The noise level of the voltage meant that a metallic transition was a possibility and we have not yet observed reliable excursions during an inductive measurement.

The magnetic susceptibility determination offers a ready means of examining superconductivity via the magnetic shielding it produces. The shielding is seen as enhanced diamagnetism, and unlike resistivity data, it depends only on proportion of superconducting material and not on whether and to what extent the superconducting particles are linked. The magnetic measurements were made on a high precision SQUID magnetometer; the calibration of the instrument and measurement techniques are as described previously (1,2). Data are given in Figure 2 for two samples prepared under slightly different temperature conditions. The magnetism indicates the onset of the transition above 90K, but the susceptibility continues to decreases further at lower temperature before leveling out. This is translated into percent superconducting material for each sample in Figure 3. In each case, a superconducting component manifests itself above 90K. As temperature is decreased further, more of the sample becomes superconducting, indicating that other species are also present with a lower critical temperature. This mixture of phases causes the gradual drop in susceptibility rather than a discontinuity. The difference between the two curves shows how sensitive the proportions of the phases are to sample preparation and annealing conditions. Magnetism and critical current measurement is further discussed below.

Structure

X-ray crystallography (3-5) reveals that the $YBa_2Cu_3O_{7-\delta}$ structure is described by a layer sequence along the a-axis, with BaO-$CuO_{1-\delta}$-BaO-CuO_2-Y-CuO_2-Y-BaO-$CuO_{1-\delta}$-BaO. A representation of the idealized structure is given in Figure 4. The large Ba^{2+} ion is located between the Cu(2) square pyramidal copper layers and the Cu(1) square planar sheets. There are two crystallographically distinct and chemically dissimilar copper sites. Cu(1) is in a square planar of oxygen ions, while Cu(2) is five-coordinated by a square pyramid of oxygens. Local charge neutrality considerations favor Cu^{3+} ions in the Cu(1) plane while Cu^{2+} are expected to dominate in the other square pyramidal (Cu(2) planes in the orthorhombic cell). For the stoichiometry $YBa_2Cu_3O_7$, the Cu ion in the Cu(1) plane is four-coordinated by oxygen ions, and CuO_4 units are corner-sharing in the b and c directions but are connected directly through Cu in chains along a. The preference of square planar geometry in d^8 ions is well established (e.g. Ni^{2+}). This is also consistent with the short O-Cu-O bonds in the c direction. In the Cu(2) plane, the Cu ion is five-coordinated and also forms corner sharing units. Here the four oxygen ions lie on the same ab plane slightly displaced in the c

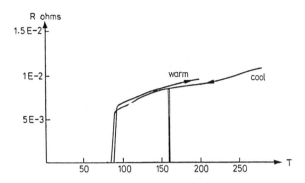

Figure 1. Resistive transition (with transient effect included).

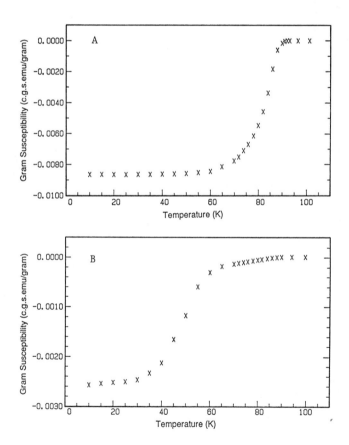

Figure 2. Gram susceptibility vs. temperature for two samples, A and B.

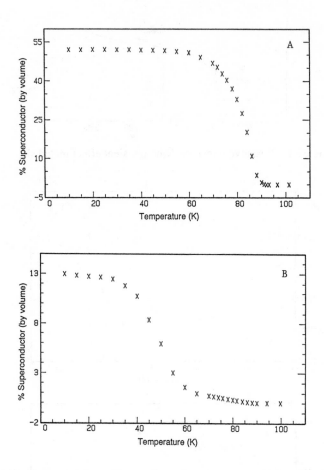

Figure 3. Percent superconductor (by volume) vs. temperature for samples, A and B.

direction from the plane of Cu ions.   This almost certainly arises from the effects of the large Ba^{2+} ion (with Goldsmith tolerance factor $t \leq 1$) and small highly charged Y^{3+} ion.   Clearly extensive distribution of Cu^{3+}/Cu^{2+} ions in both Cu(1) and Cu(2) planes is called for.   It is assumed then, that oxygen deficiency relates to the behavior of oxygen ions in the basal (Cu(1)) plane.   Such oxygen vacancies give rise to a structural distortion which changes the crystal system from tetragonal to orthorhombic.   These oxygen vacancies are randomly distributed at high temperatures, and ordered at low temperatures.   The analogous situation for oxygen defects in superconducting $LiTi_2O_4$ has been treated in detail (6).

Phase Diagram and Characterization

The phase relations among the stable phases in the Y_2O_3–BaO-CuO system (7,8), shown schematically in Figure 5, are crucial for studies aimed at isolating, identifying and studying the high T_c material, as well as for materials processing. The details of the phase compatibilities have been discussed elsewhere and only a few salient points are outlined here.   At most, three phases can coexist when the temperature and partial pressure of oxygen are fixed.   The compounds Y_2O_3, BaO, $Y_2Cu_2O_5$, Y_2BaCuO_5, $BaCuO_2$ and CuO form the boundaries of the mixed phase region outlined in Figure 5.   X-ray diffraction patterns for the phases $Y_2Cu_2O_5$, Y_2BaCuO_5 and $BaCuO_2$ are compared in Figure 6 with a characteristic pattern from the superconducting phase $YBa_2Cu_3O_{7-\delta}$.   The black $BaCuO_2$, blue $Y_2Cu_2O_5$, bright green Y_2BaCuO_5 and brown-black CuO all contain localized Cu^{2+} ions and these materials can also be identified by their characteristic EPR spectra, Figure 7. $BaCuO_2$ and Y_2BaCuO_5 both exhibit broad, anisotropic signal ($g_{||} > g_{\perp}$) with no evidence of Cu hyperfine interactions, consistent with the high Cu^{2+} content in the non-superconducting phases.   The resonance in $YBa_2Cu_3O_{7-\delta}$ is discussed further below; suffice it to say that all anisotropic feature are attributable to localized Cu^{2+} in square pyramidal configurations.

Oxygen Content: the Orthorhombic-Tetragonal Transition

Both the tetragonal and orthorhombic structures of the Y-Ba-Cu-O system are derived from a regular ABX_3 perovskite; they differ, however, in the specific oxygen vacancy arrangements and these depend critically on preparation conditions.   If this composition were referred to as a perovskite structure, its idealized composition would be $YBa_2Cu_3O_9$, which would require an extra sheet of oxygen atoms at the Y level.   Slow cooling, and extensive oxygen annealing, appears to produce superconducting orthorhombic materials with high transition temperatures.   We find, in agreement with other workers , that vacuum annealing leads to a loss in superconducting properties and a transition to a tetragonal phase. Thus, vacuum annealing of a superconducting orthorhombic $YBa_2Cu_3O_{7-\delta}$ sample produces noticeable differences in line intensities around $2\theta = 32.5°$, 47°, 58° and 68° in the X-ray diffraction patterns (Figure 8).   Similar results have recently been obtained by Kini et al (9) who produced the tetragonal phase via rapid quenching of $YBa_2Cu_3O_{7-\delta}$.   A schematic representation of the dependence of oxygen

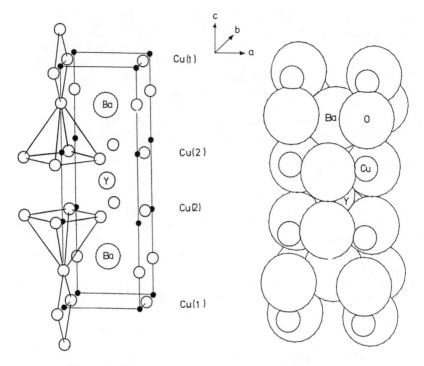

Figure 4. Structure and space filling model of $YBa_2Cu_3O_{7-\delta}$.

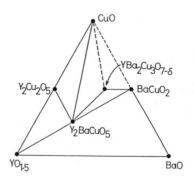

Figure 5. Phase diagram for the Y-Ba-Cu-O system.

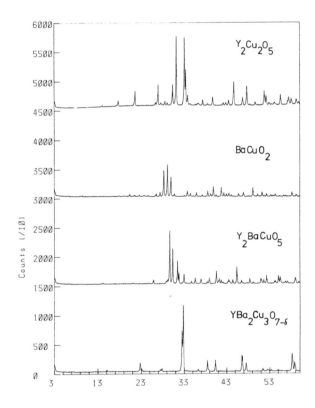

Figure 6. X-Ray diffraction patterns of $Y_2Cu_2O_5$, $BaCuO_2$, Y_2BaCuO_5, $YBa_2Cu_3O_{7-\delta}$.

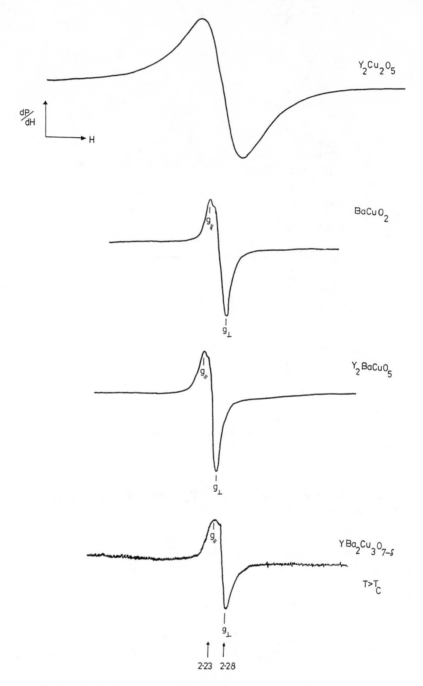

Figure 7. ESR spectra of $Y_2Cu_2O_5$, $BaCuO_2$, Y_2BaCuO_5, $YBa_2Cu_3O_{7-\delta}$.

content on reaction temperature is shown in Figure 9.   At a stoichiometry $YBa_2Cu_3O_6$ we have 1 x Cu^+ and 2 x Cu^{2+}; at O_7 we find 1 x Cu^{3+} and 2 x Cu^{2+}. For oxygen content below O_6, we expect a disproportionation into Y_2BaCuO_5, CuO and BaO (Figure 5).   Decomposition temperatures are dependent upon the vacuum/overpressure conditions ; that is, decomposition with oxygen overpressure may not proceed until ca 1050° C   (10).   However, under dynamic vacuum conditions, or oxygen gettering, this temperature would be considerably lower. Thus the orthorhombic-to-tetragonal transition is dependent both on the annealing temperature and annealing atmosphere (oxygen, vacuum, etc).   Rapid quenching from various temperatures will then freeze in the oxygen content appropriate to that temperature.   Clearly, temperatures in excess of 900° C are required to form the 1:2:3 phase, and only a relatively narrow temperature range then exists for precisely synthesizing the bulk superconducting phase.

High Resolution Electron Microscopy (HREM)

High resolution lattice images were obtained using a modified Jeol 200CX operating at 200 kV.   Computer image simulations were produced using a full numerical solution to the Schrodinger equation from the multislice method (11) and incorporating the atomic coordinates of David et al (3).   Crystallites were tilted with respect to the incident electron beam into high symmetry orientations in the selected area diffraction mode and images were recorded after aligning the illumination at magnifications of either ca 450,000 or 690,000. Astigmatism was corrected by observing the granularity of the amorphous carbon film.   Figure 10 shows lattice images obtained for 2 of the 3 possible high symmetry orientations together with computed image simulations; these results are in good agreement with those obtained by other workers (12).   The unit cell parameters determined from electron diffraction studies are in close agreement with those of superconducting $YBa_2Cu_3O_7$ from single crystal experiments.   In the samples investigated, there was no evidence for planar defects, in contrast to other studies, on materials of the same nominal composition (12,13).   This suggests that the presence of such defects may be extremely sensitive to preparation and annealing conditions.   Exposure of the sample to intense irradiation leads to the formation of small particles within an amorphous layer present at the urface of a number of crystallites.   Energy dispersive X-ray analysis of this region reveals that the amorphous layer is Ba-rich in comparison to the bulk sample.   Furthermore, optical diffractometry from the region (a) of Figure 11 suggests that the lattice spacings of these particles could correspond to those of Ba metal.   One possibility is that prolonged beam damage leads to a reduction of a Ba rich surface phase to metallic Ba, consistent with other studies in which beam induced reductions have been observed in CuO (14).   Further studies of this phenomenon are in progress.

Electron Spin Resonance Studies (ESR)

X- and Q-band ESR experiments have been carried out on both the normal and the superconducting states of $YBa_2Cu_3O_{7-\delta}$.   Both above and below T_c, a resonance

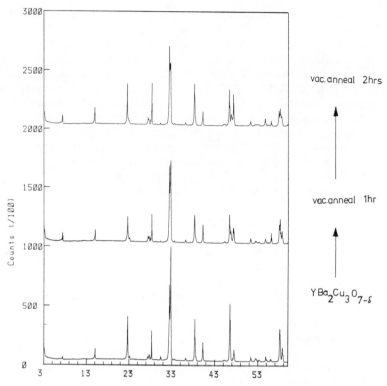

Figure 8.   Effect of vacuum annealing as characterized by X-ray diffraction patterns.

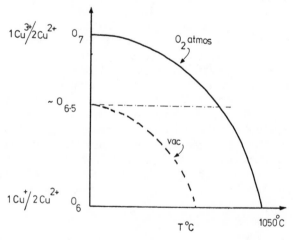

Figure 9.   Schematic representation of Oxygen concentration as a function of conditions.

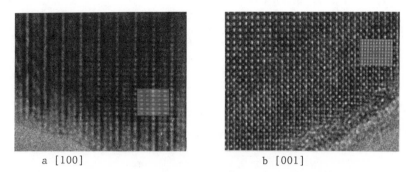

a [100]   b [001]

Figure 10.   HREM in (100) and (001) directions.

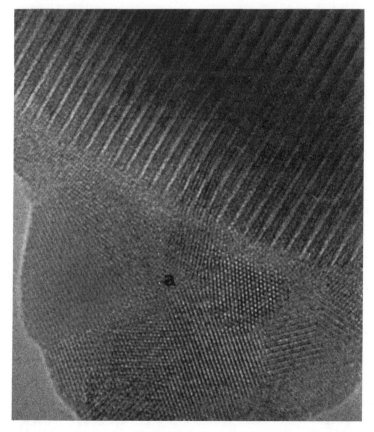

Figure 11.   HREM showing surface phase.

occurs close to the free spin region. The intensity of this signal depends critically on the preparation and annealing conditions. The intensity of the line down to T_c follows a Curie-law dependence. At X-band, the signal has approximate axial symmetry with $g_\parallel = 2.23$ and $g_\perp = 2.08$, with no evidence of Cu hyperfine structure. However, experiments at 35 GHz show a completely anisotropic signal with $g_1 = 2.20$, $g_2 = 2.28$ and $g_3 = 2.08$. This signal can be attributed to localized Cu^{2+} states. Two possibilities can be identified. First, the signal could arise from defect induced, or intrinsic Cu^{2+} sites within the 90 K superconducting phase. Second, the resonance could originate from other Cu^{2+} compounds of the phase diagram (Figure 5). A similar resonance to that in Figure 12 has been investigated by Michel and Raveau (15) in solid solution between Y_2BaCuO_5 and $Y_2BaZn_{0.99}Cu_{0.01}O_5$. Support for the assignment also come from our own studies where Y_2BaCuO_5 was prepared and exhibited similar ESR spectra, at both X and Q-band, to those observed in the 90 K superconducting material.

Cu^{2+} in Y_2BaCuO_5 exists in a distorted square-pyramidal coordination, and this is entirely consistent with the g values ordering, viz $g_\parallel > g_\perp$. Although this is strongly suggestive of an assignment to Y_2BaCuO_5, further experiments are needed. In particular, we find that the intensity of this signal increases noticeably upon vacuum annealing of a superconducting sample. The local symmetry of (potentially) localized Cu^{2+} in the $YBa_2Cu_3O_{7-\delta}$ phase is also based on a square pyramidal structure, and further experiments with internal calibrants are under way.

Below T_c, an intense line at zero field is observed. The signal increases in intensity as the temperature is reduced and Figure 12 shows a typical sequence of ESR spectra spanning the transition from normal to superconducting states. The appearance of this signal below T_c is strongly linked with the occurrence of bulk superconductivity in the $YBa_2Cu_3O_{7-\delta}$ system; intensive studies of a wide range of samples have confirmed this. It occurred to us that one possibility was the crystallization into mobile $Cu^{2+}–Cu^{2+}$ bipolaronic states below T_c. Spatially localized $Cu^{2+}–Cu^{2+}$ dimer, as formed, for example, in nonmetallic copper acetate, also gives rise to an intense zero field absorption in the ESR arising from spin transitions in the triplet electronic state. In this case the location of the dimer signal can be moved by going from X- to Q-band frequencies. No such effect was observed in the 90 K superconducting materials. We therefore attribute their absorption to r.f. surface impedance effects. It is known that r.f. surface resistance of superconducting $YBa_2Cu_3O_{7-\delta}$ decreases by more than two orders of magnitude at the transition temperature (16). The ESR technique, therefore, offers an attractive and simple method for the non-destructive screening of superconducting samples.

Magnetization and Critical Current Measurements

We have used standard ballistic magnetization apparatus to measure the magnetization curves of bulk samples of $YBa_2Cu_3O_{7-\delta}$. Figure 13 shows two magnetization curves, measured at 4.2K up to 2.2T. The outer curve is for a slab 0.5 mm thick. The second curve is the same slab crushed to a powder of about 20

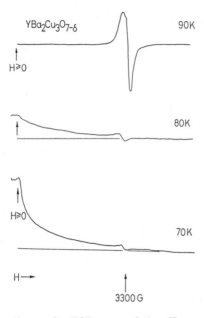

Figure 12. ESR spectra below T_c.

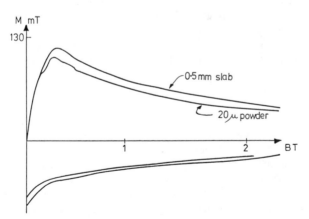

Figure 13. Magnetization curves.

μ mean diameter. It can be seen firstly that there is very little difference between the two curves and secondly that both curves are asymmetric in that the mean of the increasing and decreasing curves are diamagnetic. This allows us to estimate H_{c1} as about 0.07T at 4.2K.

The similarity of the curves confirms the results of Svenaga [17] and shows that the current is almost entirely carried within the grains. Although resistive transitions show there must be a continuous superconducting path through the sample in zero field, the critical current of this path must be very low and the connections between grains are probably driven normal by quite small magnetic fields. It is also consistent with surface impedance measurements [18]. If the current between grains is low, the current within the grains must be high to give the magnetization observed. From Figure 13, using 20 μ grains, we estimate a local current density of 10^6 A/cm^2.

Fundamental Limits to J_c. Before applying standard theories of critical currents we should consider whether these are a completely new type of superconductor. It has been suggested that they might consist of rings of surface currents connected by weak links [19]. This model, however, is not compatible with the reversible diamagnetic curves observed. A second picture is of Type II superconducting grains with weak links between them. This is certainly an acceptable model, but is not qualitatively different from a model in which vortices are weakly pinned at the grain boundaries. Effects attributed to arrays of weak links can be explained with equal validity by an inhomogeneous array of pinning centers.

It therefore seems reasonable to apply earlier theories to these materials, since they require only that the Ginzburg Landau equations be obeyed. Different pinning mechanisms give a similar order of magnitude so if we use the magnetic pinning expression $J = 0.6 S_v B_{c2}(1-b)(b^{1/2}\mu_0 K^3)$ and take the surface area per unit volume S_v as about ½a (a is the vortex spacing) we expect a reasonable value for an ideal microstructure. If dH_{c2}/dT is 0.76 T/K and T_c is 90K this gives $J = 2 \times 10^4$ A/cm^2 at 77K and half B_{c2} (5T).

Hot Isostatic Pressing

Thin films of the perovskite-structured oxides can be prepared by evaporation and by sputtering. The alternative route, best suited for monolithic components, is via powders. In this regard, hot isostatic pressing looks particularly attractive: high density is achieved quickly, shape can be controlled, and, by trading pressure against temperature, the temperature can be kept in the range that maximizes the volume fraction of the superconducting phase. This approach, in tandem with knowledge of the phase diagram (Figure 14), is an important area in the fabrication of both thin films and electrical machines. We have developed models [20, 21] for both sintering and hot isostatic pressing of the Y-Ba-Cu-O ceramic powders, and have employed computer methods for selecting the optimum pressing conditions. Programs have been written which give the final density of a compact sample as a function of temperature, pressure and processing time. Figure 14 shows the predicted densification at both 900° C and 650° C for various processing times.

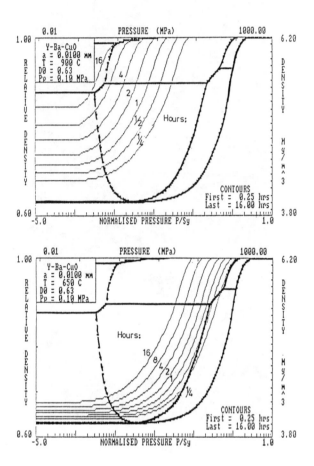

Figure 14. Phase diagram for hot isostatic pressing conditions.

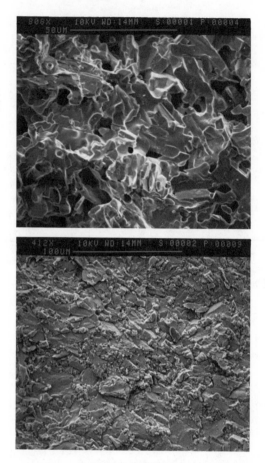

FIgure 15.   Micrograph of hot isostatic pressed phase.

We have pressed a single-phase superconducting sample at 100 MPa and 950° C for 6 h. As predicted the resulting sample was fully dense and Figure 15 shows microscope pictures both before and after processing. The resulting material was multiphasic, with both superconducting and non-superconducting fractions. One problem is that a great deal of oxygen has been absorbed by the steel encapsulation at these elevated temperatures and, consistent with this, a thin layer of metallic copper appears on the surface of the capsule. This may be prevented by working at high overpressures of oxygen, and experiments are now under way; knowledge of the variation of stiochiometry, cell symmetry and superconducting properties in $YBa_2Cu_3O_{7-\delta}$ with temperature and oxygen pressure is crucial.

Acknowledgment

Support from NSF Grant No. CHE83-00516 is gratefully acknowledged.

Literature Cited

1. O'Connor, C. J.; Deaver, B. S., Jr.; Sinn, E. J. Chem. Phys. 1979, 70, 5161.
2. O'Connor, C. J.; Sinn, E.; Cukauskas, E. J.; Deaver, B. S., Jr. Inorg. Chim. Acta 1979, 32, 39.
3. David, W. I. F.; Harrison, W. T. A.; Gunn, J. M. F.; Moze, O.; Soper, A. K.; Day, P.; Jorgenson, J. D.; Hinks, B. G.; Beno, M. A.; Soderholm, L.; Capone II, D. W.; Schuller, I. K.; Segre, C. U.; Zhang, K.; Grace, J. D. Nature 1987, 327, 310.
4. Steinfink, H.; Swinnea, J. S.; Sui, Z. T.; Hsu, H. M.; Goodenough, J. B. J. Am. Chem. Soc. 1987, 109, 3348.
5. Murphy, D. W.; Sunshine, S.; van Dover, R. B.; Cava, R. J.; Batlogg, B.; Zahurak, S. M.; Schneemeyer, L. F. Phys. Rev. Lett. 1987, 58 (18), 1888.
6. Harrison, M. R.; Edwards, P. P.; Goodenough, J. B. J. Solid State Chem. 1986, 54, 136.
7. Frase, K. G.; Liniger, E. G.; Clarke, D. R. Commun. Am. Ceramic Soc., in press.
8. Hinks, D. G.; Soderholm, L.; Capone II, D. W.; Jorgenson, J. D.; Schuller, I. K.; Segre, C. U.; Zhang, K.; Gruce, J. D. App. Phys. Lett., in press.
9. Kini, A. M.; Geiser, U.; Kao, H. C. I.; Carlson, K. D.; Wang, H. H.; Monaghan, M. R.; Williams, J. M. Inorg. Chem. 1987, 26, 1834.
10. Strobel, P.; Capponi, J. J.; Chaillout, C.; Marezio, M.; Tholence, J. L. Nature 1987, 327, 306.
11. Goodman, P.; Moodie, A. Acta Cryst. 1974, A30, 280.
12. Ourmazd, A.; Reutschler, J. A.; Spence, J. C. H.; O'Keefe, M.; Graham, R. J.; Johnson, D. W., Jr.; Rhodes, W. W. Nature 1987, 327, 308.
13. Hewat, E. A.; Dupuy, M.; Bourret, A.; Capponi, J. J.; Marezio, M. Nature 1987, 327, 400.
14. Long, N. J.; Petford-Long, A. K. Ultramicroscopy 1986, 20, 151.
15. Michel, C.; Raveau, B. J. Solid State Chem. 1983, 49, 150.
16. Hein, M.; Klein, N.; Muller, G.; Piel, H.; Roth, R. W. Superconductivity Workshop, Genoa, June 1987.
17. Svenaga, M. Proc. MRS Meeting 1987, in press.
18. Waldram, J. Nature 1987, in press.
19. Muller, K. A.; Takashige, M.; Bednorz, J. G. Phys. Rev. Lett. 1987, 58, 1143.
20. Ashby, M. F. Acta Met. 1974, 22, 275.
21. Artzt, E.; Ashby, M. F.; Easterling, K. E. Metallurgical Trans. 1983, 14A, 211.

RECEIVED July 22, 1987

INDEXES

Author Index

314

Affiliation Index

Subject Index

ACS Books Department staff working on this book included
Paula M. Bérard, Janet S. Dodd, Keith B. Belton,
Carla L. Clemens, and Robin Giroux

Elements typeset by Hot Type Ltd., Washington, DC
Jacket printed by Atlantic Research Corporation, Alexandria, VA
Printed and bound by Maple Press Co., York, PA